Eric Le Bourhis

Glass

Related Titles

McGee, T

Glass Manufacturing Engineering

2006

ISBN-13: 978-0-471-50836-6

Affatigato, M.

Advances in Glass and Optical Materials II

Ceramic Transactions. Volume 197

August 2006

ISBN-13: 978-0-470-08343-7

Doremus, R. H.

Glass Science

Second edition

1994

ISBN-13: 978-0-471-89714-1

Eric Le Bourhis

Glass

Mechanics and Technology

WILEY-VCH

WILEY-VCH Verlag GmbH & Co. KGaA

The Author

Professor Eric Le Bourhis
Université de Poitiers
Laboratoire. de Métallurgie Physique,
UMR 6630 CNRS,
SPMI-Téléport 2-Bd. Marie et Pierre Curie,
BP 30179
86962 Futuroscope-Chasseneuil Cedex
France

Library of Congress Card No.:
applied for

British Library Cataloguing-in-Publication Data
A catalogue record for this book is available from the British Library.

Bibliographic information published by the Deutsche Nationalbibliothek
The Deutsche Nationalbibliothek lists this publication in the Deutsche Nationalbibliografie; detailed bibliographic data are available in the Internet at <http://dnb.d-nb.de>.

Typesetting Thomson Digital, India
Printing betz-druck GmbH, Darmstadt
Binding Litges & Dopf Buchbinderei GmbH, Heppenheim
Cover Design WMX-Design, Bruno Winkler, Heidelberg
Printed in the Federal Republic of Germany
Printed on acid-free paper

ISBN: 978-3-527-31549-9

Contents

Glass; Mechanics and Technology. Eric Le Bourhis

ISBN: 978-3-527-31549-9

Foreword

It is a pleasure to introduce this book which approaches glass from the scientific and technological sides. This is an important contribution to the glass literature since few books address both these aspects together. Glass mechanics has appeared as a preliminary for further development of glass functionality and performance. This issue is complex and requires detailed understanding of glass science and industry. This textbook gives a comprehensive approach to glass mechanics taking constant care to keep as close as possible to the applications. The author shares his own experiences both in the academic and industrial fields resulting in an illustrative and relevant description of glass research and development. This book will be of great interest as training material for people entering the glass industry as well as future engineers and researchers in the field.

Aubervilliers, France
June 2007

René Gy

Glass; Mechanics and Technology. Eric Le Bourhis

ISBN: 978-3-527-31549-9

Preface

The project of writing this book started from lectures given to future engineers in materials science. The prime idea was to present both science and technology that participate in industrial R&D. I chose the glass industry as an example having working experiences on both industrial and academic sides. The science–technology approach and concepts developed and illustrated in the book can be utilized in other contexts and for other materials. Fracture and ageing, photoelasticity, rheology, contact resistance, diffusion and production tools issues are shared by many fields. For teaching purposes, appendices suggest exercises and experimental approaches. It is hoped to offer an overview of the background science and technology used in industry and R&D. The book focuses on mechanics, which has to be considered throughout the industrial process.

Preparing this book, I am indebted to many colleagues from industry and academe for their support, discussions and motivations. I would like to thank R. Gy, H. Arribart, J. Prieur, S. Valladeau, E. Bodiot, O. Gaume, N. Brown, J. Zhang, C. Bleuze, J.M. Poujade, P.H. Guering, M.H. Chopinet, E. Barthel, L. Lesage (Saint Gobain group), A. Alliche (ECP Paris), R. Vacher, C. Levelut, D. Cavaillé, J. Pelous (Montpellier University), M.M. Chaudhri (Cambridge University), C.R. Kurkjian (Rutgers University), T. Rouxel, J.C. Sangleboeuf, V. Keryvin (Rennes University), G. Patriarche, L. Largeau (LPN Marcoussis), P. Gadaud (ENSMA Poitiers), D. Mousserion, P.O. Renault, P. Goudeau, J. Bonneville, C. Jaouen, C. Templier, T. Cabioc'h, F. Tranchant, C. Coupeau (Poitiers University), L. Pranevicius and S. Tamulevicius (Kaunas University). I should like to acknowledge support from Wiley VCH, in particular K. Sora and W. Wüst.

The author also wishes to thank S. Valladeau for his help in carrying out again fractography and photoelasticity experiments, and R. Gy, E. Barthel and F. Mammeri for critical reading and suggestions on parts of the book. Some students worked with me to develop research and applicative studies, and their contributions have been of utmost importance. I would like to thank M. Bustarret, C Teles, M. Beaurain, N. Tournerie, D. Metayer, F. Mammeri, D. Faurie and N. Chemin who shared investigation interest during their training period.

This book is dedicated to my wife, sons and parents.

Futuroscope-Chasseneuil
September 2007

Eric Le Bourhis

Glass; Mechanics and Technology. Eric Le Bourhis

ISBN: 978-3-527-31549-9

Symbols and Definitions (units in parentheses)

a Contact radius	(m)
a Distance	(m)
a Acceleration	(m s^{-2})
a_o Atomic diffusion jump	(m)
a_c Atomic diffusion interface path	(m)
a Mark diagonal	(m)
A Cation field intensity	(m^{-2})
A Elastic anisotropy	
A_p Projected surface	(m^2)
A_c Contact projected surface	(m^2)
B Brittleness index	($m^{-1/2}$)
B, $\mathbf{B}$ Magnetic field	(T)
BON Bridging oxygen number	
$c(x, t)$ Concentration	(m^{-3})
c Radial crack length	(m)
c_l Lateral crack length	(m)
c_P Specific heat	(J kg^{-1} K^{-1})
c_R Cooling rate	(K s^{-1})
C Brewster constant	(B)
C_g Atomic packing density	
D Diffusion coefficient	(m^2 s^{-1})
E Young's modulus ($2(1 + \nu)G$)	(Pa)
E, $\mathbf{E}$ Electric field	(V m)
E_i Indenter Young's modulus	(Pa)
E_p Potential energy	(J)
E_s Specimen Young's modulus	(Pa)
f Friction coefficient	(N m^{-1} s)
F Load, force	(N)
F_c Critical load for fracture	(N)
F_m Maximum force	(N)
g Gravitational acceleration	(m s^{-2})
g(T) Growth rate	(m s^{-1})

Glass; Mechanics and Technology. Eric Le Bourhis

ISBN: 978-3-527-31549-9

Δg Energy barrier for diffusion jump ($J\ at^{-1}$)
$\Delta g'$ Energy barrier for diffusion at nucleus interface ($J\ at^{-1}$)
$\Delta g''$ Energy barrier for diffusion at germ interface ($J\ at^{-1}$)
G Shear modulus (Pa)
G' Conservation or storage modulus (Pa)
G'' Loss modulus (Pa)
$G(t)$ Relaxation modulus (Pa)
ΔG Energy barrier for diffusion jump ($J\ mol^{-1}$)
$\Delta G'$ Energy barrier for diffusion at nucleus interface ($J\ mol^{-1}$)
$\Delta G''$ Energy barrier for diffusion at particle interface ($J\ mol^{-1}$)
ΔG_v Molar enthalpy of the transformation from liquid to crystal ($J\ mol^{-1}$)
h Penetration (m)
h Total penetration (m)
h_e Elastic penetration (m)
h_l Depth of lateral crack (m)
h_m Maximum penetration (m)
h_p Plastic penetration (m)
h_v Viscous penetration (m)
H Hardness or mean pressure (Pa)
H_v Vickers hardness (0.927p) (Pa)
$H(t)$ Heaviside function ($= 1$ for $t > 1$; $= 0$ for $t \leq 0$)
I Second moment (m^4)
I_o Incidence angle (rad)
J Diffusion flux ($m^{-2}\ s^{-1}$)
J Shear compliance (Pa^{-1})
$J(t)$ Compliance (Pa^{-1})
k Thermal conductivity ($W\ m^{-1}\ K^{-1}$)
k Wave vector ($rad\ m^{-1}$)
K Stress intensity factor (SIF) ($Pa\ m^{1/2}$)
K_c Fracture toughness ($Pa\ m^{1/2}$)
L Length (m)
m Fragility
m Weibull modulus
m, M Mass (kg)
M Torque (N m)
n Optical index
n Volumic concentration (m^{-3})
n Surface concentration (m^{-2})
$n(T)$ Nucleation rate ($m^{-3}\ s^{-1}$)
p Polarisation vector (C m)
p Pressure (Pa)
P Property
P_s Probability
Q Heat (J)
r Position or distance (m)

r Radius	(m)
r_c Cation radius	(m)
r_o Interatomic distance	(m)
r_O Oxygen radius	(m)
R Radius	(m)
RD(r) Radial distribution	
S Surface, section	(m^2)
S Entropy	($J\ K^{-1}$)
t Time	(s)
t_H Thickness	(m)
T Temperature	(K)
T_f Glass fictive temperature	(K)
T_g Glass transition temperature	(K)
T_L Liquidus temperature	(K)
T_m Melting temperature	(K)
T_R Relaxation time (η/G)	(s)
ΔT Thermal gradient	(K)
tan(δ) Loss factor	
U Fractured specimen total energy	(J)
v Velocity	($m\ s^{-1}$)
v_f Free volume (per atom)	(m^3)
v_l Limit velocity	($m\ s^{-1}$)
v^* Minimal volume for atomic jump	(m^3)
V Volume	(m^3)
V_m Molar volume	($m^3\ mol^{-1}$)
w Angular velocity	($rad\ s^{-1}$)
w_c Critical energy for a particle to form	(J)
Y Yield stress or yield pressure	(Pa)
Y_R Representative flow stress	(Pa)
z Impacter (indenter) penetration	(m)
Z Atomic number	
Z Cation valence	
α Shear angle	(rad)
α_l Thermal expansion coefficient of the liquid	(K^{-1})
α_s Thermal expansion coefficient of the solid	(K^{-1})
β Phase shift	(rad)
β Angle between indenter flank and specimen surface	(rad)
γ Shear strain	
δ Loss angle	(rad)
ε Strain	
ε_1, ε_2, ε_3 Principal strains	
ζ, ζ_{cl}, ζ_{cs}, ζ_{sl} Surface energies	($J\ m^{-2}$)
η Viscosity	(Pa s)

η_R Viscosity at reference temperature	(Pa s)
θ Wetting angle	(rad)
θ Distortion	(rad)
λ Wavelength	(m)
λ_m Wavelength at maximum light intensity	(m)
ν Poisson's ratio	
ν_E Atomic transition frequency, Einstein frequency	(Hz)
ν_i Indenter Poisson's ratio	
ν_s Specimen Poisson's ratio	
ξ Reduced time	(s)
ρ Density	(m^{-3})
σ Stress	(Pa)
σ_1, σ_2, σ_3 Principal stresses	(Pa)
σ_c Core stress ($\sigma(z = 0)$)	(Pa)
σ_m Mean stress	(Pa)
σ_M Maximum stress	(Pa)
σ_s, Surface stress ($s(z = \pm t_H/2)$)	(Pa)
τ Shear stress	(Pa)
υ Reciprocal dispersion	
ϕ Shape factor	
ϕ_i Volume fraction	
$\phi(T)$ Shift factor	
Φ Heat flow	($J\ s^{-1}$)
$\psi_1(t)$ Shear relaxation function	
$\psi_2(t)$ Hydrostatic relaxation function	
Ω Atomic volume	(m^3)

Shorthand Notation

$$\nabla = \left| \begin{array}{l} \dfrac{\partial}{\partial x} \\ \dfrac{\partial}{\partial y} \\ \dfrac{\partial}{\partial z} \end{array} \right. \quad \text{Gradient operator}$$

$\partial/\partial t$, $\partial/\partial x$ Partial derivatives	
For an x only or r only space variation $\nabla = d/dx$ or d/dr	(m^{-1})
Overbar or overdot $= d/dt$ Time derivative	(s^{-1})
Double overdot $= d^2/dt^2$ Second time derivative	(s^{-2})
$' = d/dV$ Volume derivative	(m^{-3})
[ ] Concentration	($mol\ l^{-1}$ or m^{-3})

Physical Constants

g Gravitational acceleration: 9.8 m s^{-2}
N Avogadro number: 6.022×10^{23} mol^{-1}
h Planck's constant: 6.626×10^{-34} J s
R Perfect gas constant: 8.314 J mol^{-1} K^{-1}
k Boltzmann's constant ($= R/N$): 1.381×10^{-23} J K^{-1}
e Electron charge: 1.6×10^{-19} C
ε_o Vacuum permittivity ($\mu_o \varepsilon_o c^2 = 1$, $4\pi\varepsilon_o = 1/(9 \times 10^9)$)
c velocity of light: 3×10^8 m s^{-1}
m_e electron mass: 9.11×10^{-31} kg
m_p proton mass: 1.673×10^{-27} kg
m_n neutron mass: 1.675×10^{-27} kg
V_m Perfect gas standard volume: 22.4 l mol^{-1}

International System
Length: metre (m)
Mass: kilogram (kg)
Time: second (s)
Electrical current: ampere (A)
Temperature: kelvin (K)
Amount of substance: mole (mol)
Luminous intensity: candela (Cd)

Prefixes
Tera (T), 10^{12}
Giga (G), 10^9
Mega (M), 10^6
Kilo (k), 10^3

Milli (m), 10^{-3}
Micro (μ), 10^{-6}
Nano (n), 10^{-9}
Pico (p), 10^{-12}

Glass; Mechanics and Technology. Eric Le Bourhis

ISBN: 978-3-527-31549-9

Femto (f), 10^{-15}
Atto (a), 10^{-18}

Units Conversion
1 atm = 1.015 bar = 0.1 MPa
1 Å = 0.1 nm
1 eV = 1.6×10^{-19} J
1 Hz = 1 s^{-1}
1 J = 1 N m
0 K = −273.2 °C
1 l = 10^{-3} m^3
1 N = 1 kg m s^{-2}
1 P = 0.1 Pa s
1 Pa = 1 N m^{-2}
1 torr = 1.33 mbar = 133 Pa
1 W = 1 J s^{-1}
1 C = 1 A s
1 V = 1 W A^{-1}
1 Ω = 1 V A^{-1}
1 cal = 4.186 J
1 F= Ne C mol^{-1}
1 in = 25.4 mm

Useful Conversions and Numbers
k = 0.086 meV K^{-1}; for 25 °C, kT = 25 meV
N $m_p \sim$ N $m_n \sim$ 1 g mol^{-1}
Human hair diameter is 100 μm, living blood cell diameter 8 μm
Glass hardness is about 5 GPa = 50 000 atm!

Abbreviations

1D	One dimension
2D	Two dimensions
3D	Three dimensions
AFM	Atomic force microscopy
ALS	Aluminosilicate
AMLCD	Active matrix liquid crystal display
AZS	Alumina–zirconia–silica refractory
BCC	Body centred cubic
BMG	Bulk metallic glasses
BO	Bridging oxygen
CBED	Convergent beam electron diffraction
CEC	Cold end coating
CRN	Continuous random network
CSC	Composed semiconductor
CT	Compact tension
CVD	Chemical vapour deposition
DRM	Dynamic resonance method
DT	Double torsion
DMA	Dynamic mechanical analysis
ESC	Elemental semiconductor
ESP	Engineered stress profile
EXAFS	Extended X-ray absorption fine structure
FCC	Face centred cubic
FDP	Flat display panel
FEM	Finite element modelling
FIB	Focused ion beam
GFRP	Glass fibre-reinforced polymer
HEC	Hot end coating
IR	Infrared
IS	Individual section
ISE	Indentation size effect
ITO	Indium tin oxide, or tin-doped indium oxide

Glass; Mechanics and Technology. Eric Le Bourhis

ISBN: 978-3-527-31549-9

LCD	Liquid crystal display
Low-E	Low emissivity
MEMS	Micro-electro-mechanical system
MD	Molecular dynamics
MNR	Magnetic nuclear resonance
NBO	Non-bridging oxygen
NBT	Notch beam test
PDP	Plasma display panel
PER	Paramagnetic electronic resonance
PMMA	Poly(methyl methacrylate)
PVB	Polyvinylbutyral
PVD	Physical vapour deposition
RD	Radial distribution
R&D	Research and development
RT	Room temperature
SEM	Scanning electron microscopy
SENB	Single edge notched beam specimen
SIF	Stress intensity factor
SLS	soda-lime-silica
TEM	Transmission electron microscopy
TEOS	Tetraethoxysilane
TTT	Time–temperature–transformation
UV	Ultraviolet
VDW	Van der Waals
VFT	Vogel–Fulcher–Tammann
XPS	X-ray photoelectron spectroscopy
XRD	X-ray diffraction

1
Introduction

Glass, existing for millions of years, has fascinated and attracted much interest both scientifically and technologically. For long, glass was considered a 'fourth state of matter' before the realization of its 'liquid-like' structure. Glass appears on cooling down a liquid continuously until its viscosity becomes so high that it freezes to a glassy state. This happens at some range of temperatures that depends on thermal history. This lets glass be a mysterious material, since the way it is prepared may change its properties. Nowadays open questions are still raised and we shall address some of these within the present book. These theoretical considerations did not however prevent the development and use of glass over the centuries. Empirical knowledge and know-how have developed through the ages so that, nowadays, glass is a commonly available material. The application of glass science to the improvement of industrial tools, so called research and development (R&D), occurred only in the past century, with a few exceptions, as for instance regarding porcelain researches focused on developing processes in Sax and Worcester (Dr J. Wall's group by 1750, refer also to Limoges (France) and Worcester (UK) museums). In their 'Perspective on the history of glass composition' Kurkjian and Prindle (1998) noted that the age of glass science started only in 1886 with the disclosure of the work of Schott and Enrst (Chapter 4). Science and technology were then becoming more and more interrelated.

Glass has been employed in many forms to fabricate glazing and containers for centuries while it is now entering new applications that are appearing in nanotechnology (fibres and displays). Many qualities make glass attractive since it is transparent, chemically inert, environmentally friendly and keeps very strong. In fact, no other materials being mass-produced have shown such qualities over so many centuries. Nowadays glass offers recycling opportunities and allows for tailoring new and dedicated applications. The glass industry has been developing considerably in efficiency in terms of production rate and quality over the past few decades. Studying glass history in terms of raw materials and process offers an interesting perspective and understanding of how the glass industry progressively developed. These progresses were achieved because of further knowledge on materials properties and tools. While great achievements were accomplished empirically, R&D offered new opportunities. It is much accepted nowadays that an industry has to

Glass; Mechanics and Technology. Eric Le Bourhis

ISBN: 978-3-527-31549-9

progress continuously to face the strong competition that has been developing since globalization (with the emergence of low-labour-cost countries). The evolution in technology varies much according to the area of application. As far as glass is concerned, many different fields are to be considered, from mass production (glazing, containers) to nanotechnology processing (optical fibres, coatings). The glass industry encompasses several international industrial groups that have strengthened their own R&D units. New products and tools are developed and patented before being launched in production. Research products become of utmost importance for these groups that face always strong technological competition. As the glass industry covers so many different applications, this supposes important technological investment and know-how. Conventional silica-soda-lime glass industry covers large application fields comprising glazing (for transportation and buildings; Fig. 1.1) and containers, and this still represents the most important glass productions in terms of volumes. At the same time, added-value functions have been developed to sustain industrial competition. Glass R&D has allowed great progresses in the manufacturing of new functional glasses improving continuously performance and quality. Nowadays R&D is aimed at adding functions to glass objects with an interdisciplinary approach involving chemistry, mechanics and optics. New fields are being opened in electronics and displays and glass technology is expected to play an important role in nanotechnology. Different routes are being employed by manufacturers to add the designed functions, either modifying the glass substrate or using coatings. The former alternative is more expensive since up-stream operations have to be modified to adjust the tools to the changes in melt composition (high-temperature flow). The development of new coatings looks more attractive and

Fig. 1.1 Modern building with glass walls (near the author's university).

would be preferred since no up-stream operations need to be changed; however, this field is newly opened and many issues are still to be solved. In particular, one has to tailor coating strength otherwise the designed function would not resist against abrasion over time. In fact, the overall development and production of glass involve important and interdisciplinary knowledge and know-how.

This book presents an overview of background science and technology used in the glass industry and R&D. It focuses on mechanics that is to be considered throughout the industrial process. It is well known that glass is a brittle material and one understands why it is important to improve the strength of glass objects. Not so well accepted is that to achieve such a result and also to improve other performances (optical, dielectrical), one has to consider mechanics at most steps of the production. A literature survey reveals in fact strong interest in glass mechanics since 15 % of articles addressing glass science focus on its mechanical properties. This is the reason for the focus of this book. the most important technological property of glass is its viscosity. It determines the melting conditions, the temperatures for working and annealing and the upper temperature for use (devitrification rates). Glass quality is directly dependent on the way it is homogenized and this is also related to the viscous flow in the glass melt (Chapter 6). Also, proper quenching of the glass allows for the production of residual stresses that will strengthen the glass and protect it from sub-critical crack growth (Chapters 7 and 9). The dimensional adjustment of display front and back plates at the pixel resolution requires control of the thermal history and knowledge of the structural relaxation of glass substrates (Chapter 10). Contact mechanics controls the density of superficial defects generated while transferring products (including elevated temperatures; Chapter 8). The distribution and importance of the superficial defects affect the strength of the manufactured object (Chapter 7). Among the few examples listed above, one understands that glass mechanics is a very broad field while being one of the most important issues for the glassmaker.

The book is composed of 13 chapters; Chapters 2 to 4 are introductory while Chapters 5 to 12 propose a progressive route into glass mechanics and technology, with a main text that could be first read following the proposed chronology. Notes, appendices and further references (books, scientific papers, videos and web sites) are proposed throughout the text for deeper insight. Appendices recall basic concepts and also illustrate through exercises and applications parts of the book. Also, examples throughout the text allow one to employ the concepts used to describe glass behaviour while their application to glass technology is extensively discussed. Some parts with more mathematics will help in getting more insight into glass behaviour. For a first lecture these can be skipped.

2
Glass, A Ceramic Material[1)]

2.1
Four Classes of Materials

A commonly used classification of the materials separates these into four categories (Askeland, 1989; Ashby and Jones, 1991; Mozdierz et al., 1993):

i. the metals,
ii. the ceramics,
iii. the polymers, and
iv. the composites,

the last mentioned being obtained from mixing materials from the three main categories. This is illustrated in Fig. 2.1 where the three main categories of materials are shown at the centre while the composites made of two of these are shown at the periphery. This classification is based on the type of bonding and the related properties as discussed in more detail below.

This classification can be examined in view of the Mendeleev table (Fig. 2.2; see also Appendix 2). Most elements in the left-hand side of the table display a metallic behaviour while the other elements are considered as non-metals.

Metals may be pure or alloyed with other metals and also non-metals (steel being an Fe–C alloy). They are conductors of electricity and heat. Ceramics are inorganic materials and result from the combination of either (i) metals and non-metals (ionic ceramics, NaCl, MgO, Al_2O_3, TiN, ZrO_2) or (ii) only non-metals (covalent ceramics, SiO_2, Si_3N_4). Combination with oxygen, nitrogen and carbon yields oxides (SiO_2, MgO, Al_2O_3), nitrides (TiN, Si_3N_4) and carbides (WC) respectively. Ceramics show a refractory behaviour; they are electrically and thermally resistant. Polymers are formed by organic chains ($-CH_2-$) and show thermal and electrical resistances. Composites are formed by two or more materials pertaining to two different classes of materials. For instance, a polymer matrix can be reinforced by ceramic fibres to form a new material with properties combining those of its constituents. The proposed

[1)] In this chapter, and also in this book, we consider mostly inorganic glasses.

Glass; Mechanics and Technology. Eric Le Bourhis

ISBN: 978-3-527-31549-9

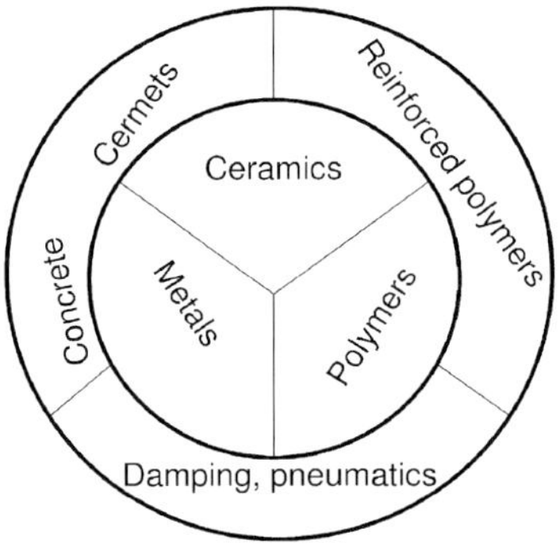

Fig. 2.1 Materials classes.

classification is mostly based on the type of bonding between the constituents of the related materials. It suggests that the structure at the nanoscale determines the properties of the material. This is true for many properties (elasticity, thermal expansion; see Chapter 7 and Appendices 1 and 3) while defects happen to play a major role in others (diffusion, fracture, plasticity; see Chapters 6 and 7, and Appendices 3 and 10). We generally distinguish four types of bonding listed in Table 2.1. Ionic and covalent bonds show the highest binding energy. Such bonds are formed in covalent and ionic ceramics. The lowest binding energy is obtained for van der Waals (VDW) bonds. Theses bonds are formed between polymer chains in thermoplastics. VDW and covalent bonds are found in thermosets that are stronger than thermoplastics.

Therefore, the related properties are expected to differ drastically from one group to another (for instance from polymers to ceramics). The main properties of the four categories of materials are listed in Table 2.2. Subjective criteria are suggested

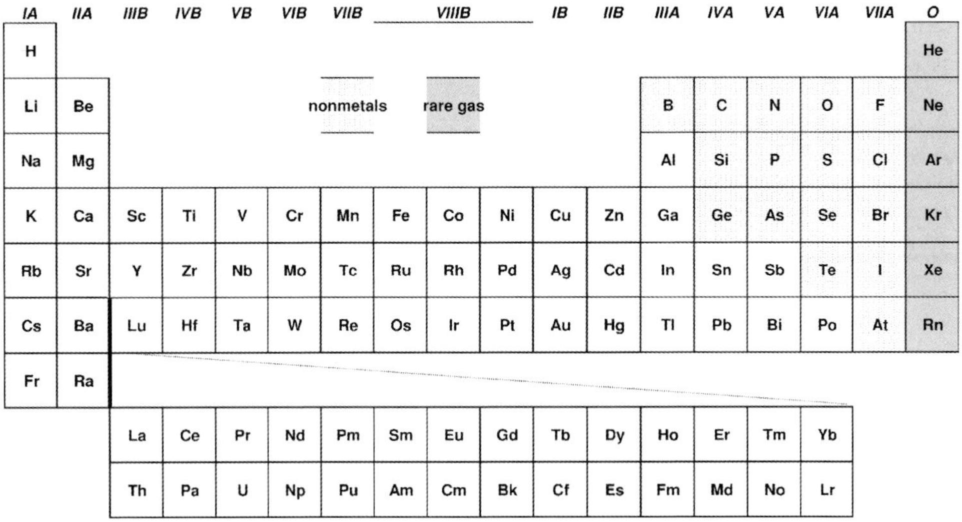

Fig. 2.2 Metals and non-metals in the Mendeleev table.

Tab. 2.1 Binding energies for the four bonding types (after Askeland, 1989).

Bond	Binding energy (kJ mol^{-1})
Ionic	600–1600
Covalent	500–1300
Metallic	100–840
Van der Waals	<45

here considering separately advantages and drawbacks. Most ceramics present large stiffness and elevated yield stress allowing for small elastic and reversible deformations until failure. They present elevated chemical and thermal resistances. Unfortunately, most ceramics are brittle with low toughness and thermal shock resistance. Like metals, they show an elevated density that is in favour of developing lighter composites made from a polymer matrix.

Most glasses pertain to the ceramics category since the bonds are either ionic or covalent. The related properties of the glass in fact fit well to the ceramics group as illustrated below using properties charts. While ionic or covalent ceramics present generally a long-range order, glasses are characterized by a short-range order. Table 2.3 shows the different states of matter in view of their respective distance of order.

Tab. 2.2 Materials categories with their respective advantages and drawbacks.

Material	Advantages	Drawbacks
Metals	Rigid $E \approx 100$ GPa	Low yield stress for pure metals
	Ductile, complex shape production	$Y \approx 1$ MPa $\rightarrow$ Alloys
	Good thermal and electrical conductivity	Hardness: $H \approx 3Y \rightarrow$ Alloys
	($K_c > 50$ MPa m$^{1/2}$)	Corrosion $\rightarrow$ Coatings
	Elevated melting point ($T_m \approx 1000$ °C)	Elevated density
	Good thermal and electrical conductivity	
Ceramics	Rigid $E \approx 200$ GPa	Low toughness ($K_{ic} < 2$ MPa m$^{1/2}$)
	Elevated yield stress $Y > 3$ GPa	Low thermal shock resistance
	High hardness	($\Delta T \approx 200$ °C)
	Elevated melting point ($T_m \approx 2000$ °C)	Difficulty in complex shape
	Corrosion resistance	production
	Good thermal and electrical resistance	Elevated density
Polymers	Ductile, complex shape production	Low stiffness ($E \approx 2$ GPa)
	Corrosion resistance	Low yield stress ($Y \approx 2$–1000 MPa)
	Low density	Low transition temperature
		($T_g \approx 100$ °C) $\rightarrow$ Creep
		Low toughness ($K_{ic} \approx 1$ MPa m$^{1/2}$)
Composites	Stiffness ($E > 50$ GPa)	Complex shape production difficulty
	Elevated yield stress ($Y \approx 200$ MPa)	Cost
	Toughness ($K_{ic} > 20$ MPa m$^{1/2}$)	Creep (for polymer matrix)
	Fatigue resistance	
	Corrosion resistance	
	Tailored and low density	

Tab. 2.3 Matter state as a function order distance (r_o is the interatomic distance). Perfect gas means that interactions between molecules can be neglected. Perfect crystal is a concept used to describe crystal structure with a repeated unit cell. In fact, all real crystals have defects.

Matter state	Distance of order: 0	3-10 r_0	>>100 r_0	∞
Gas	Perfect gas			
Liquid		Liquid ↓		
Solid		**Glass** →	Crystal	Perfect Crystal
	No order	Short-range order	Long-range order	Perfect order

Ordering distance increases from gas to crystal. In a sense, glass looks closer to a liquid than to a crystal. In fact, glass can be considered as a frozen liquid for which viscosity becomes so high that the atomic motions have slowed to the extent that characteristic relaxation time exceeds the observation period (Chapter 6). Many properties of glass can in fact be related to its liquid-like structure. Freezing happens at glass transition that sets a boundary between the glassy state and the liquid state. The 'fictive' path from the crystal state to the glass state will be illustrated in Chapter 5 in order to get insight into the complex glass structure. Both transitions, i.e. glass to crystal and liquid to glass transitions (arrowed in Table 2.3), correspond to real instability of the glassy state. Both phenomena are discussed extensively in Chapters 5 and 6 respectively.

Oxide glasses, the most common one being silico-soda-lime glass, are used to produce glazing or containers. As discussed in more detail in Chapter 5, such a glass is formed by an Si–O network that is modified by Na^+ and Ca^{2+} ions. This means that such oxide glasses are formed by covalent Si–O bonds while modifying ions interact with non-bridging oxygen (NBO) atoms through ionic bonds (Chapter 5). This allows for the strong bonding characteristic of the ceramics group. It is to be noted that modification with Ca^{2+} leads to less viscosity change than with Na^+ since Ca^{2+} may link two NBO atoms.

Organic glasses belong to the polymer category since they are formed by polymer chains interacting through VDW (thermoplastics) or VDW and covalent bonds (thermosets).

Let us consider the energy–distance curve related to a single bond. Interatomic spacing r_o is the equilibrium distance between atoms forming a bond and results from the balance between the repulsion predominant at short distances and attraction predominant at longer distances. The two contributions shown in Fig. 2.3 result in a potential energy curve that shows a minimum value reached for the atoms at

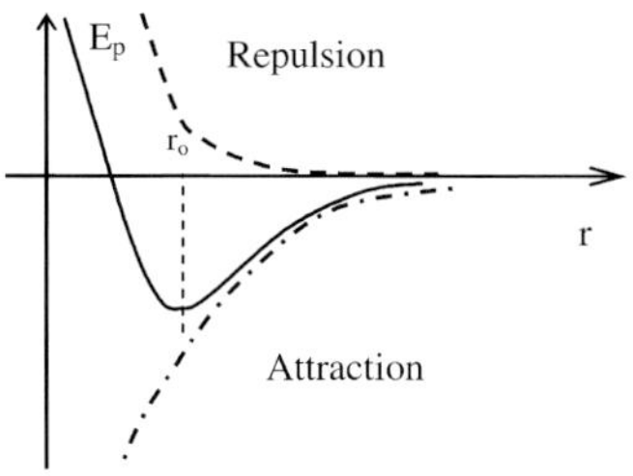

Fig. 2.3 Potential energy as a function of distance resulting from repulsion predominant at short distance and attraction predominant at long distance.

equilibrium distance r_o. This can be functionalized in the form of two terms corresponding to repulsion and attraction (see also Appendix 2):

$$E_p = E_{rep} + E_{attr} = \frac{\alpha}{r^m} - \frac{\beta}{r^n}$$

Since repulsion is predominant at short distance we have necessarily $m > n$. Attraction is responsible for the bond formation while electron repulsion happens when atoms overlap (the Pauli exclusion principle of quantum mechanics). The distance r_o allows for determining atom radius (half of the bond distance between identical atoms). This concept is very important in the modelling of material structure and we shall use it in Chapter 5. The hard-sphere model supposes that atoms pack as hard spheres, allowing for establishing relationships between atom radii and crystal parameter and quantifying packing density of glasses (Chapter 5). The minimum energy of the bond is defined as the binding energy and is the energy required to break the bond since it is the energy required to separate the atoms or ions to an infinite distance (Appendix 5). Consequently, materials having a high binding energy like oxide glasses also have a high strength and a high melting temperature. Fig. 2.4 plots

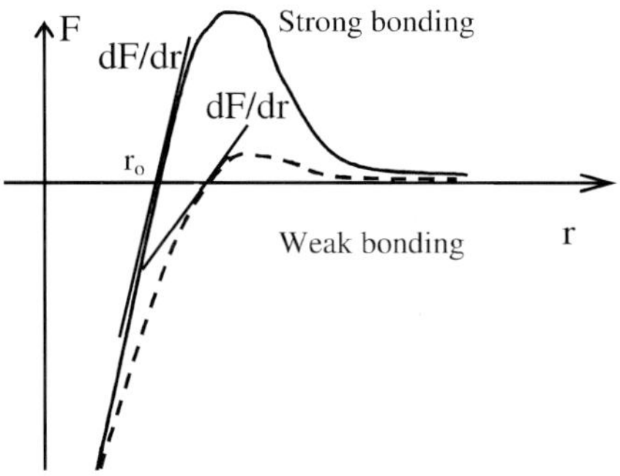

Fig. 2.4 Applied force as a function of distance for two types of bonds (weak and strong) showing the relationship between bonding strength and elastic modulus. Steep slope (dF/dr) results in high modulus (see also Appendix 5).

the applied force as a function of distance, the equilibrium being obtained for a resulting force equal to zero. The elastic modulus of a material which is the amount that it will stretch when a force is applied to it, is related to the slope of the force–distance curve at equilibrium ($r = r_o$, Fig. 2.4; Appendix 5). We shall see that, in fact, elastic moduli do not depend only on the interatomic bonding strength but also depend on the short-range ordering (possible formation of ring chain, atomic packing density; Chapter 5). The curves are shown for two materials, one with a strong bond and maximum force superior to that of a weak bond. A steep slope at $r = r_o$ that correlates with an elevated binding energy and hence an elevated melting point, signifies that a greater force is necessary to stretch the bond. Therefore, the material has an elevated elastic modulus. In a similar but more complex manner, the thermal expansion coefficient can be related to the energy–distance curve (Appendix 3). When the material is heated, the atoms or ions separate because of the additional energy supplied to the material. When the binding energy is large enough, the additional energy causes only small changes in the separation distance between ions or atoms as compared to what happens for a material showing a lower binding energy. Therefore, materials showing a large binding energy also show a small thermal expansion (Appendix 3).

2.2 Materials Properties

The next step in the classification of materials consists in fabricating the properties charts that show the classification of the materials according to one or several performances (Askeland, 1989; Ashby and Jones, 1991). In these charts the first three groups mentioned at the start of Section 2.1 (ceramics, metals and polymers) show different trends as expected from the different bond strengths. However, the different groups show a large range of response with partial overlapping of the properties between groups of materials. Fig. 2.5 shows three important diagrams from the mechanical point of view, where yield stress Y, Young's modulus E and fracture toughness K_c are reported. These properties will be defined in more detail in the following chapters. Here, we mention that these properties characterize the plastic, elastic and brittle resistances of a material corresponding to three most important modes of deformation (compaction may become an important deformation mode in a silica-rich glass network; see Chapters 5 and 8).

The trends mentioned in Table 2.2 are clearly observed in Fig. 2.5. For instance, the ceramics exhibit elevated yield stress and Young's modulus while their toughness is relatively low. Therefore, ceramics are expected to deform elastically up to brittle failure. Brittle behaviour gives origin to the sudden fracture of glass once critical stress is reached at a defect present in the glass. This will also be discussed in detail later in the book (Chapter 7). Instead, polymers show low stiffness and yield strength. They will deform dramatically elastically and plastically until fracture.

Fabricating a composite consists in mixing materials and thereafter in mixing the properties of each component. According to the scale of the phases, nanocomposites

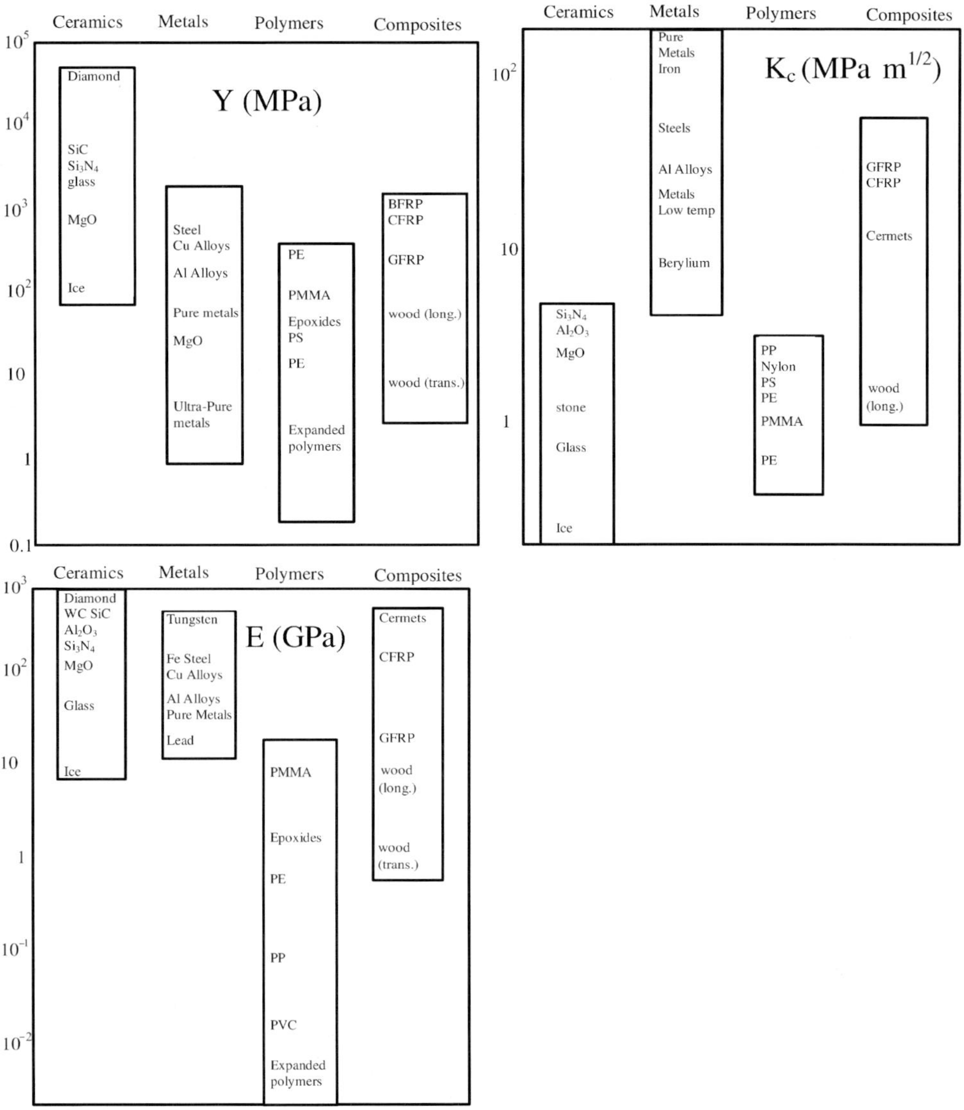

Fig. 2.5 Four classes of materials (according to Ashby and Jones, 1991).

and microcomposites are formed. Sol–gel techniques allow the fabrication of inorganic–organic materials with a mixing at the atomic scale (see also Chapter 12). Observing the high strength achieved by glass fibres, engineers incorporated them into polymers and succeeded in obtaining great reinforcement. This is now well developed and used in several fields (laminated structures, boats, car front ends and truck cabs). We can observe that indeed the composite domain covers parts of the ranges covered by the materials used for their fabrication. Changing the

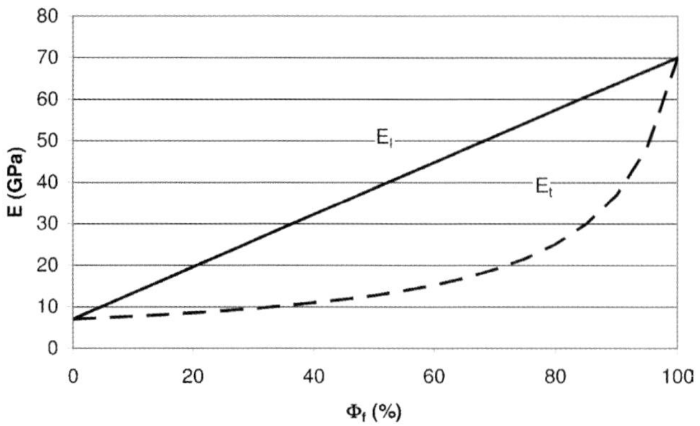

Fig. 2.6 Longitudinal E_l and traversal E_t moduli of a fibre-reinforced composite as a function of fibre volume fraction ϕ_f (all fibres are supposed to be aligned along one direction, namely longitudinal; 70 and 7 GPa moduli are taken for the fibre and matrix respectively).

concentration of the reinforcing component allows for tailoring the composite strength. The way that properties P are mixed is a complex issue. In the simplest case a rule of mixture may be employed. In its simplest form, it is written as a linear law $P = \phi_1 P_1 + \phi_2 P_2$, where ϕ_i are the volume fractions of the components ($0 < \phi_i < 1$ and $\phi_1 = 1 - \phi_2$) and P_i the respective properties. It is observed that when ϕ_1 approaches 0 or 1 then the property P approaches the properties of the components P_2 or P_1 respectively. Let us consider for instance the elastic resistance of a fibre-reinforced matrix. For unidirectionally distributed fibres in a matrix, the Young's modulus along fibres is $E_l = \phi_1 E_1 + \phi_2 E_2$ while it is $1/E_t = \phi_1/E_1 + \phi_2/E_2$ along the normal direction (so called model of springs in parallel or series respectively weighted by the volume fraction of the respective phases; see Chapter 6). The resulting moduli are shown in Fig. 2.6. Using the mixing rules allows tailoring the composite properties between the component properties, adjusting the volume fractions as well as the orientation of the fibres. It is well known so far that the simple linear rule presented above does not apply for all properties of technological importance. While a rule of mixture applies quite well to density and elasticity, it fails to predict the composite material toughness K_c because adhesion and interface issues are of utmost importance in the response and hence in the performance of a composite.

2.3 Selecting Materials

Generally not only one performance is to be optimized but several. An industrial product is then optimized to yield the best compromise between the required properties. For instance, a windscreen is designed with optical transparency, with strength to protect the driver from wind and weather, and with new functions introduced into the

Tab. 2.4 Attractive glass properties for different applications.

	Transparency	Strength	Rheology	Chemical inertness
Glazing	++	++	++	++
Containers	+	++	++	+++
Optical glass	+++	+	+	
Glass wool		++	++	++
Fibres for reinforcement			++	++

laminate (solar control, antenna, etc.). By the way, to achieve such a product, the manufacturing tools have to work in given ranges of properties, for instance viscosity and diffusion. The development of a new product supposes a compromise to be met between all these parameters. This is referred as co-selection. Table 2.4 lists the main properties of glass according to various applications.

Transparency is obviously of utmost importance for glazing and optical glasses. This performance combined with strength, rheology and cost make glass very competitive as compared to polymers. In fact, rheology is a very important issue as regards the process. Glass even though not showing exceptional performances in all fields presents a combination of desired properties that makes it very attractive.

Let us consider its transparency according to its cost (or price that takes into account the cost of raw materials, process, marketing and labour; see also Section 12.1), as shown in Fig. 2.7. In this figure transparency is plotted as a function of price per volume from the materials data base elaborated by Ashby's team (CES 4, Granta Design; Ashby, 2001). The optical performance is separated into four categories (opaque, translucent, transparent, optical quality). When considering transparency, the glassmaker is interested into the two last categories. Only dedicated optical devices require optical quality (i.e. extremely low optical loss). From the transparency–price chart we already observe that glass is a very good compromise, since only polymers compete. Glass is generally preferred because of its strength, UV protection (in containers), chemical protection, chemical inertness (allowing for conservation of foods and liquid, a property that may remain at elevated temperatures) and tailored optical properties. Developments of glazing encompass tailoring optical properties with transparency at visible wavelength while being opaque to IR and UV (Lehmann, 2005; see also Chapter 4). A chart showing so many parameters requires several dimensions; instead, Ashby (2001) proposes different methodologies to select the optimized materials or compositions either by a step-by-step selection or by using a weighted combination of the important properties that are of interest (weight-factors method). It is out of the scope of this book to detail all these methods and the interested reader should refer to dedicated books cited in the bibliography. As a preliminary, we shall see in the next section how several properties can be combined in a performance index and hence how to select a material in view of a combination of 3–4 properties using graphical methods.

The example of selection according to transparency is an archetype. Let us consider a more realistic selection of a glass according to more quantitative properties. As

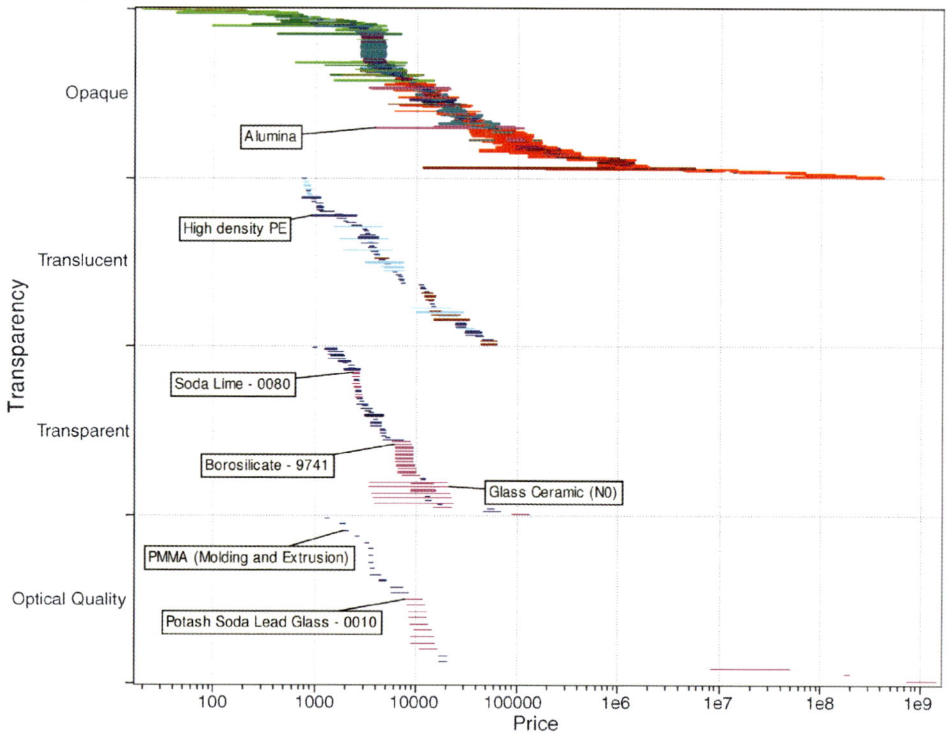

Fig. 2.7 Transparency–cost chart (obtained from CES 4, Granta Design, Ashby, 2001).

discussed in detail in Chapter 6, viscosity is one of the most important properties from the process point of view since it controls the stresses and strains to be applied to form the glass. The selection will then be according to the process requirements that set such a parameter (the viscosity in this case) at a given temperature. Research on glass has revealed that in limited domains empirical relations exist between the compositions of oxide glasses and the properties (Priven, 2004). Notably, glass composition can be continuously varied since no stoichiometric laws prevail, in contrast to crystalline solids. This allows for adjusting and approaching the desired performance. The optimization of the composition for process optimization can then be made using these empirical laws to define a range of compositions likely to yield a glass with the required properties. The estimated ranges of compositions are obviously as accurate as the empirical correlations used for the procedure. For more accuracy in the prediction, further experiments can be made for a selected domain of compositions. One will never ensure that such a range of compositions is indeed a solution to the problem, but it indicates in what range of compositions the costly experiments are probably worth undertaking. The use of designed industrial experiment plans (Sado and Sado, 1991) allows one to minimize the number of fabricated and tested specimens and to predict at minimum costs the selected domain.

2.4 Performance Indices

We have seen that selecting materials requires many parameters to be taken into account (co-selection). In fact, an extensive survey of the technological applications has been made by M.F. Ashby's team and they proposed different performance indices according to the target applications that are combinations of relevant properties of the materials.

Let us for instance consider the thermal shock resistance, also known as thermal endurance, defined as the maximum sudden change in temperature which a glass surface can withstand without breaking. Thermal shock happens, for example, when an object is moved abruptly from or into a furnace. These are complex thermo-mechanical loading conditions that are of utmost importance for the glass (and other) industry since heat treatments are extensively used to produce glazing either during forming or tempering steps (as detailed in Chapter 6). Stresses generated under a thermal gradient are proportional to thermal strains (Hooke's law; see also Appendix 9) while thermal strains are proportional to the thermal expansion coefficient of the material and to the temperature variation amplitude (ΔT; see also Chapter 6). The resulting stresses σ are written

$$\sigma = C\alpha E\Delta T \tag{2.1}$$

where α is the thermal expansion coefficient, E Young's modulus and C a coefficient dependent on the Poisson ratio ($C = 1$ for axial constraint, $(1-\nu)^{-1}$ for biaxial constraint or normal quenching and $(1-2\nu)^{-1}$ for triaxial constraint, where ν is the Poisson ratio; see also Chapter 6 and Appendices 3 and 9). It should be said that the actual value of the stresses developed by a given thermal shock would depend in fact upon many other factors, such as the thermal conductivity of the glass, the thermal conductivity of the fluid that applies the thermal shock and the thickness of the glass (these three factors change the magnitude of ΔT). Such factors are neglected here. Ashby (2001) suggested plotting α versus E in a log–log diagram (Fig. 2.8), since in such a plot αE contours are shown as parallel lines: αE constant implies $\log \alpha = -\log E +$ constant, which for a log–log plot is represented by a straight line of slope –1. On a same line, the value of αE product is then a constant and is defined as the performance index as regards thermal shock stresses. Observing that αE is expressed in MPa K^{-1}, the materials shown on the same line are submitted to identical thermal stress amplitude under the same thermal variation amplitude. For instance, along the line indexed 1 MPa K^{-1}, all materials would experience thermal stresses of 100 MPa amplitude when submitted to instantaneous (this hypothesis allows neglecting of thermal conductivity mentioned above) thermal shock of amplitude 100 °C. Fig. 2.8 shows that the different groups of materials locate at separated places even though partial overlapping is observed. Overlapping of composite and ceramics groups is not unexpected. Interestingly most important oxide glasses, that is soda-lime-silica (SLS) and borosilicate glasses match the ceramics group. As regards all the materials, one observes that αE index shows a very large variation ranging between 0.01 and

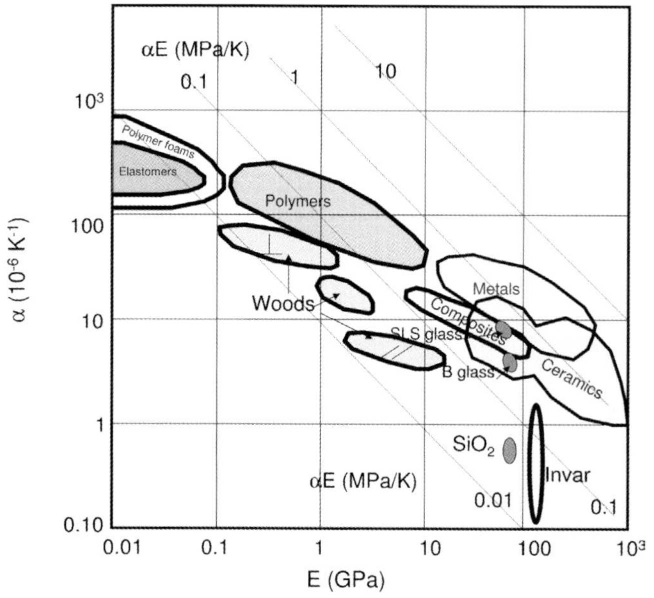

Fig. 2.8 Thermal expansion coefficient α as a function of Young's modulus *E* (according to Ashby, 2001).

10 MPa K^{-1}. This map actually allows for a wide materials selection according to what is required for the application (either high or low thermal stresses). Metals show the highest αE index values, while woods and elastomers show minimum values. Regarding glasses, silica is observed to be below 0.1 MPa K^{-1} while SLS glasses could be submitted to more elevated thermal stresses (about 0.5 MPa K^{-1}). In fact, for applications requiring thermal shock resistance, silica-rich and borosilicate glasses have been developed, so-called Nonex (for nonexpanding, invented in 1912 by E.C. Sullivan and W.C. Taylor to reduce lantern globe breakage by 50%) and Pyrex developed by Corning (Kurkjian and Prindle, 1998; see also Chapter 3). On the contrary, when tempering is considered elevated αE indices are required. This is in favour of utilizing compositions of standard silico-soda-lime type that are actually produced as tempered glazing.

When considering the risk of breakage, the next step consists in considering another important property of the material in this context, that is, its ultimate stress σ_f at rupture or, say, the stress at which the material will break. As we shall see in Chapter 7 ceramics (and indeed glass) show statistical fracture. We consider here an average or representative stress at which such articles break. Then under thermal shock, breakage will happen when the thermal stresses reach the ultimate stress. This condition is written

$$\sigma_f = C\alpha E\Delta T \tag{2.2a}$$

or forming members without units:

$$\sigma_f / E = C\alpha\Delta T \tag{2.2b}$$

where σ_f/E is the strain at fracture for a brittle material (see Section 7.2).

Fig. 2.9 plots σ_f/E as a function of thermal expansion coefficient α again using logarithmic scales for both axes. This presentation allows one to show resistances at a given thermal shock amplitude ΔT as straight and parallel lines. Again materials shown on the same line have the same thermal shock resistance. For instance, the materials shown on the 100 °C straight lines resist a thermal shock of amplitude less than or equal to 100 °C. The materials showing larger resistance are those appearing on the left and upper part of the figure. This figure shows also that the different groups of materials locate at separate places even though partial overlapping is observed. Again, overlapping of composite and metals groups is not unexpected. Oxide glasses match the ceramics group again. As expected from the α–E chart, materials show a very broad range of resistance. Ceramics cover a 10–1000 °C domain while glasses cover a reduced domain ranging from 50 to 300 °C.

Different performance indices (combining relevant properties) allow optimization of materials selection for a given application employing Table 2.5. An important parameter is the cost which can be combined with density.

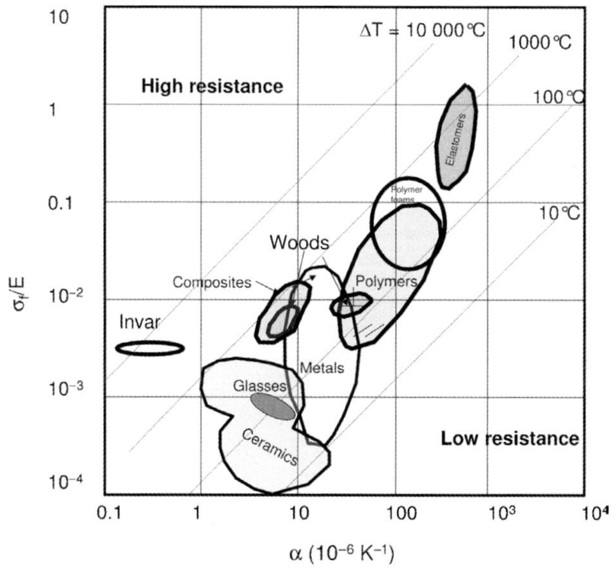

Fig. 2.9 Chart for thermal shock resistance showing σ_f/E as a function of thermal expansion coefficient α (according to Ashby, 2001).

Tab. 2.5 Performance indices for mechanical design (after Ashby, 2001). Parameter *C* is the cost per kg (see also Chapter 12).

Function, objective and constraint	Index
Window, maximum thermal shock resistance	$\sigma_f/E\alpha$
Tie, minimum weight, stiffness prescribed	E/ρ
Beam, minimum weight, stiffness prescribed	$E^{1/2}/\rho$
Beam, minimum weight, strength prescribed	$Y^{2/3}/\rho$
Beam, minimum cost, stiffness prescribed	$E^{1/2}/C\rho$
Beam, minimum cost, strength prescribed	$Y^{2/3}/C\rho$
Column, minimum cost, buckling load prescribed	$E^{1/2}/C\rho$

2.5 Shape Factors in Mechanical Design

We shall see that, besides choosing the right material, it is also important to select the right shape to produce the manufactured object. Consider for instance a paper sheet placed vertically: it shows poor buckling resistance when compressed. Instead, when folded its resistance dramatically increases. Also, plastic chair legs are fold shaped. Fig. 2.10 shows a glass roof supported by long glass beams with rectangular section set in such a way that their larger side is vertical. As discussed in detail a rectangular section allows a much better mechanical performance than a square section. It is important to note that it is the elastic failure and not the brittle failure (Chapter 7) that limits the present structural application. In fact a single tempered beam with its larger side set vertically is able to support 12.5 tons under four-point bending before buckling happens (R. Gy, personal communication).

Fig. 2.10 Glass atrium at the top floor of the building at 53–55 Av. Georges V (Paris). The glass roof is supported by tempered and laminated glass beams. These have $2 \times 19\,\text{mm}^2$ sections, the larger side being set vertically. The longer beam is approximately 7.1 m long. Architects: J.L. Robert, J.W. Wilmotte, P. Granger. The atrium at the 8th floor was designed to allow maximum daylight. (Courtesy R. Gy, Saint Gobain Recherche.)

Both shape and materials selections should be carried out together since shaping depends on the type of material, e.g. ceramics are more difficult to shape than metals. Let us concentrate on mechanics, which is the focus of the book, and more precisely now on the mechanical design that has been extensively reviewed by Ashby (2001). Several performance indices are listed in Table 2.5 according to the application. At this stage, these indices do not yet consider shape. Let us take an example to determine the performance index and determine how shape influences this performance. This will allow us to introduce the shape factor and hence understand how shape contributes to object strength.

Let us consider a beam under flexure (Figs. 2.11 and 2.12). We prescribe a given stiffness with minimum weight. The following expression holds (Fig. 2.12a):

$$\frac{F}{\delta} = k\frac{IE}{l^3} \tag{2.3}$$

where F is the load, δ the displacement of the beam centre, l the distance between supporting points (all fixed by the customer), E Young's modulus and I the second moment ($I = \int_{\text{section}} y^2 \mathrm{d}A$, where y is the distance from axis and $\mathrm{d}A$ the related surface at distance y). For a square section of area b^2, the second moment is

$$I = \int_{\text{section}} y^2 \mathrm{d}A = \int_{-b/2}^{b/2} y^2 b\mathrm{d}y = \frac{b}{3}\left[y^3\right]_{-b/2}^{b/2} = \frac{b^4}{12}$$

and Eq. (2.3) is written

$$\frac{F}{\delta} = \frac{k}{12}\frac{b^4 E}{l^3} \tag{2.4}$$

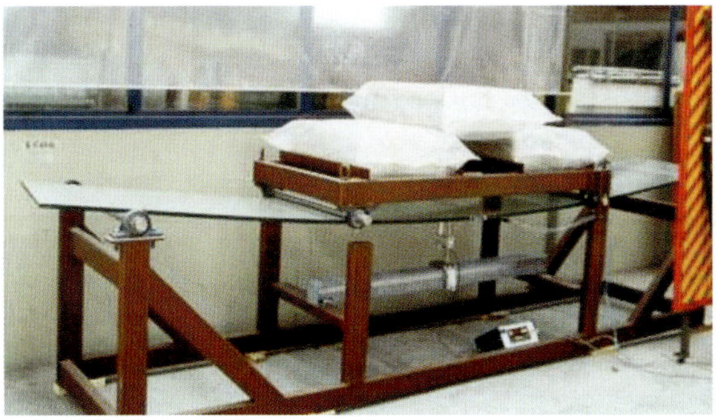

Fig. 2.11 Tempered glass beam under flexure (four-point bending). For the experiment here, the larger side is set horizontally. This set-up was used to determine the delayed elastic response (Gy, 1999; see also Chapter 6. Courtesy R. Gy, Saint Gobain Recherche.)

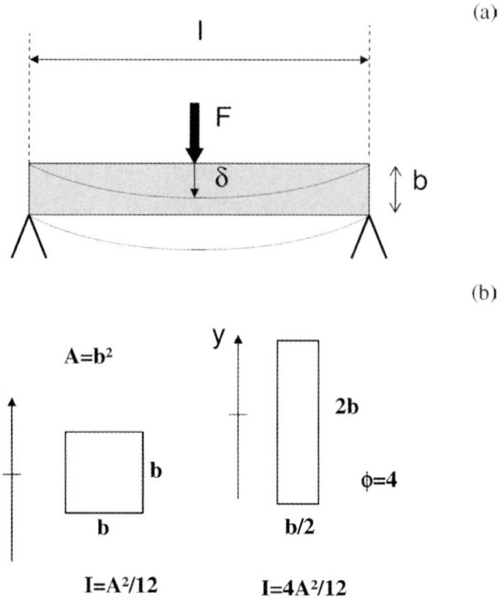

Fig. 2.12 (a) Beam under elastic flexure; (b) two beams with identical section area but different shape.

The mass of the beam is simply

$$M = \rho l b^2 \tag{2.5a}$$

and with Eq. (2.4) is

$$M = \rho l \left(\frac{F}{\delta}\frac{12}{k}\frac{l^3}{E}\right)^{1/2} \tag{2.5b}$$

This can be rewritten, separating the geometrical characteristics and the material properties, as

$$M = \left(\frac{F}{\delta}\frac{12}{k}l^3\right)^{1/2}\frac{\rho}{E^{1/2}} \tag{2.5c}$$

For a given application, mass is then minimized when

$$\frac{E^{1/2}}{\rho} \tag{2.6}$$

is maximized. This ratio is called the performance index. Note again that shape is not considered until now (or assumed to be the same up to this point).

As we shall see now, shaping the beam allows one to improve further the performance. Let us still consider a beam under flexure but now consider two different but simple shapes for the beam (Fig. 2.12b) with the same section b^2. One of the beams has a square cross-section (already studied) while the second has a rectangular cross-section, its smaller side being set horizontally. In fact, when the beam is set on its larger side then the flexure performance is poorer; this can be easily shown using the method described now or experienced in everyday life. The second moments are

$$I = b^4/12 \tag{2.7a}$$

and

$$I' = \int_{\text{section}} y^2 \mathrm{d}A = \int_{-b}^{b} y^2 \frac{b}{2} \mathrm{d}y = \frac{b}{6}\left[y^3\right]_{-b}^{b} = \frac{b^4}{3} \tag{2.7b}$$

for the square and rectangular section respectively. Equations (2.3) with (2.7b) show that the rectangular shape offers stiffer response. Calling

$$\phi = I'/I \tag{2.8}$$

the shape factor ($\phi = 4$ in the present case) and using the same procedure as before shows that the performance index transforms into

$$\frac{(\phi E)^{1/2}}{\rho} \tag{2.9}$$

This shows a square dependence on shape factor (in the present case performance is then twice as high for the elongated beam). Therefore, the shape of the beam is expected to play an important role in mechanical design. The performance can be considerably improved when using an elevated shape factor. As mentioned before, shaping has materials constraints. Then, one has to consider all required performance factors (co-selection). Shape should be considered at different scales: (i) at the macroscopic level (glazing, glass beam shape, arrangement and placing to be optimized for different applications (buildings, vehicles)); and (ii) at the microscopic scale (the arrangement of glass fibres in glass wool, in a composite or in a mat is to be optimized). In fact, there is still R&D effort towards optimizing glass wool recovery. When glass wool is produced, it is packed and stored under different environments. Glass wool recovery on unpacking is essential to allow maximum insulation and hence performance and this recovery will depend on the final arrangement of glass fibres. The study of the mechanical response of entangled glass fibres is complex and still the subject of ongoing research (Durville, 2005; Bergonnier et al., 2005).

Ashby (2001) proposed the use of charts to predict shape influence and determine the gain that can be achieved. Let us consider again the loaded beam under flexure as an illustrative example. The method is the following. A simple mathematical operation consists in dividing both parts of Eq. (2.9) by ϕ so that

$$\frac{(\phi E)^{1/2}}{\rho} \tag{2.10}$$

transforms into

$$\frac{(E/\phi)^{1/2}}{\rho/\phi} = \frac{E^{*1/2}}{\rho^*} \tag{2.11}$$

Then the charts can be used to select 'shaped materials' that are placed on the chart as artificial ones at coordinates $E^* = E/\phi$ and $\rho^* = \rho/\phi$. A shaped material is positioned relative to the unshaped one on the chart as shown in Fig. 2.13. As expected a shaped material offers a better performance than an unshaped one. In fact, on this chart, the performance index $E^{1/2}/C\rho$ that corresponds to a beam with minimum weight and stiffness prescribed is represented by lines of slopes +2 (as usually the performance is identical on the same line). It is then observed that the shaped material is positioned towards the left and bottom side as compared to the unshaped one. The magnitude of

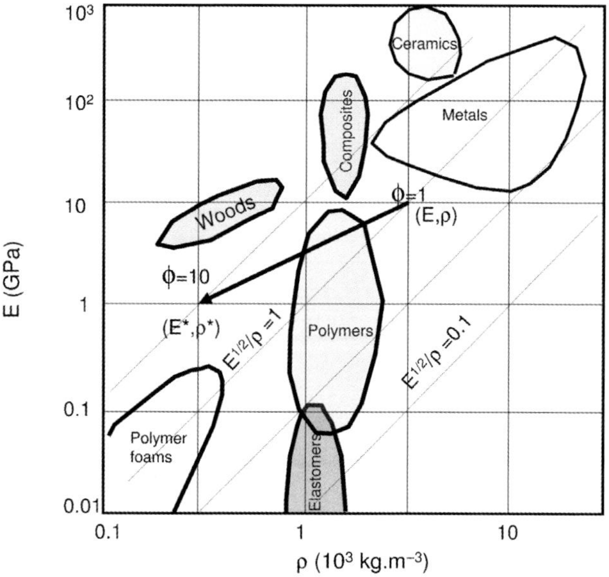

Fig. 2.13 Young's modulus–density chart (according to Ashby, 2001).

the achieved improvement can be predicted plotting the (E^*, ρ^*) position. Interestingly, composites like wood having natural microshapes show the best performance for the designed beam.

Shaping engineered materials allows one to approach the domain covered by natural or biomaterials as illustrated in Fig. 2.13. In fact, bio-inspired materials have become an important field of research (so-called biomimetic materials) since great gains in performance are expected.

3 Glass Prehistory and History

3.1 Natural Glasses

Natural glass has existed since the beginnings of time, formed when rocks melt as a result of high-temperature phenomena such as volcanic eruptions, lightning strikes or the impact of meteorites, and then cool and solidify rapidly enough so that a liquid-like structure can be frozen in glassy state (see Fig. 5.15). Rocks can be classified according to the way they form either from sediments or eruptions (Dejean-Arrecgros, 1978). The compositions of volcanic rocks do not differ significantly from sedimentary ones contrary to their respective thermal histories. It is well known that glass formation is a kinetic phenomenon (Chapter 6; Tammann, 1925).

Most well-known natural glass is obsidian that is believed to be one of the first glasses that appeared on earth at least 40 million years ago. Such an impressive age demonstrates that the vitreous state may be resistant in the long term against devitrification and water corrosion (Fig. 3.1; more images can be viewed at http://www.musee.ensmp.fr/gm//lettres.htm, refer also to Sedgwick Museum of Earth Sciences collection, Cambridge U.K.).

Glass utilization would have started when stone-age people became aware of the strength and sharpness that could be obtained from obsidian. Stone-age humans (75 000 BC, long before humans learned how to make glass) are believed to have formed (here form means mechanically, not thermomechanically) and used cutting tools and arrowheads made of obsidian.

The formation of natural glasses may happen under three main conditions, the first two corresponding to a compromise between the glass composition (viscosity regime) and the cooling conditions (as discussed in more detail in Chapter 5):

i. Magmatic formation: the silicate (silica-rich) melt being rejected from earth's mantle at the surface by volcanic activity cools down at rates scaling from 1 °C per minute to 1 °C per month (Chapter 5).
ii. Metamorphic formation: these conditions correspond to extremely fast heating and cooling of rocks and melts (1 °C per second) during impact, friction or lightning.

Glass; Mechanics and Technology. Eric Le Bourhis

ISBN: 978-3-527-31549-9

Fig. 3.1 Obsidian. (Copyright Ecole des Mines de Paris.)

iii. Sedimentary silica gel may be formed after drying over large periods of time scaling from years to thousands of years. Also natural biological glass forms sponge (Fig. 3.2, refer also to University Museum of Zoology, Cambridge U.K.) and mollusc skeleton and shell (Sundar et al., 2003). These mild synthetic conditions have inspired sol–gel science (see Chapter 12).

Figure 3.3 shows the composition ranges of natural glasses together with human-made glasses according to the relative concentrations of formers, modifiers and amphoteric ions (see Chapter 5 for precise definitions). The classification is based on the way the ions participate in the glass network (forming the network or modifying it, amphoteric being intermediate). Such a diagram is very instructive since it shows that the chemical domain occupied by industrial glasses separates dramatically from the one occupied by natural glasses. The relation between network modifiers,

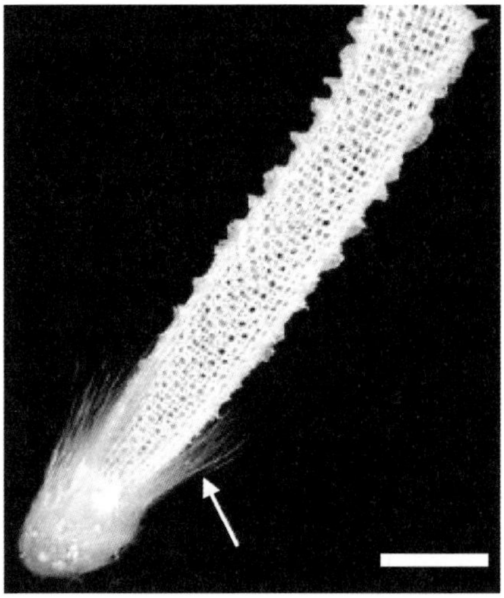

Fig. 3.2 Silica glass sponge; scale bar is 5 cm. (Reprinted from Sundar et al., 2003, with permission from Macmillans Publishers Ltd.)

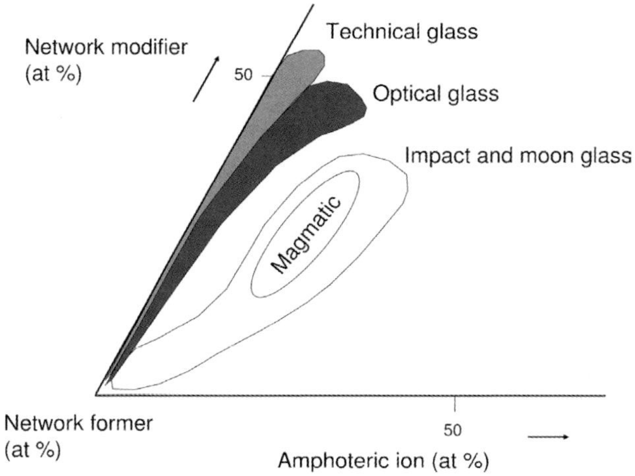

Fig. 3.3 Compositional ranges of manufactured and natural glasses. Network modifiers: Li, Na, K, Ba, Ca, Mg, Fe^{2+}, Pb, Zn; network formers: Si, P, B; amphoteric or intermediate ions: Al, Fe^{3+}. (According to Heide et al., 2003.)

network formers and intermediate oxides is indeed much different since industrial glasses have more network modifier than natural ones. The majority of natural glasses are peraluminous ($Al_2O_3 > Na_2O + K_2O + CaO$) whereas human-made glasses are peralkaline ($Al_2O_3 < Na_2O + K_2O + CaO$) in terms of geosciences.

Magmatic glasses (obsidian; Table 3.1) tend to be silica-rich and contain less than 10 % of alkaline. Their dark colour is due to iron oxides. Their purity and elevated viscosity allowed for the natural production of these glasses. Notably on the Moon, eruptive glasses form more easily because of the absence water, which leads to the induced elevated viscosity. Magmatic glasses can be found in Iceland, the Canaries and Wyoming. Premise stones have compositions close to that of obsidian, while

Tab. 3.1 Typical glass compositions of natural and manufactured glasses. Natural (magmatic: obsidian; metamorphic: australite, lunar glass) (weight percent, E and S glass fibre for composites).

Glass	SiO_2	B_2O_3	Na_2O	K_2O	CaO	MgO	Al_2O_3	Fe_2O_3
Obsidian	74.6		4.0	4.7	0.8	0.08	14.2	1.1
Australite	70.4		1.6	2.5	2.9	2.2	14.3	5.0
Lunar glass	54.5		0.1	<0.1	0.2	23.0	0.1	21.4
Egypt (1500 BC)	67.8		16.1	2.1	3.8	2.9	3.2	0.9
Italy (9000 BC)	77.8		6.4	8.7	2.1	0.7	2.2	0.8
Container glass	67–70		13–18	6–7	6–8	0–1	3–4	0–1
Glazing	71–73		13–15.5	0–1	6.5–12	2–4.5	0–2	
Glass wool	65	4.5	17	1	8	2.5	2	
Nonex	73	16.5	4.2					
Pyrex	80.5	12.9	3.8	0.4			2.2	
E glass (ALS)	54	10.0			17.5	4.5	14.0	
S glass	65					10.0	25.0	

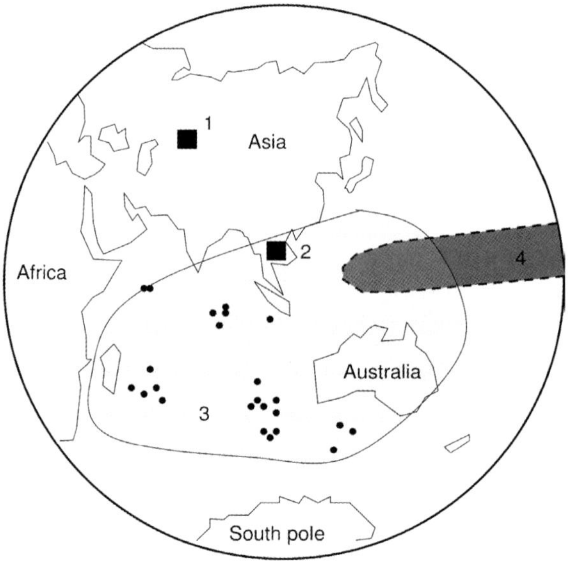

Fig. 3.4 Occurrence of australite and possible Zhamanshin strewn field (according to Heide et al., 2003): 1, Zhamanshin impact crater; 2, suggested Cambodian source region; 3, microtektite strewn field; 4, bottle-green spherule strewn field.

after ejection they release the enclosed gas forming a porous material. So called foam glass (or porous glass or cellular glass) can be produced industrially in a similar manner, by gas ejection at an appropriate stage during manufacture (Jones, 1956).

When formed by impact, metamorphic glasses show chemical compositions dependent on the chemical composition of the impact area. For instance, compositions of metamorphic lunar glasses differ drastically from earth ones (compare australite and lunar glasses for instance in Table 3.1). Silica-rich metamorphic glasses are found as small fragments in Arabia at Wabar impact and in the Libyan desert. Their formation process is not so far known in detail (Frischat et al., 2001)[1]. Some metamorphic glasses have been observed to be distributed over very large areas scaling with several continent sizes (Fig. 3.4; e.g. australite). They are believed to be generated by large meteorites impacting earth under inclined trajectories so that a cloud of silicate drops is formed and extended by aerodynamic transport. The mass of such tektites would vary from a gram to several kilograms.

Formed by different processes, natural glasses show different ages and compositions. From a scientific point of view they are of great interest since they allow for studying long-term resistance against devitrification and water corrosion (see Chapter 5). Most magmatic glasses formed 65 millions years ago (as dated from ^{40}K 40Ar transformation), while the ages of metamorphic glasses range between 1 (australite) and 65 (Chicxulub on the Mexican coast) million years.

[1] Frischat et al. (2001) entitled their paper 'Mystery of the Libyan desert glasses'.

The chemical resistance against corrosion of natural glasses (see also Chapter 5) over large periods of time is an attractive field of research. It offers a large perspective on glass evolution with time under different environments. It reveals that some glasses can resist either devitrification or corrosion over million of years. This was unfortunately not the case for some early human-made glasses as discussed in more detail below. This resistance of natural glasses is also of utmost importance from a technological point of view. For example, our modern society produces today radioactive wastes that require long-term confinement. Nuclear plants produce very dangerous wastes that may remain active over very long periods of time scaling over 1000–10 000 years. During this time it is of utmost importance to isolate these complex and dangerous wastes from the biosphere. One way of doing so is to embed the nuclear wastes into a glass matrix. Therefore, the long-term stability of thc glass matrix used to enclose the wastes is demanded. As discussed above, human-made glasses differ dramatically from natural ones and our knowledge of the durability of industrial glasses is much reduced as compared to that obtained observing natural glasses. Therefore, extrapolation for human-made glasses is unfortunately very hazardous over large periods of time. Today, more information on the formation and resistance of natural glasses and early glasses is researched while further knowledge on industrial glass stability is being developed. Early human-made glasses offer, however, the opportunity to study periods of time scaling with centuries (see also Section 3.2). This study is also very attractive since weathering depends on the location (climate, humidity), on the relevant atmosphere (car traffic, industrial pollution) and on the exposed time (medieval windows, cathedrals; Garcia-Valles et al., 2003). Interestingly, some early glasses were sometimes removed and stored before the industrial revolution so that the influence of the industrial atmosphere (coal, fuel combustion) can be measured. This allows for studying glass stability and evolution in view of several environmental parameters.

3.2 Early Glasses

Most ancient manufactured glasses have been discovered in tombs in Egypt (3500 BC). Also, Phoenician merchants are believed to have discovered glass in the region of Syria around 5000 BC. Their glass discovery is so far rather speculative. In AD 77, Pliny the Elder mentioned for the first time glass technology and how it was discovered by chance (Ernout et al., 1947). Natron blocks (a sodium carbonate used in the process of mummification) were used for resting cooking pots. With the intense heat, the blocks would have eventually melted and mixed with the sand of the beach to form an opaque liquid which would have formed a glass on cooling down slowly after the fire was stopped. Once glass fabrication was discovered, the first glassmakers were motivated to produce decorative objects, like vessels and maybe to simulate gems and semi-precious stones using silica with appropriate colorants such as copper, manganese and iron salts. Notably, the decorative application for glass is still attractive. Baccarat jewellery and decorative objects represent 24 % of its business, with black crystals being a great attraction and success today (François-Baillet and Marchi, 2006). In fact, because of impurities in the glass, the glass would have been transparent only when care was taken in the raw material selection and firing. At that time there was no demonstrated interest in transparency which was developed later as discussed below. The most ancient glass workshop was discovered in Egypt at Magatta palace and would have been in use by 1500 BC. The first so-called glass factory dates back to 1250 BC as revealed from the survey of Qantir Pi-Ramses site in the Nile delta region by Rehren in 2006. Parallel developments of ceramic and metallurgical processes allowed for glass to be used to produce foils and glazes on pots and vases (Fig. 3.5). The Romans formed glass foils by pouring the glass melt on a metallic plate which was covered by sand to prevent glass adhesion to the plate. The thinning of the foil was obtained by stretching it before the glass transition was passed. The glass transparency might have been then important since glass foils could have been used for windows. After the fall of the Roman Empire, the technology was dispersed (Doremus, 1994) with new centres being developed in Byzantium and in the Middle East by the Arabs. Glazes were formed over ceramics, allowing the protection of the ceramics from liquids (water, wine) and decorating the ceramic bulk (Fig. 3.5).

The first glassmaking 'manual' dates to around 650 BC. Instructions on how to make glass are contained in tablets from an Assyrian library (650 BC). Blowing was invented in Syria. An iron blowpipe some 1.5 m long was used, one extremity being flared to sustain the glass gob. This end was warmed in the furnace before the operations so that glass would adhere better to the blowpipe. The glassmaker then slowly twisted the blowpipe into the glass melt to collect the gob (or gather). From this point, the blowpipe was constantly rotated to prevent the glass from sagging and to get symmetrical glassware. When the glassmaker blew through the blowpipe a large bubble could be formed and was further extended on blowing and under its own weight (Fig. 3.6). The glassmaker had to control carefully its development and the glass distribution all over the glass object. Whenever thickness was decreasing more rapidly in one part of the glass item, in this area the glass would cool more rapidly. Then, as compared to the other areas of the glass, the thinnest parts presented larger viscosity and hence more resistance to glass flow. This was of great help to the glassmaker since this phenomenon would prevent further thinning and rupture of the area and allowed for glass distribution over the entire article to be controlled (Barton and Guillemet 2005). In the last century BC, the Romans began to blow glass inside moulds, which allowed a great increase in the variety of glass object shapes and topography. By AD 200, blowing allowed the formation long bottles which were separated from top and bottom to form long cylinders. Such cylinders were then cut lengthwise (Fig. 3.6), placed on a rock slab and annealed at a relevant temperature where they could be opened and flattened to produce flat glass. A similar process is still employed to produce coloured stained glass (Saint Gobain, Saint Just, refer to http://www.saint-gobain-glass.com/saint-just/indexflash.asp). Blowing appeared to offer a great variety of shapes which could be further transformed when re-annealed.

Early glasses in the western world were almost all soda-lime-silica composition but varied with the availability of raw materials (Table 3.1). Alkaline (soda) was used to

Fig. 3.5 Islamic ceramics with glazes in (a, b) plane view and (c) in cross-sectional view. (After Le Bourhis and Le Guiner 1986.)

lower the viscosity and allow the fabrication of glass at temperatures obtained from burning wood. The liquidus temperature also decreases, which facilitates glass production since glass transformation occurs at a higher viscosity regime preventing crystallization appearing. Little attention was paid to alkaline earth elements by the Romans and this continued into the Middle Ages, although they are known to be of utmost importance to prevent water diffusion and corrosion of glass (Chapter 5). Fortunately, beach sand and crude sources of alkaline were used as ingredients for glass fabrication and both components contain enough lime and magnesia to yield chemical durability. In the Middle Ages much glass was still dark green or dark brown as a result of the impurities that were present in the raw materials. Although still little was known about glass technology at that time, it appears that more attention was

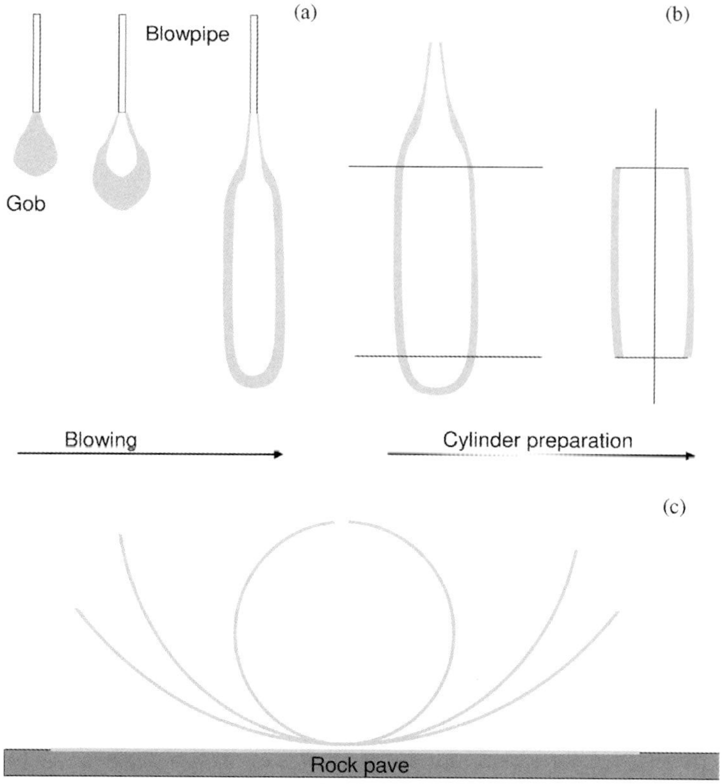

Fig. 3.6 (a) Blowing, (b) cutting and (c) flattening.

paid to raw materials selection. The raw material selection procedure resulted in the rejection of alkaline earths and alumina yielding glass with poor resistance to moisture (Kurkjian and Prindle, 1998). This resulted in deteriorating glass articles. Intentionally added CaO might date back to the 16th century, when Bohême crystal was invented. In fact, glass manufacture was mostly developed empirically. Scientific studies of glass were not carried out until much later (Faraday, 1830, 1857; see also Cable and Smedley 1989; James, 1991). Faraday investigated very important properties from the fabrication point of view: (i) melt conductivity that is of utmost importance when electrical furnaces are considered (Chapter 10) and (ii) glass colour, where he realized that ruby glass colour could be due to small gold particles (Chapters 4 and 5; Doremus, 1994).

3.3 First Optical Glasses

The history of optical glasses starts with flint glasses that were first produced in the 17th century. They were called flint glasses since they were based on high-purity silica from flint nodules and calcined lead oxides. Adding PbO into silica allowed for lower temperature production than soda-lime-silica glasses while limiting the content in K_2O yielded resistance to water corrosion. These lead glasses took advantage of high refractive index and ease of cutting and polishing. Lead flint was used to produce goblets, bowls and vases taking advantage of its elevated refractive index and dispersion. In the 17th century, there was a growing interest in scientific instruments for microscopic and astronomic observations. Galileo and Kepler made considerable discoveries thanks to improvements in optics using crown soda-lime-silica glasses. The crown process was a major breakthrough in glassmaking. A long thin metal tube was used to blow the glass in a large bubble later opened and spun into a circular disc by centrifugal forces and allowed the fabrication of flat glass. By that time colour and transparency of glass had become of utmost importance. The quality was still poor and optical instruments suffered from achromatic aberrations.

This poor optical quality spurred the development of mirrors (e.g. in reflecting telescopes). Around the 17th century, in France, a new process was developed for the production of plate glasses, principally for use in mirrors. In fact, mirrors had poor optical qualities until then and improvement was searched for. The molten glass was poured onto a special table and rolled out flat. After cooling, the plate glass was ground on large round tables by means of rotating cast iron discs and increasingly fine abrasive sands, and then polished using felt discs. The result of this 'plate pouring' process was flat glass with very good optical transmission qualities. When coated on one side with a reflective and low-melting metal, high-quality mirrors could be produced. This allowed for the development of reflecting telescopes with always a more reflective surface.

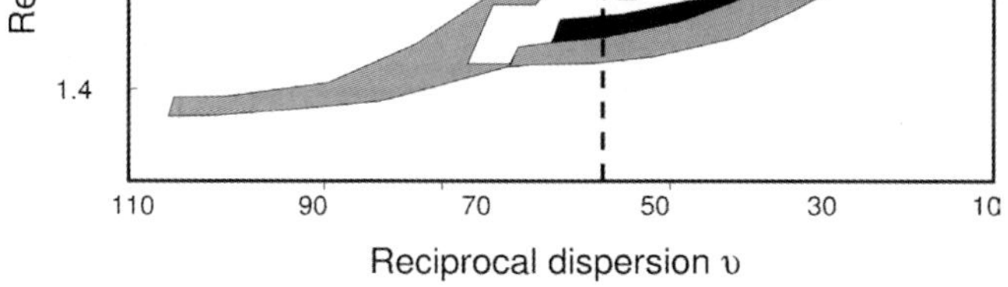

Fig. 3.7 Refractive index as a function of reciprocal dispersion (see Appendix 1) showing historical development of optical glasses. Grey area represents modern glasses; white area represents earlier glasses (1880–1934, Schott and others); black area represents glasses before 1880. (According to Kurkjian and Prindle, 1998).

It was also realized that achromatic lenses could be obtained by forming a doublet with a flint and a crown element (with high and low refractive index respectively; Fig. 3.7). This discovery allowed considerable improvement in the quality of the lenses and this was in favour of conventional refracting telescopes. Moreover, optical glasses were considerably improved by the discovery of Abbe and Schott allowing for the extension of the refractive index versus dispersion domain (white domain in Fig. 3.7). Nowadays the relevant domain has considerably extended either in terms of refractive index or dispersion as illustrated in Fig. 3.7.

3.4 Modern Glasses

3.4.1 Soda-Lime-Silica Glasses

Soda-lime-silica glasses used in containers and flat glass have changed little over the past centuries. Great improvements in glass quality were achieved by changes in the raw materials, the compositions (slight changes actually) and the melting and forming operations (Fig. 3.8). During the 17th century Deslandes from Saint Gobain added up to 6 % of lime to increase the resistance of plate glass to moisture. Around 1880 Schott suggested a few percent of alumina to improve chemical durability and resistance to devitrification. These composition refinements were of utmost importance for improving glass quality and resistance. While standard glass composition was progressively fixed, the industrial revolution yielded new instruments and machines to automate glass production and then to increase production, yield and quality. In 1854, the Siemens brothers transformed coal into CO gas that was further burnt in a furnace while they added regenerators that alternately stored and released heat wastes. The fabricated glass was thus preserved from coal dust contamination, improving its quality, while regenerators, still used nowadays (see

Fig. 3.8 Glassmakers at Saint Gobain in 1824. (Copyright Saint-Gobain.)

Chapter 10), saved energy and allowed much better production yields. By 1822, the Young factory proposed new silica-lime-based refractory to build furnaces instead of clay-based ones. These achieved much lower thermal expansion and hence the thermomechanical resistance and the fatigue resistance of furnaces were much improved. The furnace life could be increased accordingly.

Towards the end of the 19th century, container fabrication was greatly changed when Ashley (UK) and Boucher (France) proposed semi-automatic machines to produce bottles. At the same time an American engineer, Michael Owens (1859–1923), invented an automatic bottle blowing machine. It was a successful process, which arrived in Europe after the turn of the century and considerably transformed the containers industry. A further improvement of the process is due to Hartford Company which invented the feeder to provide the forming machines with a thermally conditioned glass gob. Up to this time the glass parison was taken from the glass melt inducing thermal heterogeneities. The feeder worked continuously like an extrusion machine allowing for the production of a gob of a given size at a given frequency. The feeders are still employed nowadays and with the invention of individual section (IS) machines by 1925, they deliver several forming machines (see also Chapter 10).

As in the containers industry, the glazing process was progressively and strongly automated. Up-stream conditioning of the glass melt benefited as well from refractory developments and thermal waste regeneration. By 1903, Window Glass (USA) had developed the Lubbers process which used automatic blowing. After 20 minutes of blowing, a cylinder 1 m in diameter could be produced. Later operations (cylinder separation and flattening) remained manual. Two processes (Fourcault in 1913 and Pittsburgh in 1925) were then developed for producing a glass ribbon continuously directly from the glass melt. This was a great improvement in terms thermal and thickness homogeneity. Both processes are illustrated in Fig. 3.9 and encompass a drawing tool (or bar) in contact with (or immersed in) the glass melt, cooling pieces above and metallic rollers drawing the ribbon upwards. The drawing tool and bar were used to avoid the thinning of the ribbon at its base that would result otherwise from surface forces. The tool used in the Fourcault process was in contact with the ribbon surfaces and thereafter damaged its quality and strength (Chapters 7 and 8). This drawback could be avoided using a refractory draw bar immersed 10 cm below glass melt surface. This was a great advantage in favour of the Pittsburgh process. Glass sheets of good optical quality could be produced with widths of 4 m. For a 2 mm thickness, the drawing speed could reach 150 m h^{-1} (Barton and Guillemet 2005).

In 1909 Benedictus invented laminated glazing while Saint Gobain patented tempered glazing in 1932. These inventions are still utilized nowadays since glass being a brittle material needs to be strengthened against fracture (Chapters 6 and 7). Laminated glazing is formed by two sheets of glass bonded together by a polymer. Whenever one of the glass sheets breaks the other one keeps the glazing in one single piece. Tempering of the glass allows for compressive residual stresses to build at the surface and hence preventing flaw extension and breakage of the glazing (Chapters 6–8). The float process was developed after the Second World War by Britain's

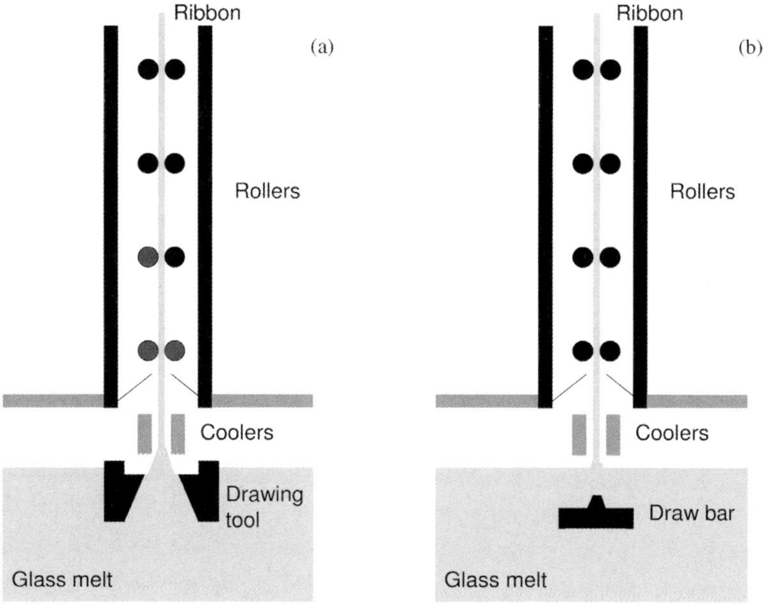

Fig. 3.9 (a) Fourcault and (b) Pittsburgh process for flat glass production. Note the drawing tool contacting the glass surface in the Foucault process (According to Barton and Guillemet, 2005).

Pilkington Brothers Ltd (Pilkington, 1969). It was a revolution in flat glass production since polishing of the plates was no longer necessary. It was a big astonishment since many glass manufacturers had focused their research and technology on improving glass polishing while this new process allowed excellent surface quality and transparency without this operation. In 1959, after seven years of experimentation and an investment of £7 million Pilkington Ltd introduced this economical means to produce distortion-free glass. Nowadays almost 90 % of flat soda-lime-silica glass is exclusively produced in this way (Palucka, 2006). The process is based on the idea of using a flat surface of molten metal to support a continuous ribbon of glass. The metal to be used has to be liquid over the temperature range 600–1050 °C corresponding to the processing of glass. This requirement is met by seven metals, namely Li, Ga, In, Sn, Ta, Pb and Bi (Appendix 2). Li is not dense enough to support glass while Ta, Pb and Bi evaporate above 1050 °C. Of the remaining candidates, Sn is cheap and available. Oxidation problems were solved by sealing the bath compartment with a reducing atmosphere of nitrogen and hydrogen. Molten glass, when poured across the surface of a bath of molten tin, spreads and flattens. As the ribbon progresses along the bath it is progressively cooled and then drawn horizontally without damage towards the annealing lehr (Chapter 10). It was realized that after holding the ribbon at high enough temperature the ribbon thickness was always about 6 mm, which was attributed to equilibrium between gravity and surface tension. Fortunately, this thickness corresponded to most of the company's

(a) (b)

Fig. 3.10 (a) Flat tempered glazing in the popular 2 CV and (b) tempered windscreen fabricated before the introduction of annealed and laminated ones. Note the photoelastic signature left by the tempering process (Chapters 6 and 11). These signatures are better observed at dawn. Observing old cars offers surprises. One may see windshields showing a circular zone approximately facing the driver. This was obtained using differential cooling rates. On breakage the circular zone remained in one piece (being less tempered) while the rest fragmented in several pieces. The vision of the driver was then preserved through this circular area. Nowadays laminated windshields are used instead and no fragmentation happens.

production needs. Tailoring ribbon thickness was to require further developments (guide bars, top rolls; Chapter 10). The contact with the molten tin allowed the excellent surface quality of the glazing produced. The next production steps (namely forming, tempering and laminating) require annealing the glass sheet again in different production lines (Fig. 3.10).

3.4.2 Borosilicate and Aluminosilicate Glasses

Abbe and Scott established that borosilicate glasses had superior resistance than soda-lime-silica glasses to chemical attack and had better thermal shock resistance (see also Chapters 2 and 6) because of their lower thermal expansion. Corning Glass Works developed glasses named Nonex (for non-expanding) in 1912 and Pyrex in 1915 that were immediately a sales success for ovenware. In 1930, aluminosilicate E glass fibres were introduced with excellent electrical resistivity and resistance to moisture. Fibre drawing developments allowed for the production of flawless fibres achieving very high strength close to the ultimate strength (Chapters 7 and 10). Thereafter, they were revealed to be excellent candidates for strengthening polymers (Chapter 2) and were progressively used to produce composites. Also aluminosilicate glasses could be chemically strengthened to achieve compressive residual stresses well above those achieved with thermal tempering (Chapter 6). In 1960, the fusion draw process was invented (Chapter 10). Its success was to come later with the development of displays since this process allows for very thin borosilicate flat glasses to be produced. This process is now being widely employed to produce panels for displays (Chapters 10 and 12).

4
Applications of Glass

Glass is used for a very large range of applications. The most well-known applications (reviewed in this chapter) are glazing, containers and glass wool which correspond to the largest industrial volumes. There are many more applications, some of them appearing just recently: glass containers for long-term conservation of industrial wastes (see also Chapter 3), hard disk drives (see Zou and Toratani, 2001; from Hoya corporation), displays (see also Chapter 12), glass ceramics for cooking tops, amorphous semiconductors for photocopiers and metallic glasses for golf clubs and cutting tools (Zandonella, 2005; metallic glasses allow for large elastic energy storage since they present both elevated strength and elastic modulus as compared to their crystalline counterparts; see also Chapter 7). The applications are driven by one or several properties that make the use of glass attractive (Chapter 2). Table 4.1 lists the different glasses used in their relevant fields. Obviously many applications concern optical properties. Glass fibres are employed to fabricate and strengthen composites, while borosilicate glass chemical inertness has proved to be very attractive for medical purposes.

4.1
Glazing

Soda-lime-silica glazing is the most well known application and is attributed to economical reasons. Glazing can be produced as flat glass with excellent surface quality, using the float process developed by Pilkington in the 1960s (Chapters 3 and 10). This good surface quality encompasses improved optical and mechanical performance (since flaw population is much reduced; Chapter 7). For safety reasons, in particular in buildings and transportation, glazing are either tempered (strengthened) or laminated, that is, formed by two substrates bonded by a polyvinylbutyral (PVB) film. Tempered glazings offer more mechanical resistance and eventually break into small harmless pieces (Chapters 7 and 8). For very high strength levels required for airplanes and trains, chemically tempered glazings are preferred. Laminated glazing are preferred for roofs since, once breakage happens, roofs remain in one single piece thanks to the PVB bond. Optical transparency may be tailored by sanding,

Glass; Mechanics and Technology. Eric Le Bourhis

ISBN: 978-3-527-31549-9

Tab. 4.1 Glasses and applications.

Glass	Application
Soda-lime-silica	Glazing, packaging
Borosilicate	High temperatures (cookware, laboratory glassware), pharmacy
Aluminosilicate	Fibres for reinforcement
Lead glasses	Lead-crystal tableware ('crystal' glass), protecting panels
Silica	Optical fibres
Chalcogenide	IR optics
Vitroceramics	Cooking tools, optics
Bioglasses	Medical

etching or rolling the glass (the latter process at elevated temperature). This is very popular in interior architecture (Dubois Petroff, 2001; Fig 4.1).

Insulated glasses allow for reducing heat transfer which occurs by radiation, conduction and convection. Through the use of insulated glass units with a low-conductivity gas fill between two or three panes of glass, conductive and convective heat loss can be significantly reduced (Fig. 4.2).

Radiative heat loss requires tinted or coated glazing. In fact, while visual transparency is in most cases required, tailored properties have been developed in infrared (IR) and ultraviolet (UV) domains to reduce heat transfer (Arbad et al., 2005; Lehmann 2005). Transparency is important in the visible spectrum with a minimum transmission for visible light (LTA) through windshields of 70 and 75 % in the USA and Europe respectively. In contrast, IR transparency might be a problem (see also Appendix 1). In a hot climate, IR transparency means heat flows from the outside into the building or the car. In a cold climate the opposite situation happens. Therefore, it is important to consider the related IR spectra. The IR solar radiation presents a maximum intensity at $\lambda_m = 0.5\ \mu m$ in the visible range (Wien law assumes a black

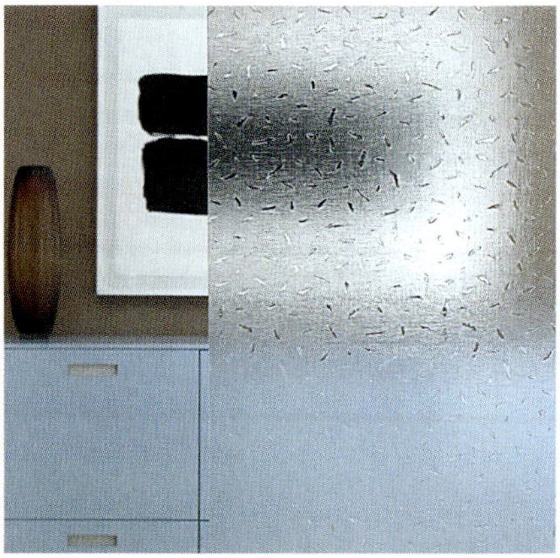

Fig. 4.1 Photograph of windows with one ordinary window glass (left) and marked glass (right). (Copyright Saint-Gobain.)

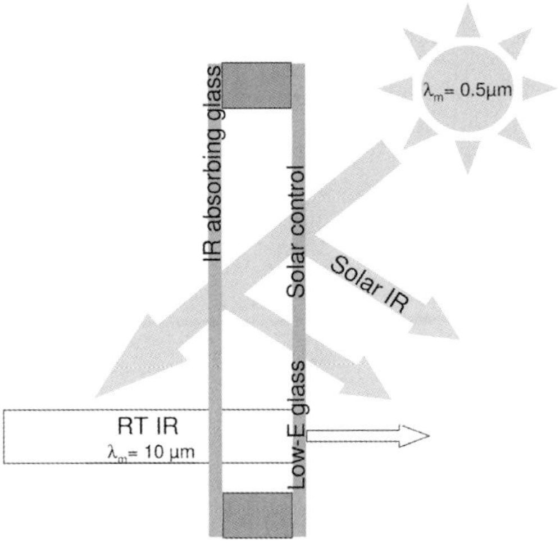

Fig. 4.2 Window unit with relevant functions.

body approximation, $\lambda_m T = 3000\ \mu m$ K; Ngô and Ngô, 1988). Because of the need to let through as much natural daylight as possible, transparency in the near IR may lead to excessive temperatures in modern buildings and vehicles because of the large glass surface. In fact, the IR region of sunlight contains 53 % of the total energy while the visible and UV regions contribute 44 and 3 % respectively. IR radiation spectrum for a 20 °C building shows instead a maximum intensity at much higher wavelength, $\lambda_m \sim 10\ \mu m$, that is, much shifted to the IR. Transparency in this wavelength range would lead to heat loss through building windows during winter. Therefore, new glass compositions have been developed to obtain the best compromise between optical transparency and IR insulation (Fig. 4.3; Saint Gobain group code for these

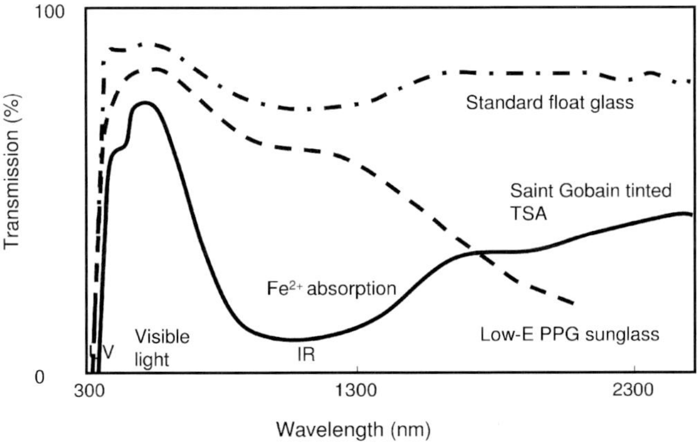

Fig. 4.3 Light transmittance of standard float, tinted and low-E (F:SnO_2) coated glasses. (After Arbad et al., 2005; Lehmann, 2005.)

glazing is TSA; Lehmann, 2005). Light-green coloured glass has been produced for such purpose not without technological challenge. In fact, as heat is primarily delivered to the top of the glass batch and glass melt (unless electrical power is used) the spectral properties of the final product have a corresponding impact on the heat-transfer efficiency through the depth of the glass melt and thus affect the convective flows and fining of the glass (Chapter 10). In order to gain weight, glazings become thinner and this forces the glassmaker to produce always darker glasses, with new challenges. For this reason, and others, it is of interest to develop new coating technologies (Chapter 12). Glazing can be designed to reflect IR radiation instead of absorbing it. Hence, instead of changing the composition of the glazing its surface can be treated using coatings. Low-emissivity coatings (low-E, based on $F{:}SnO_2$) have been developed to reflect thermal (20 °C) IR radiation for insulation purposes. In contrast, a bare glass absorbs and reemits in this region. Its emissivity, that is, the ratio of the energy radiated by the surface to energy radiated by a black body at the same temperature predicted by Planck's law, is 0.84 compared to 0.2 achieved with $F{:}SnO_2$ coatings. In general, higher conductance of the coatings results in a lower emissivity of the product. For anti-solar control, Ag-based multilayer coatings have been developed to reflect solar IR radiations. The Saint Gobain group codes for these coatings are Planitherm (low-E) and CoolLite (anti-solar), while PPG codes are Sungate. Coatings allow more industrial flexibility and have become a major value-added market for the flat-glass industry (see also Chapter 12).

In fact, changing the composition of the glass is a complex procedure and moreover the glass production during the composition transition is rejected. This is one of the reasons why glassmakers have involved their R&D in the field of

Fig. 4.4 Photograph of windows taken from indoor looking out with one ordinary window glass (left) and self cleaning glass (right). (Copyright Saint-Gobain.)

coatings. Nowadays, multilayered coatings allow great efficiency in radiation selection. Technologies have developed to produce such multilayer coatings at the surface of glass panels using CVD and PVD methods (Chapter 12). PVD techniques allow the production with a high reliability of coatings with 20 different sub-layers. Because of the flexibility of these techniques, the efficiency of the industrial process and improved performances, these functional glazings have become widespread. These technologies are also used for automotive applications as discussed in the next

Fig. 4.5 Examples of flat glass applications: glazing for (a) building, (b) phone box and (c) vehicle.

paragraph. More recently, glazings have been functionalized by titanium oxide layers in order to generate a self-cleaning function. This is of much interest for buildings that present more and more surfaces of glass at high altitudes and positions that are not easily reached. Using glazing with coatings functionalised by titanium oxide allows the dust to be photo-catalytically cracked. This photo-catalytic property when combined with a hydrophilic property allows for cracking dust later removed by rain or water. These treatments allow the glass transparency to be retained over long periods of time (about one year) while an untreated glazing has to be cleaned (Lehmann, 2005; Fig. 4.4).

For vehicles, glazings with complex shapes are now being fabricated. In fact, glazing is an integral part of the design of new cars. As for building applications, the surface area of glazing in vehicles (and particularly in cars) has been increasing continuously. Side and back windows are tempered while laminated windscreens are now required for safety reasons (see also Chapters 6 and 7). Moreover, glass roofs have become more and more attractive. With increasing glazing surface area, anti-solar coatings are required (Chapter 12). Such coatings are detected on the windscreen since they reflect blue (Renault vehicles) or pink (Pontiac vehicles) colours. As mentioned for building applications, IR insulation is now necessary to prevent extreme temperatures in vehicles during summer and at the same time to reduce gas consumption resulting from air-conditioning that equips most vehicles having large glazing areas. As for flat glass in buildings (Fig. 4.5), the use of coatings allows more flexibility in the production of IR insulating windows (Chapter 12). For achieving complex shape and/or tempering flat glass needs to be re-annealed at temperatures close to transition temperature that is in favour of strong CVD coatings (Chapter 6).

4.2 Containers

Containers include packaging (bottles, cans, jars, flasks) and glassware (plates, dishes, bowls, glass, moulds) that are formed at large scale by pressing and/or blowing a glass gob into a forming mould (Chapter 10). Packaging and glassware (Fig. 4.6) utilize glass optical transparency and/or UV protection (see Appendix 1) combined with chemical inertness (for contained food or liquid). As a result of its durability, glass is also useful as a container for corrosive fluids, or liners for chemical reactors.

For pharmaceutical packaging, borosilicate glasses are preferred to standard silico-soda-lime since they are more chemically resistant. In fact, standard glass is not completely inert since mobile alkaline ions may be leached into the contained product (Chapter 5). Therefore, low alkaline concentration glass containers have been developed. In the field of elevated temperature applications, low thermal expansion glasses are used. Pyrex glass is the most well known. Some manufacturers also tried to develop tempered containers from standard glass for the same purpose. These resist unless thermal shock stresses exceed the built-in residual stresses and then brittle fracture happens generating several pieces of glass (Chapter 7). Glass

(a) (b)

Fig. 4.6 (a) Containers and (b) fragrance flask. The fragrance bottle colour results from gold aggregates; an earlier use of this phenomenon was the ruby glass obtained by Murano glassblowers. (Copyright Saint-Gobain.)

containers were much utilized going back some decades since they could be reused. Glass presents several advantages as a packaging material (Murphy, 2005):

i. chemical inertness,
ii. ability to be recycled,
iii. non-polluting nature on ultimate disposal,
iv. ability to be manufactured from abundant raw materials,
v. UV filtration (amber and green glass), optical and transparency qualities,
vi. low gas permeability (Appendix 13) and
vii. high intrinsic strength (Chapter 7).

Despite these advantages, glass-packaged products face significant competitive pressures from products made of alternative materials. Nowadays, plastic, paper and related material, and metallic (aluminium and steel) containers share a large part of the container market, although some of them cost sometimes more than they contain (Fig. 4.7; see also Chapter 12). This has forced the glass industry to produce lighter containers (this has led to the development of tools for achieving uniform glass distribution all over the article) for short-term conservation purposes while offering new attractive designs and promoting glass recycling. (Metallic can manufacturers are developing environmentally friendly technology which allows for recycling as well.) For long-term conservation (e.g. liquors, wines, champagne) glass tends to be preferred to other materials because of its low permeability to gas (Appendix 13) and

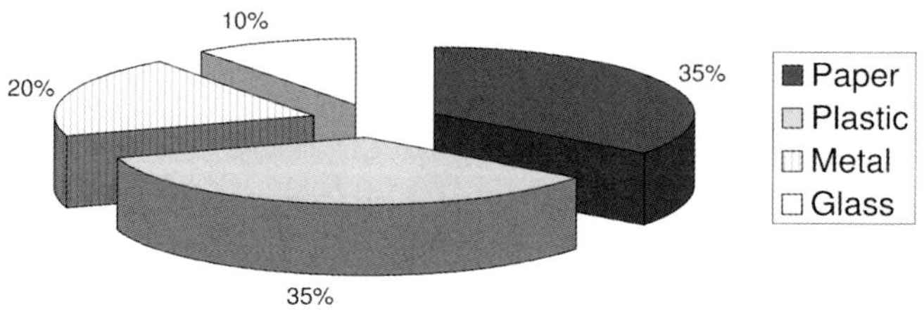

Fig. 4.7 Distribution of container industries (according to Murphy, 2005).

because of its chemical stability. By the way, marketing is a complex issue: glass still provides superior taste, health, visual appeal and value (see also Chapter 12). One expects champagne to be contained in a heavy glass bottle! In terms of R&D there are opportunities for sol–gel science (Chapter 12) and chemistry to produce new colorations and new design technology being environment friendly. When considered for short-term conservation containers, these new functions are to be produced at similar industrial costs otherwise other materials compete dramatically.

The situation is much different for cosmetics and fragrance packaging where design plays a very important place in the marketing of the product. Dedicated compositions of glass are being employed, for instance coloured (with transition metal Mn, Co, Cr and oxides) and opal glasses (so called milk glass). Also, lead crystal glassware offers great flexibility as revealed by art glass development and industrial tools being adapted to the low-viscosity regime. Blowing, flameworking and hot-glass sculpting are however still employed and developed in this area.

4.3 Optical Glass

Optical glasses can be produced with improved transparency in a designed range of wavelengths (see also Appendix 1). This is the case for silica lenses and fibres used in optoelectronics technologies which have become widespread. These are used for transatlantic communication cables, telecoms and cable TV. In fact, silica-based fibres achieve extreme transparency over tens, or even hundreds of kilometres, as a window glazing over a few millimetres. The mechanical reliability of optical fibres is as important as their optical properties. Optical waveguides can be obtained when a refractive index gradient is tailored (following $n_p \sin i_p = n_{p+1} \sin i_{p+1}$, Descartes' refraction law) as shown in Fig. 4.8. These fibres can be formed with an internal core the refractive index of which is greater than that of the outside so that total reflection can be obtained at the interface. Another way to propagate an electromagnetic signal is to produce a continuously changing refractive index and light follows a path similar to the popular mirage effect at the proximity of a hot road during summer. The first type of fibre is produced by drawing a glass preform containing in its core the high-index glass (Chapter 10). The second type of fibre can be produced by chemical diffusion (Chapters 6 and 10 and Appendix 13). These fibres can be protected from

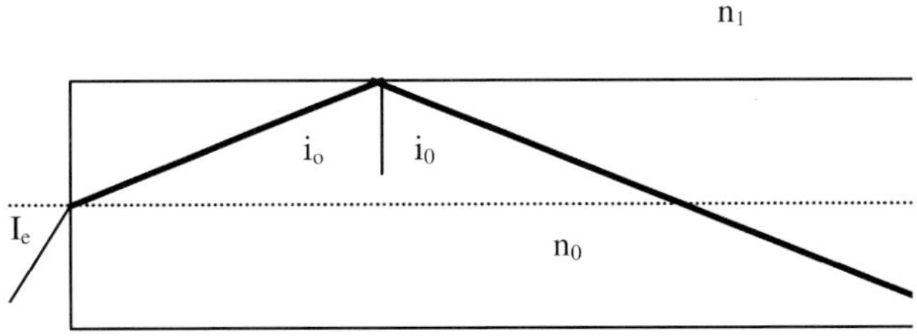

Fig. 4.8 Optical fibre principle (I_e is the entrance aperture) with discontinuously changing index ($n_0 > n_1$). The light beam follows Descartes' law with reflection at the refractive index interface.

moisture and contact by hermetic coatings which are deposited during the sizing step (Chapter 10). As a matter of fact, fibres offer exceptional stxrength and optical performance that have yield such an impressive development in communications. One drawback of glass when considered for lenses is its density, and load gains are accomplished by controlling better glass distribution (Section 4.2). It should be noted that nowadays spectacles are made of organic glasses. These allow load gains and comfort. Organic glasses are polymers formed by long organic chains (similarly, chalcogenide glasses are constituted of Se chains, their applications being described below). One single spectacle lens is composed of several layers, some of these being

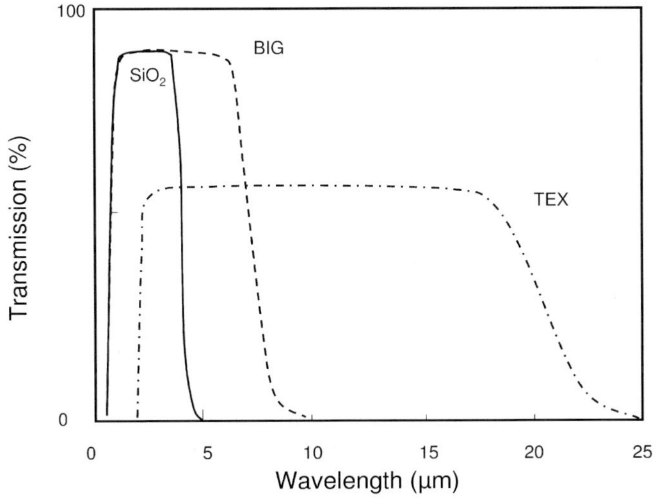

Fig. 4.9 Transmission of 'infrared' glasses compared to silica. BIG is a multicomponent fluoride glass based on BaF_2, InF_3 and GaF_3. TeX is a multicomponent chalcogenide glass based on the combination of Te, Se and I_2 (according to Lucas, 1999). Observe the huge changes in the IR window.

only protective because of the poor scratch resistance of polymers as compared to glass (Chapter 8).

More recently chalcogenide glasses have been proposed for application in IR vision (they are also used for DVD-RW taking advantage of their laser-induced glass–ceramic transition). In fact these glasses formed with group VI atoms associated with either group IV or V atoms (Appendix 2) allow for improved transparency in the IR domain (Appendix 1; 8–12 μm wavelengths, Fig. 4.9). This leads to application for night observations, detection of thermal loss in buildings and *in situ* medical observation (Lucas, 1999). These glasses while showing 'plastic' behaviour remain very brittle at room temperature (Guin et al., 2002; see Chapters 7 and 8). Hence, there is an important demand for improving the contact resistance and hence extending their future applications.

4.4 Glass Fibres for Insulation and Reinforcement

The glazing, packaging and optical applications of glass are the most well known since they refer to glass transparency. Glass is also used for thermal and acoustic insulation or to reinforce plastics (Figs 4.10 and 4.11) and concrete (CEM-FIL®) because of their low cost and high strength. In the latter case, compositions incorporating ZrO_2 allow to resist concrete chemical attack. Polymers reinforced by glass fibres are fabricated as composites and used in transportation applications (automotive front ends, vehicle furniture, plane cockpits, etc.). As discussed in Chapter 2, composite properties are a mix of those of the constituent parts, sometime in a complex way. In the simplest case a rule of mixture of properties may be employed in a linear form: $P = \phi_1 P_1 + \phi_2 P_2$, where ϕ_i are the volume fractions of each component (glass and polymer in the present context). It allows

Fig. 4.10 (a) Glass wool (copyright Saint-Gobain) and (b) fibre-reinforced composite used for the front end of an automobile.

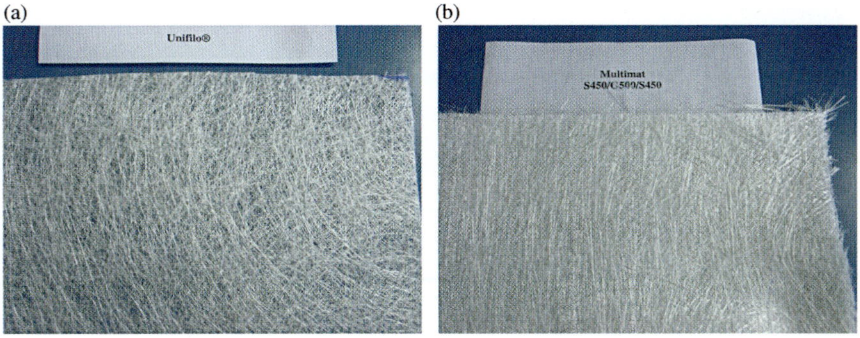

Fig. 4.11 Glass mats: (a) Saint-Gobain Unifilo® and (b) Multimat formed by E-glass fibres.

tailoring stiffness (Fig. 2.6) and load gains that are attractive for transportation and in particular for airplanes since, for instance, saving a mass of 1 kg on commercial planes increases the travelled distance by about 1 mile (Gay, 1997). Moreover as discussed later in the book (Chapter 7), fibres can be produced with much higher mechanical strength than any glazing. This makes the fibres very attractive for improving composite strength. Glass wool is produced for buildings and vehicles and forms an air–fibre composite. Notably, the glass composition is tailored accordingly for different applications (Table 4.1; Chapter 5). Wool fibres have been optimized to allow maximum recovery when unpacked and set (Chapter 2). Also, their dissolution *in vitro* has been of utmost importance since the problems generated by asbestos (see water diffusion and hydrolysis of siloxane bonds, Chapters 5 and 7). Biodegradable glass fibres have been developed accordingly (Chartier, 1997).

4.5 Abrasive Tools

Abrasive applications are not so well known. Glass can be used as a bond between abrasive grits to fabricate cutting tools. Glass composition is then tailored to obtain grit wetting and a strong grit–glass interface (Alliche and Le Bourhis, 2000). Figure 4.12 shows different grinding wheels composed of grits (alumina, silicon carbide) that are bonded together by a glass matrix. A scanning electron micrograph shows different grey contrast for the grits and bonds (Fig. 4.13). Note the elevated porosity of such a wheel. In fact, porosity is required for cooling and evacuating the metal chips generated when tooling a piece (Coes, 1971; Malkin, 1989; Tawakoli, 1993; Alliche and Le Bourhis, 2000). For high-speed grinding operations, wheels are made of a metal core (with high thermal conductivity) and abrasive segments fixed at the wheel circumference.

Fig. 4.12 Vitrified bonded abrasive wheels. (Copyright Saint-Gobain.)

Fig. 4.13 Scanning electron micrograph of a vitrified bonded abrasive.

4.6
Glass Manufacturers

Some well-known glass manufacturers are listed below, subdivided into the applications reviewed above.

i. Flat glass: Saint Gobain (France), Guardian (USA), Pilkington (UK), PPG (USA), Glaverbel (USA), Asahi (Japan), Schott (Germany).
ii. Packaging: Saint Gobain (France), BSN Glasspack (USA), Anchor Glass (USA), Owens Illinois (USA), Rexam, Vitro (Mexico), Gerresheim (Germany, USA), Pochet (France), Rocco Bormioli (Italy), Wheaton (USA).
iii. Optical glass: Heraeus (Germany), Schott (Germany), Corning (USA).
iv. Glass wool: Saint Gobain (France), Amstrong (USA), Johns Manville (USA), Owens Corning (USA), Rockwool (EC), Uralita (EC).
v. Reinforcing fibres: Saint Gobain (France), Johns Manville (USA), PPG (USA), Owens Corning (USA), Nottobo, Porcher.
vi. Abrasives: Saint Gobain[1)] (France), Asahi (Japan), Noritake (Japan), 3M (USA), SAIT (Italy), SIA (Switzerland), Tyrolit (Austria).

Note that none of the manufacturers are present in all six main domains. As a matter of fact, each domain requires specific know-how and tools.

1) Winter, Unicorn and Norton are now part of Saint Gobain group.

5
Glass Structure

5.1
Introduction

Glass structure is out of equilibrium. It is obtained when a liquid is cooled down in such a way that on passing the melting temperature or liquidus temperature (T_L), 'freezing' happens instead of crystallization (Tammann, 1925). The final temperature must be so low that molecules or atoms move too slowly to rearrange to the more stable crystalline form. In Chapter 6 we shall discuss extensively the atomic mobility in terms of viscosity. In particular, we shall see that glass structure may depend upon thermal history. For instance, a liquid will yield denser glasses when the cooling rate decreases. This chapter shows the prominent characteristics of glass disordered structure. Alternatively, glass can be built up onto a substrate, either by deposition or chemical reaction from the vapour, the substrate being cold enough that rearrangement cannot take place. In fact, although they are named differently, amorphous silica (a-SiO_2) films and silica glass bulk are the same materials having the same chemical composition and a random network. Ryu and Tomozawa (2006) suggest that the concept of fictive temperature can be used to describe a-SiO_2 film structure as well (see also Chapter 6). However, amorphous layers have been achieved with a limited number of substances as compared to the usual route (Doremus, 1994). Many definitions of glass have been proposed: we shall use the most commonly used one that defines it as an amorphous solid showing a glass transition (Zarzycki, 1982).

The glass is frozen in a structure showing only short-range order that makes its characterization quite difficult as compared to crystalline structure that presents long-range order (periodical structure with the repetition of a unit cell; see also Table 2.3). As discussed below, diffraction experiments on glasses yield results that most closely resemble those of liquids (see also Chapter 2).

Glass; Mechanics and Technology. Eric Le Bourhis

ISBN: 978-3-527-31549-9

Let us start with silica and related glasses because of their practical importance and the well-documented literature.

5.2
Silica Glass and Related Glasses

5.2.1
Glass Network

Silicon (Si) is in group IV of the Mendeleev table (Appendix 2) and tends to bond to four different oxygen (O) atoms. Note that what is described here applies as well to germania glasses changing Si into Ge. Hence, each oxygen atom is shared by two silicon atoms. In three dimensions (3D), Si and O atoms arrange to form oxygen tetrahedral elements with a Si atom at the centre (Fig. 5.1). This is in terms of chemistry the resulting bonding through four sp^3 hybrid orbitals of silicon and two 2p orbitals of oxygen atoms (see also Appendix 2). Then the structure can be thought of as built up by the sharing of corners by pairs of these oxygen tetrahedra. Two tetrahedra meet at only one corner, but all corners of all tetrahedra are actually shared. This is shown in Fig. 5.2 for the crystalline and glass form of SiO_2. Interestingly, a very open 3D structure may then be built in many ways, in which the Si–O–Si lines may or may not be straight. One observes rings of atoms of different sizes when a disordered structure is generated (this will be more obvious in 2D). In fact, it is helpful to project the structure on a surface where tetrahedra become triangular elements. One can also consider the projected glass as an imaginary 2D oxide Si_2O_3 (Jones, 1956). In such a 2D projection of the network we shall shift from a perfectly ordered structure to a disordered one (Fig. 5.3).

Figure 5.3a shows again the crystalline projected structure of SiO_2 (quartz) while Fig. 5.3b shows the continuous random network (CRN), or a network without periodicity illustrated now in 2D and first proposed for silica glass by Zachariasen (1932). It is observed in both structures that oxygen atoms are bonded to two silicon atoms forming Si–O–Si chains. Figure 5.3a reveals a long-range order characteristic of a crystalline solid (repetition in space of a unit cell). In fact, any atom's position can be predicted in this ordered structure. Figure 5.3b shows instead only a short-range order that is in great contrast with the 2D projection view of Fig. 5.3a. In fact, it reveals no long-range order any more while the structure still respects the stoichiometry

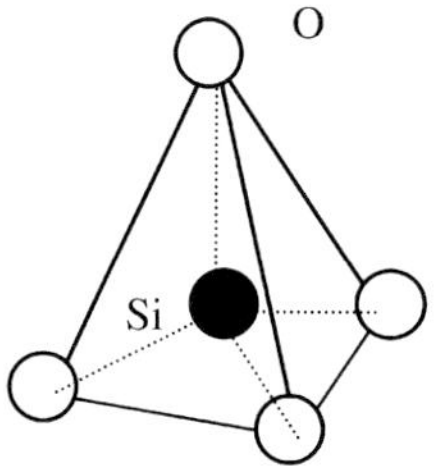

Fig. 5.1 Elemental tetrahedron, Si (group IV) bonds to four O.

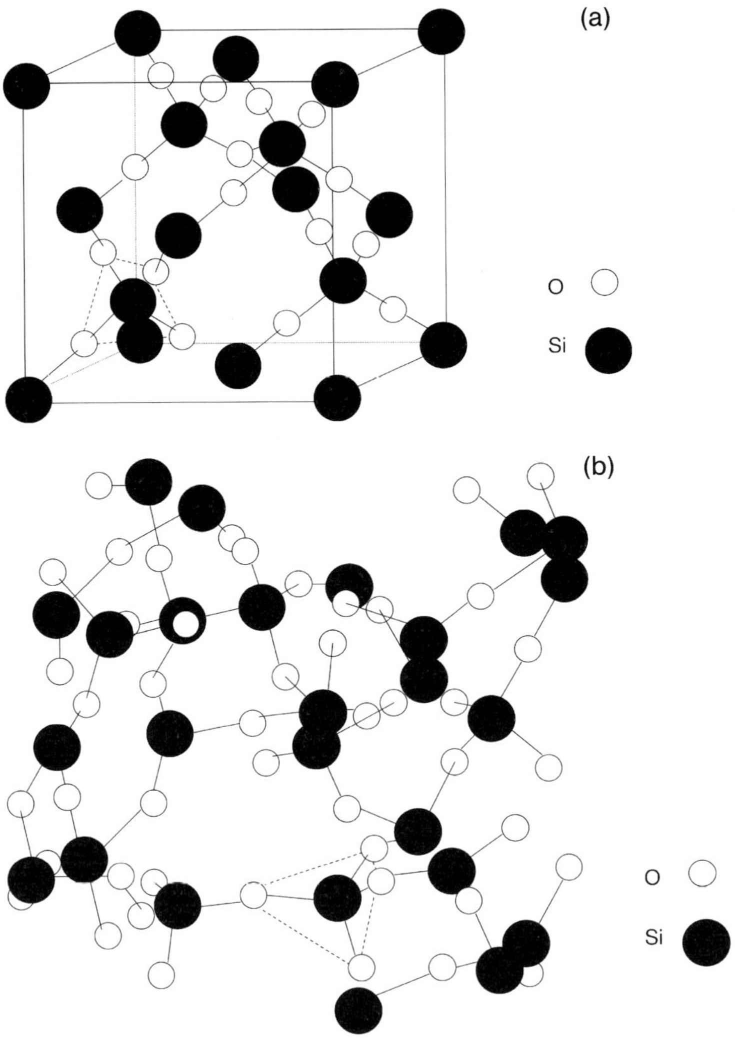

Fig. 5.2 SiO_2 structure: (a) crystal form; (b) glassy form. An elemental tetrahedron is shown in each case (dashed lines).

since all oxygen atoms are bonded to two silicon atoms forming Si–O–Si bonds. The atomic arrangement cannot be predicted any more in the latter case. One further observes the existence of rings with a different number of bonds. It seems from modelling that smaller rings would appear on cooling glass rapidly or, say, so-called network polymerization would be affected by thermal history (Ito and Taniguchi, 2004). In fact, the glass structure is known to be determined by its past thermal history and this phenomenon will affect the details of the structure and the structural relaxation as discussed extensively in Chapter 6.

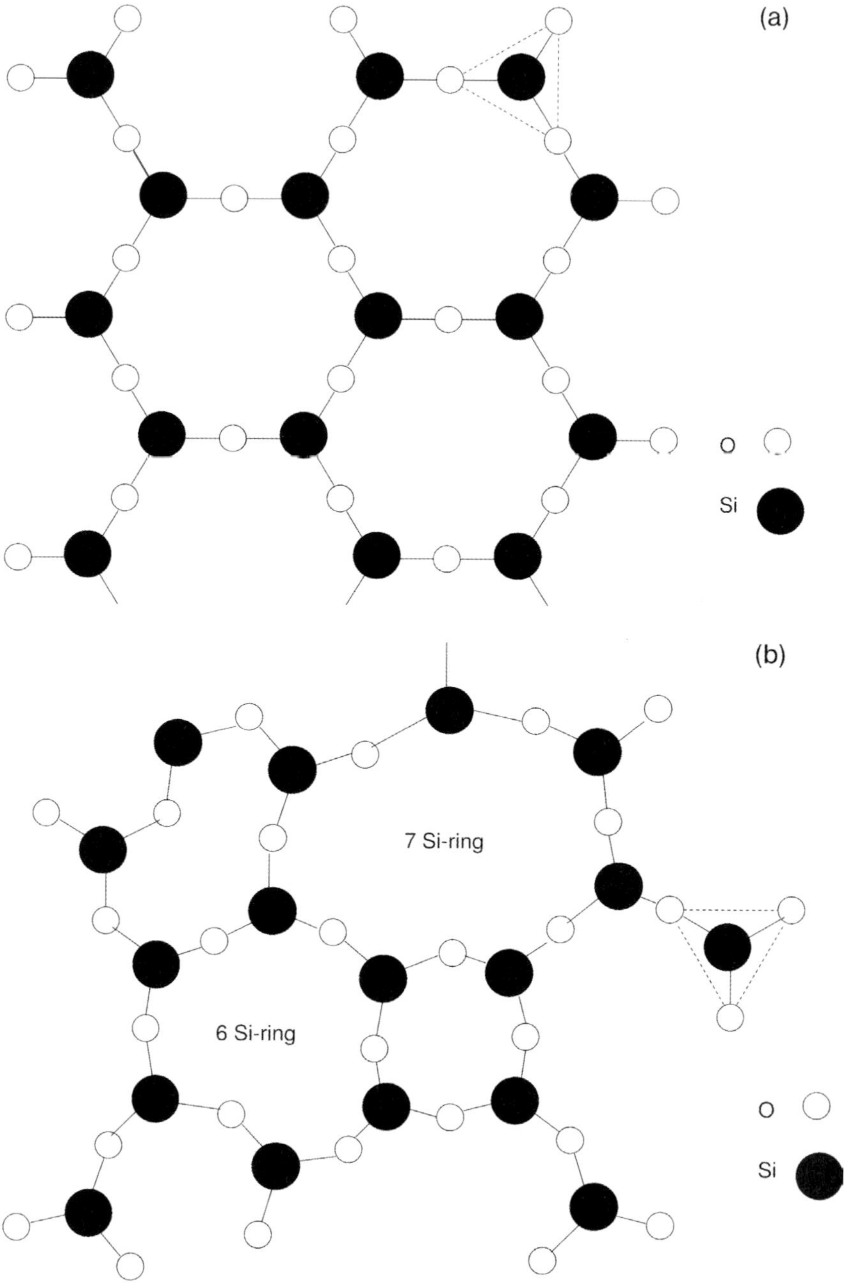

Fig. 5.3 2D projections of structures showing the path from crystalline to glassy structure: (a) crystalline form of SiO_2; (b) glassy form of SiO_2; (c) glassy form of SiO_2 modified by Na^+ ions. Note the presence of one NBO per Na^+. The projection of a tetrahedron is shown in each case (dashed lines).

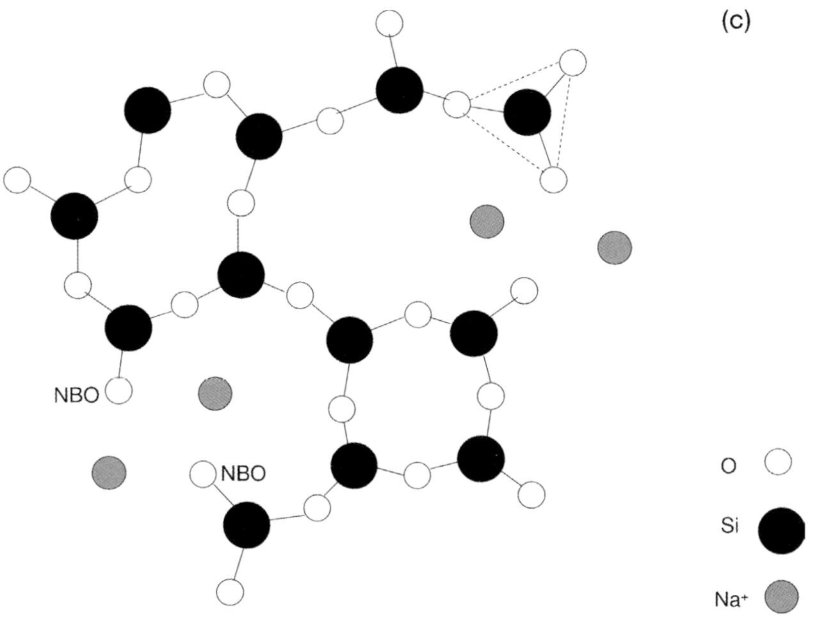

Fig. 5.3 *(Continued)*

5.2.2
Glass Network Modification

In Fig. 5.3c the CNR is shown this time when sodium oxide is added to the structure (or more generally an alkaline or an alkaline earth oxide). Na^+ is called a modifier since when added to the glass it modifies the Si–O network. Na^+ ions are charged with an elemental positive charge $+e$ and each of them are compensated by the formation of one non-bridging oxygen (NBO) atom that holds a negative charge $-e$. Therefore, in a glass containing modifying ions, NBO atoms appear (as opposed to bridging oxygen (BO) atoms which form the network). Figure 5.3c shows additionally that two Na^+ ions are necessary for compensating the charges generated by the rupture of one O–O bond (one Na^+ ion compensates one NBO). The two tetrahedra still interact electrostatically. In fact, electrostatic interactions exist between the two NBO atoms and the two Na^+ ions. As revealed experimentally however, glass network modification induces a viscosity decrease or a greater facility for viscous flow. CRN model accounts quite well for the observed change in properties as the glass composition is modified. Modifiers disrupt the network and are used in fact to lower the viscosity of the glass melt and hence to facilitate its fabrication at lower temperatures (Chapter 6). In fact, silica glass is difficult to process and instead of standard melting procedure (Chapter 10) it is produced following ceramic routes. The melting temperature of silica is as high as 1713 °C corresponding to cristobalite–liquid equilibrium. Instead, adding 25 % of soda to silica lowers the liquidus temperature to only 793 °C (Jones, 1956). Alkali metal ions like Na^+ are mobile and allow for electrical conductivity, ion migration. In contrast, alkaline earth ions like Ca^{2+} ions (one alkaline earth ion is compensated electrically by two NBOs) are relatively immobile and can hinder the diffusion of other ions, in particular alkali, and hence improve the chemical resistance of the glass (see also Section 5.8). It is then clear why most important commercial glass, SLS glass, is based on a mixture of silica (network former), sodium and calcium oxides (alkaline and alkaline earth modifiers respectively).

As expected then, the relative concentration of bridging and non-bridging oxygen atoms has an important influence on the properties of glasses (Kurkjian and Prindle, 1998; Dériano, 2002). Let us call BON the number of bridging oxygen atoms in a tetrahedral element; we can anticipate the following rules:

i. For BON = 4, that is [O]/[Si] = 2 since each BO is shared by two silicon atoms, the network is a three-dimensional one with all four corners bridging.
ii. For BON = 3, [O]/[Si] = 2.5, the network is two-dimensional with three corners bridging (note that some tetrahedra may be linked to four others and some therefore to less than three, the said number being the average value over the network).
iii. For BON = 2, [O]/[Si] = 3, the network is formed by one-dimensional chains with one corner bridging.
iv. For BON < 2, [O]/[Si] > 3, the network is composed of individual SiO_4 tetrahedral elements, some of these being bound together.

Using Q^x terminology with x the number of bridging oxygen atoms on a tetrahedron, we refer to (i) Q^4, (ii) Q^3, (iii) Q^2 and (iv) Q^1, Q^0 structures respectively. The stiffness of the network decreases gradually by replacing the bridging oxygen atoms by non bridging ones until only individual isolated tetrahedra remain.

Consider a modified oxide glass of general composition $(A_2O)_x(SiO_2)_{1-x}$ with A the alkali metal. The concentration of BO and NBO varies with x as shown in Fig. 5.4[1]. For $x = 0$, that is for pure silica, only BO exist (Q^4), the relative concentration being then 2, while increasing alkali concentration yields a dramatic decrease of BO and an increase of NBO accordingly.

Dietzel (1942) proposed classifying the oxides according to their ability to form or modify the glass network in view of their cation field. He defined the cation field intensity A as a charge-to-distance squared ratio:

$$A = \frac{Z}{(r_c + r_o)^2} \tag{5.1}$$

where Z is the cation valence and r_c and r_o are the cation and oxygen radii respectively (Appendix 2). The Dietzel field amplitude varies within a relatively broad range of 1.2 to 0.1 (Table 5.1) and is indicative of the magnitude of the electrostatic attractive force

[1] Introducing x A_2O yields $2x$ NBO. The total number of oxygens is $x + 2(1 - x)$. Hence introducing xA_2O leaves $x+2(1-x)-2x=(2-3x)$ NBO. $[NBO]/[Si]=2x/(1-x)$. $[BO]/[Si]=(2-3x)/(1-x)$.

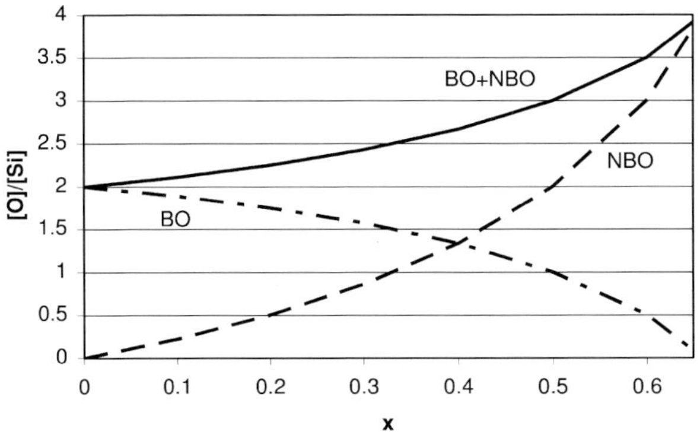

Fig. 5.4 Relative concentrations of bridging oxygen (BO) and non-bridging oxygen NBO as a function of x in a modified silica glass of general formula $(A_2O)_x(SiO_2)_{1-x}$.

between the considered element and oxygen (Appendix 2). The ordered sequence of oxides is relatively convincing, since for $A > 1$ we find the oxides that form the glass network while for $A < 0.35$ the oxides (alkali in particular) are known to modify the network. Between these two extreme behaviours, intermediate oxides may show both behaviours according to the glass composition (Table 5.1). Dietzel's approach suggests that both smallness and high charge of the cation favour the formation of glassy oxides. Too much importance must not be attached to one criterion or another (for instance single bond strength to heat of dissociation ratio). These are mostly empirical and break down when structures are unusual, as in As_2O_3 and PbO for instance (Jones, 1956). Also the situation is very complex when weak VDW bonds coexist with strong covalent bonds as in chalcogenide glasses (Section 5.4).

Besides decreasing the viscosity, modifiers allow for the liquidus (or melting) temperature to decrease to a more viscous domain and hence favour glass formation

Tab. 5.1 Dietzel field intensity and role of cations in glass ($r_0 = 1.4$ Å).

Role	Cation	Z	r_c (Å)	A
Network formers	Si	4	0.4	1.23
	B	3	0.25	1.10
	Ge	4	0.53	1.07
Intermediates	Ti	4	0.74	0.87
	Al	3	0.53	0.80
	Zr	4	0.86	0.78
	Be	2	0.41	0.61
	Mg	2	0.86	0.39
	Zn	2	0.88	0.38
Modifiers	Ca	2	1.14	0.31
	Pb	2	1.33	0.27
	Li	1	0.90	0.19
	Na	1	1.16	0.15
	K	1	1.52	0.11

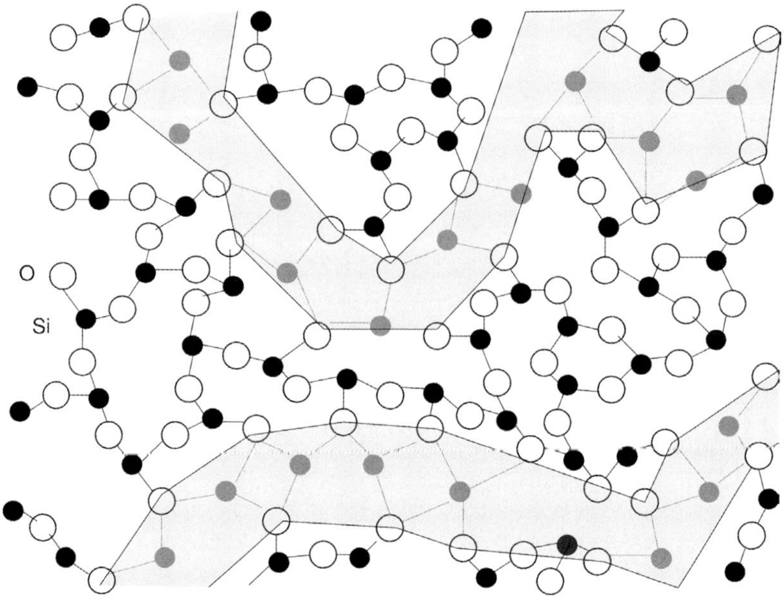

Fig. 5.5 Modified random network as proposed by Greaves (1985) showing Si-rich zones connected by modifier-rich percolating channels marked in grey.

(Barton and Guillemet, 2005). Alkali metal ions (Li, Na, K, valence 1) are more efficient than alkaline earth ones (Ca) for inducing both effects. Pb, while being classified as a modifier, allows fabrication of glass associated with only 8 wt% of silica. This supposes Pb atoms are forming the network while bridging two oxygen atoms. This question is still open and under debate. Alumina modifies the silica network (decrease of the viscosity); when introduced with alkali metal, Al may participate in network formation with a related increase of the viscosity.

More recently Greaves (1985) introduced a modified random network where modifiers form zones connecting silica-rich volumes as shown for a modified silica glass in Fig. 5.5. This model offers interesting predictions and suggestions as regards the plastic deformation of silicate glasses. It can be anticipated that such glasses would preferentially flow through modifier-rich zones (see also Chapter 8 and Appendix 10).

5.2.3
Short-Range Order

Short-range order can be investigated using techniques like X-ray and neutron diffusion, some results being discussed in more detail below. X-ray diffraction was among the first techniques allowing for detailed insight into glass structure. It remains a tool of utmost importance allowing for the detection of residual crystallites. Mathematical analysis of the glass response allows extraction of the radial atomic distribution of a given atom as detailed below. Neutron diffraction gives the same information as X-ray diffraction but allows for

Tab. 5.2 Experimental tools and associated information.

Experimental tool	Information
X-ray or neutron diffraction	Radial distribution and/or crystal phase
IR and Raman spectroscopies, MNR	Local symmetry, coordination
EXAFS	Local symmetry, radial distribution, coordination
X-ray photoelectron spectroscopy (XPS)	Surface arrangement (Section 5.8)
X-ray reflectometry	Surface arrangement (Section 5.8)

penetrating deeper into specimens. Neutron diffraction requires a nuclear reactor. More recently, extended X-ray absorption edge fine structure (EXAFS) has been extensively used to extract coordination numbers and interatomic spacings. The power of this technique is that by selecting the X-ray energy, the structure around a single kind of atom can be explored rather than getting composite information involving all kinds of atoms as for X-ray diffraction. Also, infrared (IR) and Raman spectroscopies (Chapter 8), magnetic nuclear resonance (MNR, including 'magic angle spinning' improvements), paramagnetic electronic resonance (PER) and Mössbauer spectroscopy (Zarzycki, 1982; Williams et al., 1997) are employed to get further insights into local symmetries, coordinations and ion environments. The last two techniques listed in Table 5.2 are very important as regards surfaces. In fact, a glass surface is transformed by water and alkali diffusions depending on the environment and the atmosphere. As a result, glass surfaces may show very different structures and properties as compared to the glass bulk. With the nanotechnology breakthrough, the surface structure and its related properties become of utmost importance and have attracted much scientific and technological interest.

Let us consider X-ray or neutron diffraction. The results of such experiments are usefully presented in terms of a radial distribution where we select an atom at random and consider a series of concentric spheres around it (Fig. 5.6).

From the experimental spectra, the radial distribution (RD) can be extracted in the form (Fig. 5.7)

$$\mathrm{RD}(r)\mathrm{d}r = 4\pi r^2 \rho(r)\mathrm{d}r \tag{5.2}$$

where $\rho(r)$ is the density at distance r. $\mathrm{RD}(r)$ corresponds to the probability of finding a neighbouring atom at the distance r from another atom the position of which is

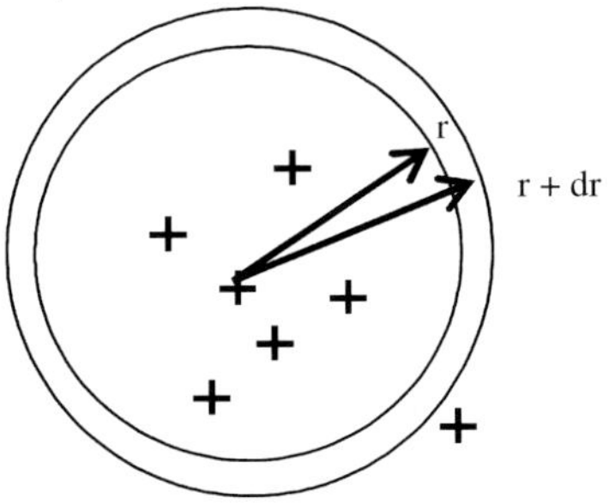

Fig. 5.6 Schematic glass structure centred on a given atom.

considered arbitrarily as the reference ($r = 0$). For a crystal (Chapter 2), the radial distribution would exhibit sharp peaks at periodic distances. Instead in a glass one observes: (i) at small distances diffuse maxima with expected deviation in bond angles and bond lengths, the area under the peak being indicative of the coordination and (ii) at long distances ($r \rightarrow \infty$) RD(r) converges to the mean distribution, that is, $4\pi r^2\rho_o$, where ρ_o is the average density of the considered glass[2]. This is indeed observed in Fig. 5.7 since glass RD converges towards a parabolic evolution at long distance with less and less characteristic features being observed. At short distance, RD peaks are observed corresponding to the short-range order of the glass. For instance, in silica (Fig. 5.7a) we observe that silicon atoms are bonded to oxygen atoms at a distance of about 0.16 nm while the distance between two silicon atoms (pertaining to two different tetrahedra) is about 0.32 nm, that is, about twice the previous value, indicating that Si–O–Si lines are not very far from collinear. In between, the O–O band is detected at about $\sqrt{8/3}$ times the Si–O distance, as expected in a tetrahedral arrangement. After about five interatomic distances, order almost vanishes, confirming the view that no long-range order prevails in a silica glass.

The radial distribution of soda-silica glass shows similar information for Si–O and O–O distances being almost unaffected (Fig. 5.7b). A new peak appears corresponding to the Na–O distance (Jones, 1956). This suggests that the Na ions occupy the free space left in the silica open structure. This view is supported by the fact that the density of silica increases when soda is incorporated, while gas permeability decreases (at least by 5 decades in soda-rich glasses; Jones, 1956; see also Appendix 13). Obviously, we do not suggest that Na^+ ions simply fill network holes, for at the same time we must consider a change in the number of silicon and oxygen atoms.

The short-range order of the glass will control the density or the atomic packing density C_g of the glass that has an important role in glass properties (elastic modulus, Poisson ratio, glass flow). As for a crystalline structure the atomic packing density C_g is defined as the ratio between the minimum theoretical volume occupied by the ions and the corresponding effective volume of the glass. For a glass of composition A_aB_b, C_g is written

$$C_g = \frac{av_a + bv_b}{\left(a\frac{M_A}{N} + b\frac{M_B}{N}\right)/\rho} = \rho N \frac{av_a + bv_b}{aM_A + bM_B} \tag{5.3a}$$

Here ρ is the glass density, N the Avogadro number, M_i the molar mass and v_i the theoretical ion volume:

$$v_i = \frac{4}{3}\pi r_i^3 \tag{5.3b}$$

where r_i is the ion radius (Appendix 2).

[2] For a gas of non-interacting point particles (structureless) the curve would be a parabola with its vertical axis and its apex at the origin. The probability of finding a second atom at a given distance from a first one is dependent only on the volume available, and therefore varies with the square of the distance. For a perfectly ordered crystal, it consists of a number of vertical lines of infinite height, one at each of the possible interatomic distances (depending on the structure type). A hump above the parabola is indicative of a more than random probability of finding a second atom within a given range of distances from the chosen atom that is a degree of order.

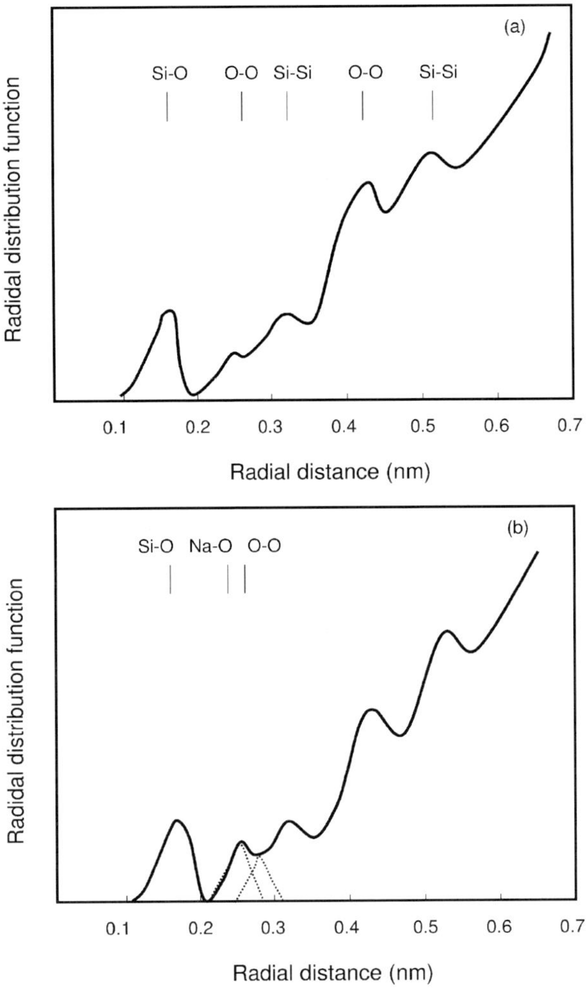

Fig. 5.7 Radial distribution curves of (a) vitreous silica and (b) Na_2O–SiO_2 extracted from X-ray diffraction data (according to McMillan, 1979).

Silica glass packing density C_g is 0.45 and is lower than that of standard float glass ($C_g = 0.52$; Rouxel, 2006) both being below that of a crystalline structure where the highest atomic packing density is obtained for a face centred cubic (FCC) structure and is 0.74. As discussed below, metallic glasses show the highest atomic packing density greater than 0.7. The Poisson ratio increases accordingly from 0.2 for silica glass to 0.4 for metallic glasses (Appendix 9). Hence, silica and related glasses show excess volumes that have important consequence on the properties (transport, mechanical; see Chapter 6). Metallic glasses as compared to other glasses show the lowest excess volumes.

5.3
Borate Glass and Related Glasses

Boron (B) is in group III of the Mendeleev table (Appendix 2) and is a network former with strong Dietzel field (Table 5.1). It tends to bond to three different oxygen ions adopting a triangular coordination. Tetrahedral coordination is also found. As regards the three coordinated B atoms, it has been suggested that these form boroxyl groups B_3O_6. These groups are planar and linked to form a three-dimensional network by boron–oxygen–boron bonds (Fig. 5.8). Such a structure is in good agreement with X-ray diffraction results (Doremus, 1994). Such a structure is then much different from that of silica. When adding alkali metals to borates the induced changes are opposite to those generated in silica; this is called in the literature the 'boron oxide anomaly'. In fact, the viscosity may increase and the thermal expansion decrease when adding alkali metal, contrary to what is generally observed in silica-based glasses. It is believed that such a trend could be related to a change in the coordination of boron atoms.

Borosilicate glasses play an important role in glass manufacturing since they are corrosion resistant to water and allow for low thermal expansion. They are composed of Si and B atoms which form the network, Si being four coordinated and B three coordinated. The resulting network is softer than silica with lower viscosity and lower melting temperature. Contrary to what happens when modifiers are introduced in silica, the bond potential remains symmetrical in borosilicates inducing low variation in the mean distances with increasing temperature, that is, a low thermal expansion (see also Appendix 3). In borosilicate glasses, alkali metal ions like Na^+ ions will either modify the network introducing NBOs or change the boron coordination (from 3 to 4). The latter phenomenon is believed to be responsible for the 'boron oxide anomaly' where viscosity may increase with increasing alkali metal concentration as stated for borate glasses.

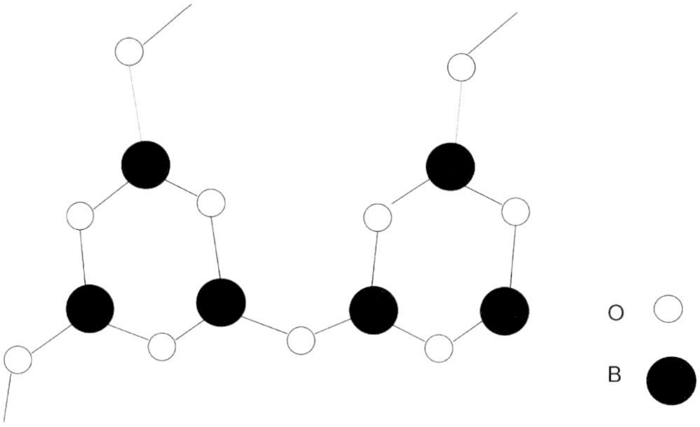

Fig. 5.8 Boroxyl groups sharing a common oxygen atom.

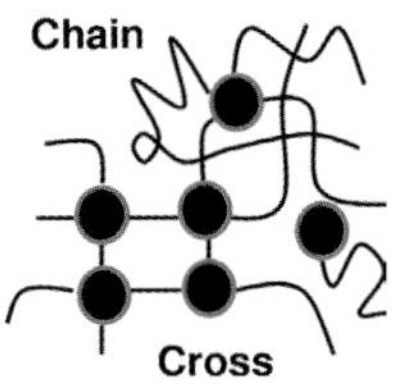

Fig. 5.9 Schematic structure of organic or chalcogenide glasses with chains and cross-linking.

5.4 Organic and Chalcogenide Glasses

Organic glasses are used to produce optical lenses and to protect glass articles (cold-end coatings, fibre sizing; see also Chapters 4 and 12). They are formed by organic chains either cross-linked or not (Fig. 5.9). In thermoplastics only weak bonds (VDW) are formed between chains. In thermosets, cross-linking is obtained when covalent bonds form between the organic chains. The simplest polymer known as polyethylene is formed by long chains made of $-CH_2-$ moieties. The length of the carbon–carbon chain depends on the way the polymer is fabricated. Moreover, side group functions can be added to the carbon chains. When cooled from a melt, the polymer will generally present a mixture of very small crystalline regions embedded in a glassy matrix. The crystalline regions consist of oriented chains. Only when quenching is rapid enough, the chains cannot rearrange and a glassy phase is obtained. The chains form a glass three-dimensional network with strong (covalent) and/or weak (VDW) bonding between chains with related properties. Strong cross-links result in higher viscosity regime for instance.

Interestingly, chalcogenide glasses can be described in analogy to organic glasses, the term polymerization being sometimes employed in the glass context as well. A chalcogenide glass is formed from group VI and IV atoms (GeSe for instance; Chapter 2, Appendix 2). Group VI atoms form long chains (Se chains for instance) interacting through VDW bonds and cross-linking due to the covalent bonding between the group IV atoms (Ge for instance). Note that Ge can bond four different chains (Guin et al., 2002) inducing strong network stiffness.

5.5 Metallic Glasses

Metallic glasses have been fabricated only recently. Duwez's team pioneered in the 1960s elevated cooling rates ($10^6\ K\,s^{-1}$) to fabricate binary glass compositions like AuSi glass (Klement et al., 1960). More recently more complex alloys of 4 to 6 different elements have been produced at much lower cooling rates ($1\ K\,s^{-1}$) offering the possibility of producing larger specimens (so called bulk metallic glasses (BMGs); Johnson, 1996; Inoue, 2000). As a matter of fact, the actual structure of the alloys made with so many elements is complex (Fig. 5.10) as reviewed by Wang et al. (2004) and is still debated (Sheng et al., 2006). These glasses did not yield a technological breakthrough in terms of technological applications, even though they present more

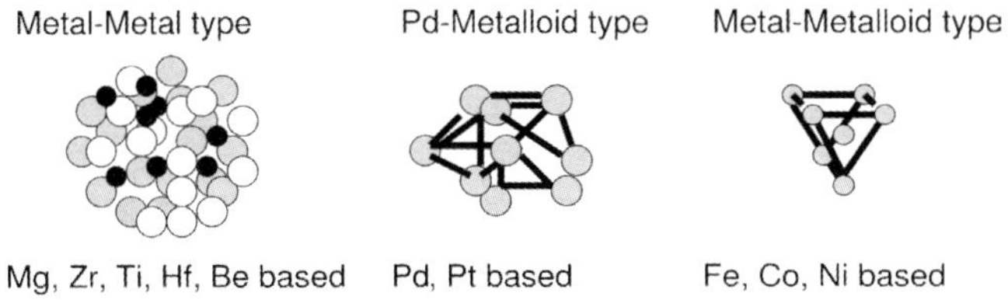

Fig. 5.10 Different atomic configurations of three types of BMGs (according to Inoue and Takeuchi 2002).

elevated elastic limit and corrosion resistance than crystalline metals. They have played an important role in the past decades in the study and understanding of glass mechanical properties since they can be plastically deformed as bulk (under compression) well below their glass transition (Appendix 10). This has allowed great insight into plasticity and flow in glass (Chapters 6 and 8 and Appendix 8).

BMGs were found to have a new type of glassy structure with a high degree of dense randomly packed atomic configurations with much smaller excess volume and higher packing density ($C_g > 0.7$) than in oxide glasses. Density measurements show that the density difference between BMG and the fully crystallized state is in the range 0.3–1.0% which is much smaller than the reported range of about 2% for ordinary amorphous alloys (Wang et al., 2004). This indicates that BMGs have higher density randomly packed atomic configurations. Inoue classified the BMGs into three types: metal–metal type, metal–metalloid type and Pd–metalloid type alloys. The configurations are different according to the three types of BMGs shown in Fig. 5.10.

Metal–metal alloys present icosahedral local order which after annealing may yield icosahedral quasicrystalline phase. Some authors have questioned the stability of these glasses during a stay at elevated temperatures when observing superplasticity (Nieh et al., 2002; Appendix 10). Metal–metalloid-type glassy alloys present network atomic configurations consisting of trigonal prisms which are connected with each other through 'glue' atoms comprising Zr, Nb, Ta or lanthanide metal. Pd-based BMGs consist of two large clustered units of a trigonal prism capped with three half-octahedra for the Pd–Ni–P and a tetragonal dodecahedron for the Pd–Cu–P region, as shown in Fig. 5.10.

5.6 Avoiding Crystallization

5.6.1 Nucleation and Growth of Crystallized Phases

As mentioned above, glass forms when a liquid is cooled down in such a way that 'freezing' happens instead of crystallization. In fact, the cooling rate must be large enough so that no appreciable amount of crystalline material is formed. It cannot be excluded that some glasses have crystalline regions. This issue is much debated today for metallic glasses (Appendix 10). The detection of crystalline phases is dependent on the

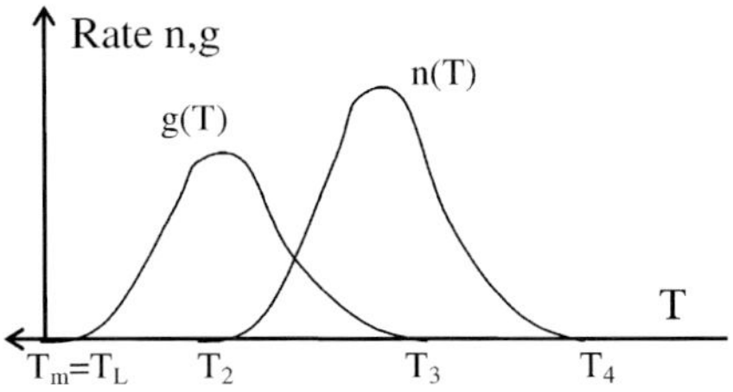

Fig. 5.11 Nucleation n and growth g rates as a function of temperature; note the qualitative graph with arbitrary units (differing for n and g: $n(T)$ is in $m^{-3}\,s^{-1}$, $g(T)$ in $m\,s^{-1}$).

tool employed (optical observation, electron microscopy, X-ray diffraction). If some crystals grow in the interior of a glass, the optical properties of the glass for instance will be changed. Light scattering will deleteriously affect the long-distance transmission of an optical fibre for instance. Figure 5.11 shows the nucleation and growth rates which both depend on the temperature. Growth (g) starts at the melting or liquidus temperature T_L. Nucleation (n) starts at a lower temperature (T_2) resulting from the energy barrier w_c that we calculate below. There is no barrier for growth, so it occurs at any undercooling. Hence, crystalline phases may appear in the critical zone situated between T_2 and T_3. Therefore, glass formation depends on the overlapping of growth (g) and nucleation (n) curves and on their respective kinetics as compared to thermal history (Zarzycki, 1982; Ashby and Jones, 1991). As will be illustrated below the viscous regime at these temperatures plays a very important role and the cooling rate amplitude will determine whether crystallites form or not. Modifiers allow for the liquidus temperature to decrease to more viscous domains hence favouring glass formation. The curves shown qualitatively in Fig. 5.11 can be described using a simple analytical approach described below. We shall consider a very simple glass constituted by only one type of atoms. Modelling becomes very difficult in a more realistic case (Fokin et al., 2006). The methods allowing for reliable measurements of nucleation rates are also not straightforward (Doremus, 1994). The physics captured here will help obtain an insight into glass formation.

5.6.2
Nucleation of Crystallized Phases

Crystallite nucleation is controlled by thermodynamics.

i. When atoms aggregate, a new surface is created between the emerging crystals and the embedding liquid requiring energy, while phase transformation from liquid to crystal provides energy to the process. We shall define w_c as the energy necessary to form a particle that further develops into a crystallite. Below the critical energy w_c, atoms aggregate and later separate. We shall develop in more detail below the expression of this energy barrier in simple configurations obtained for a monatomic melt in either homogeneous or heterogeneous nucleation regime.

ii. The former phenomenon (i) will take place if particles can reach the nucleus passing through the diffusion barrier $\Delta g'$.

Correspondingly, nucleation rate can be expressed as a function of two terms showing energy barrier w_c and diffusion barrier $\Delta g'$ respectively:

$$n(T) = n\nu_E \exp\left(-\frac{w_c}{kT}\right)\exp\left(-\frac{\Delta g'}{kT}\right) = n\nu_E \exp\left(-\frac{Nw_c}{RT}\right)\exp\left(-\frac{\Delta G'}{RT}\right) \quad (5.4)$$

where n is the concentration of atoms, ν_E the atomic transition frequency in the liquid (order of magnitude $10^{13}\,s^{-1}$), T the temperature and N the Avogadro number. The first term $\exp(-Nw_c/RT)$ regards the thermal activation related to the energy barrier w_c for a particle to form as stated above. The second term $\exp(-\Delta G'/RT)$ is the diffusion contribution. In the case of a unique atom type, the vicinity of the particle always presents structural units to grow the particle. Therefore, no diffusion on long distances is necessary and the diffusion barrier is that appearing for the viscous flow in the melt ($\Delta G' = \Delta G$; Chapter 6).

We shall consider now two nucleation configurations to determine w_c. The first situation is when nucleation happens homogeneously within the liquid while the second situation (the most realistic one) is when nucleation happens at heterogeneities (at impurities or surfaces; Fig. 5.12).

5.6.2.1 **Homogeneous Nucleation**

Considering first the case of homogeneous nucleation, the energy of a particle w is the sum of two contributions:

i. the liquid to crystal formation w_f
ii. the interface energy w_i.

These two terms can be expressed as

$$w_f = \frac{V}{V_m}\Delta G_v = \frac{4\pi r^3}{3V_m}\Delta G_v \quad (5.5a)$$

where r is the particle radius (assumed to be spherical), ΔG_v the molar enthalpy of the transformation from liquid to crystal and V_m the molar volume of the crystal, and

$$w_i = 4\pi r^2 \zeta_{sl} \quad (5.5b)$$

where ζ_{sl} is the surface energy (at the interface between the solid s and the liquid l).

Figure 5.12 plots $w = w_f + w_i$ as a function of particle radius r, and shows a maximum value for a particle radius of

$$r_c = 2\zeta_{sl} V_m/\Delta G_v \quad (5.6a)$$

The maximum energy is then determined to be

$$w_c = \frac{16\pi\zeta_{sl}^3 V_m^2}{3\Delta G_v^2} \quad (5.6b)$$

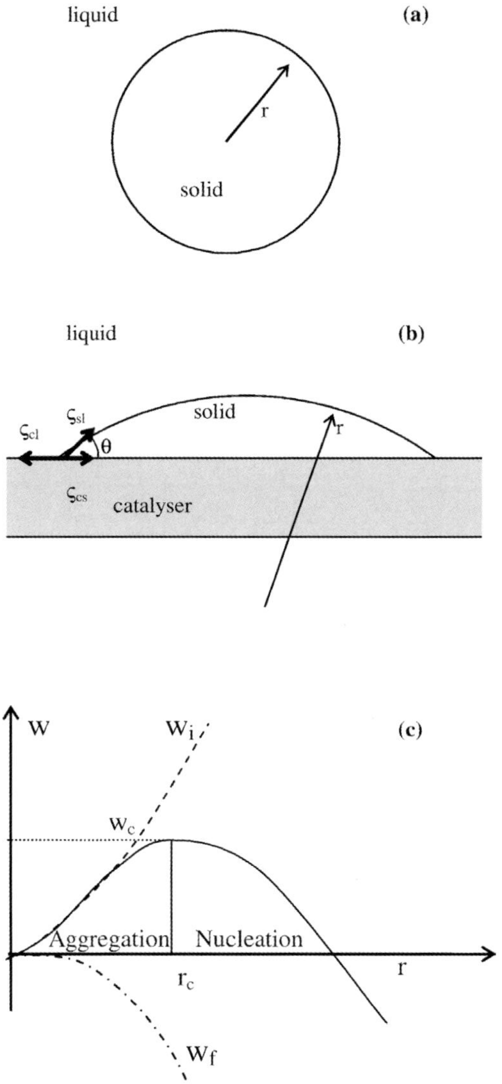

Fig. 5.12 Schematic (a) homogeneous and (b) heterogeneous nucleation representations. (c) $w(r)$ plot showing r_c, w_c maximum.

This maximum in energy w_c sets a limit on size r_c. When an aggregate reaches this critical size, it will develop further. Instead, below this critical size the atoms aggregate and further separate. Let us consider now the Stokes–Einstein relation to express the diffusion term. The diffusion coefficient is written

$$D = \frac{kT}{3\pi a_o \eta} = \nu_E a_o^2 \exp\left(-\frac{\Delta G}{RT}\right) \tag{5.7}$$

where a_0 is the distance of the atomic jump (taken as twice the particle radius) and η the viscosity. Then, we get for the nucleation rate

$$n(T) = n\nu_E \exp\left(-\frac{\Delta G}{RT}\right)\exp\left(-\frac{Nw_c}{RT}\right) = n\frac{D}{a_0^2}\exp\left(-\frac{Nw_c}{RT}\right)$$
$$= n\frac{kT}{3\pi a_0^3\eta}\exp\left(-\frac{Nw_c}{RT}\right) \tag{5.8}$$

We observe that there is balance between the contribution of w_c decrease (ΔG_v increases with decreasing temperature, it is zero at T_L) and viscosity increase when temperature decreases. This corresponds qualitatively to the plot shown in Fig. 5.11 with an observed maximum between T_2 and T_4.

5.6.2.2 Heterogeneous Nucleation

We consider now a more realistic situation where heterogeneous nucleation originates from solid impurities in the glass melt. Heterogeneous nucleation is to be prevented on forming a glass article. Instead, vitroceramics are fabricated introducing metallic impurities in the glass melt to induce heteronucleation as detailed in Section 5.7. In such a case, nucleation may happen much more easily as we will show below. We shall refer to impurities that help crystallization as catalysers. Let us calculate the energy barrier w_c.

In the case of heterogeneous nucleation three interfaces have to be considered instead of one. The mechanical equilibrium of a particle on the catalyser is given by the Young Dupré equation (Fig. 5.12; see also Chapter 12):

$$\zeta_{cl} = \zeta_{cs} + \zeta_{sl}\cos\theta \tag{5.9}$$

The surface of the particle is the sum of its surface in contact with the liquid S_{sl} and its surface in contact with the catalyser S_{cs}:

$$S = S_{sl} + S_{cs} \tag{5.10}$$

with

$$S_{sl} = \int_0^\theta 2\pi r \sin\theta' r\, d\theta' = 2\pi r^2(1-\cos\theta) \tag{5.11}$$

$$S_{cs} = \pi r^2 \sin^2\theta = \pi r^2(1-\cos^2\theta) \tag{5.12}$$

while its volume is

$$V = \int_0^\theta \pi(r\sin\theta')^2 r\sin\theta'\, d\theta' = 2/3\pi r^3(1 - 3/2\cos\theta + 1/2\cos^3\theta) \tag{5.13}$$

Again the energy of the particle is the sum of two contributions (volume w_f and surface w_i):

$$w_f = V/V_m \Delta G_v = 2/3\pi r^3 (1 - 3/2\cos\theta + 1/2\cos^3\theta)\Delta G_v/V_m \quad (5.14)$$

and

$$w_i = \zeta_{sl} S_{sl} + \zeta_{cs} S_{cs} - \zeta_{cl} S_{cs} = \zeta_{sl} S_{sl} + (\zeta_{cs} - \zeta_{cl}) S_{cs} \quad (5.15a)$$

Substituting the mechanical equilibrium equation allows one to write

$$\begin{aligned} w_i &= \zeta_{sl} S_{sl} - (\zeta_{sl}\cos\theta) S_{cs} = \zeta_{sl}\pi r^2 [2(1 - \cos\theta) - (1 - \cos^2\theta)\cos\theta] \\ &= \zeta_{sl}\pi r^2 (2 - 3\cos\theta + \cos^3\theta) \end{aligned} \quad (5.15b)$$

It follows that

$$w = w_i + w_f = (1 - 3/2\cos\theta + 1/2\cos^3\theta)\{2/3\pi r^3 \Delta G_v/V_m + 2\zeta_{sl}\pi r^2\} \quad (5.16)$$

Therefore, the maximum of the function is found for the same critical radius as for homogeneous nucleation:

$$r_c = 2\zeta_{sl} V_m/\Delta G_v \quad (5.17)$$

while

$$\begin{aligned} w_{c\,het} &= (1 - 3/2\cos\theta + 1/2\cos^3\theta) w_{c\,hom}/2 \\ &= (1/2 - 3/4\cos\theta + 1/4\cos^3\theta) w_{c\,hom} \end{aligned} \quad (5.18)$$

The ratio $w_{c\,het}/w_{c\,hom}$ is plotted as a function of angle θ in Fig. 5.13 and this figure illustrates why heterogeneous nucleation has to be carefully considered since even for wetting angles as high as 30° the $w_{c\,het}/w_{c\,hom}$ ratio is less than 1 %.

Therefore, 'dangerous' catalysers are such that they allow wetting of the solid ($\theta \to 0$) and a related decrease of the energy barrier enhancing heterogeneous nucleation. It should be noted that for a given volume, heterogeneous nucleation leads to much larger particle radius since the radius ratio is

$$r_{het}/r_{hom} = [(1 - 3/2\cos\theta + 1/2\cos^3\theta)/2]^{-1/3} \quad (5.19)$$

5.6.3
Crystal Growth

The nuclei being formed, they will then grow. Growth rate g can be calculated assuming a thermally activated transfer of atoms from the liquid phase to the solid

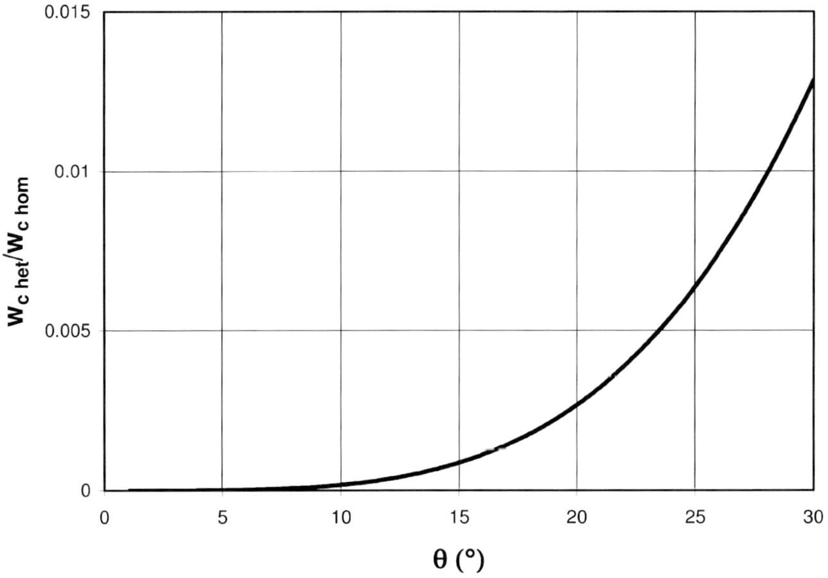

Fig. 5.13 Plot of $w_{c\ het}/w_{c\ hom}$ as a function of angle θ.

phase and from the solid phase to the liquid phase (Fig. 5.14):

$$g(T) = fa_c(\nu_{ls} - \nu_{sl}) \tag{5.20}$$

where f is the fraction of surface site ($0 < f < 1$), a_c the solid–liquid interface thickness and ν_{ls} and ν_{sl} the frequencies for respective transfers (Fig. 5.14).

We have

$$\nu_{ls} = \nu_E \exp\left(-\frac{\Delta G''}{RT}\right); \quad \nu_{sl} = \nu_E \exp\left(-\frac{\Delta G_v + \Delta G''}{RT}\right) \tag{5.21}$$

with ν_E the atomic transition frequency in the liquid, T the temperature and $\Delta G''$ and $\Delta G_v + \Delta G''$ the respective barriers. ΔG_v is the molar enthalpy of liquid–crystal transformation. One can write

$$g(T) = fa_c\, \nu_E \exp\left(-\frac{\Delta G''}{RT}\right)\left[1 - \exp\left(-\frac{\Delta G_v}{RT}\right)\right] \tag{5.22}$$

The interface transport can be described by a diffusion coefficient. In the general case we do not know the energy barrier for diffusion through the interface. Long-distance diffusion is to be considered. As a first approximation and since only one type of atom is considered here, we can suppose the diffusion barrier is the same

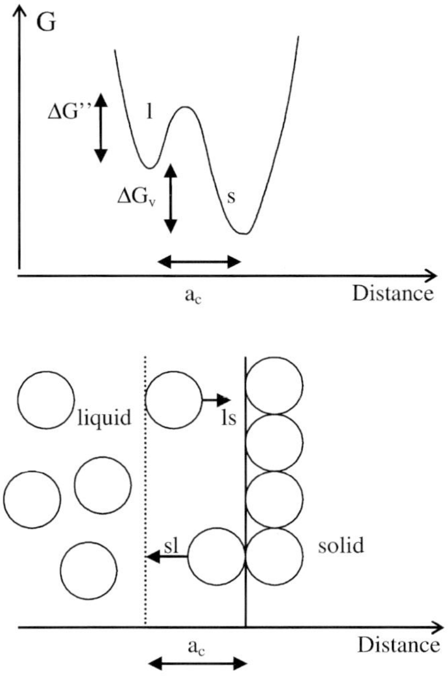

Fig. 5.14 Liquid–solid interface modelling. Note we consider only one type of atom and hence the same composition for the liquid and the solid. This is rarely the case (l, liquid; s, solid).

as that appearing in viscous flow, $\Delta G'' = \Delta G$, and take again the Stokes–Einstein relation

$$D = \frac{kT}{3\pi a_o \eta} = \nu_E a_o^2 \exp\left(-\frac{\Delta G}{RT}\right)$$

and express the growth rate supposing that $a_c = a_o$ (twice the particle radius) as

$$g(T) = f\frac{kT}{a_o^2 3\pi\eta}\left[1 - \exp\left(-\frac{\Delta G_v}{RT}\right)\right] \tag{5.23}$$

Note that at $T = T_L$, ΔG_v is null so that $g = 0$, while ΔG_v amplitude increases when temperature decreases. However, viscosity increases as well and progressively balances the ΔG_v contribution. This also describes well qualitatively the growth (g) curve shown in Fig. 5.12 with a maximum being observed between T_L and T_3. The role of viscosity in inhibiting nucleation, growth and hence crystallization should be emphasized. This result applies to general glass compositions. In fact, glass formation is made possible using compositions allowing for a viscous regime around the liquidus temperature. In Chapter 6 we shall further discuss the importance of this property when glass is to be frozen in. Regarding crystallization, it is well known that adding alkali to silica allows a decrease in viscosity. As a matter of fact, at very high content of alkali, say that corresponding to the composition of orthosilicate, glasses are never obtained since the rates of nucleation and crystallization become too high. Also, devitrification will depend on the viscosity regime (Section 5.6.5). As mentioned at the beginning of this section, once glass is not composed of a single atom type (which indeed is unrealistic) crystallization becomes complex to model (Fokin et al., 2006). In general the crystalline phase presents a different composition from the original glass, so that glass composition in the surrounding of the growing crystalline phases changes with implication for the viscosity regime. Also, a depleted zone may appear for the elements present at higher stoichiometry in the crystalline phases.

5.6.4
TTT Diagram

The simple monatomic models presented above allow an insight into glass formation even though we made rough approximations. The practical need however is to form a realistic glass without detectable crystalline phases. In this regard temperature–time–transformation (TTT) diagrams have to be elaborated theoretically and experimentally for more complex glass compositions (Fig. 5.15). The TTT diagram shows in a temperature–time plot the crystalline and glass domains.

When forming a glass, the cooling route must, in no case, pass through the crystalline domain (Fig. 5.15). Then the critical cooling rate allowing for the production of glass is that tangent to the 'nose' (T_n, t_n) of the limiting front:

$$\frac{dT}{dt_c} = \frac{T_m - T_n}{t_n} \tag{5.24}$$

The critical cooling rate is 10^{-2}–10^{-3} K s^{-1} for silica glass while pure metals require cooling rates as high as 10^6 K s^{-1} (Fig. 5.16). Some metallic alloys (BMGs) can now be formed as bulk glasses at moderate rates (Inoue et al., 1996; Lu et al., 2003; Wang et al., 2004). While pure metals can be produced by melt spinning, alloys can be produced by arc melting and drop casting at cooling rates as low as ~1 K s^{-1}.

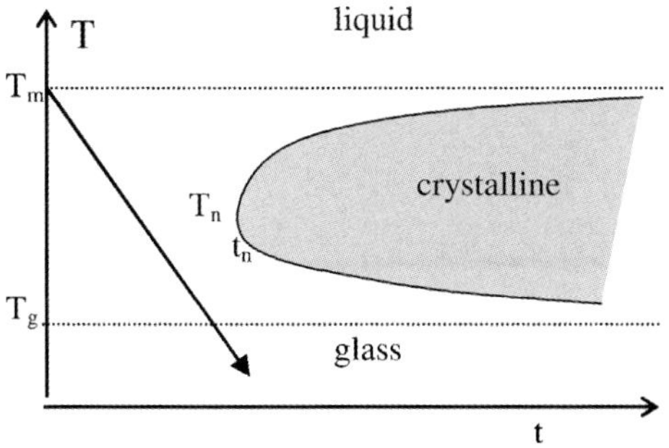

Fig. 5.15 Schematic TTT diagram and one thermal history for producing glass (arrow).

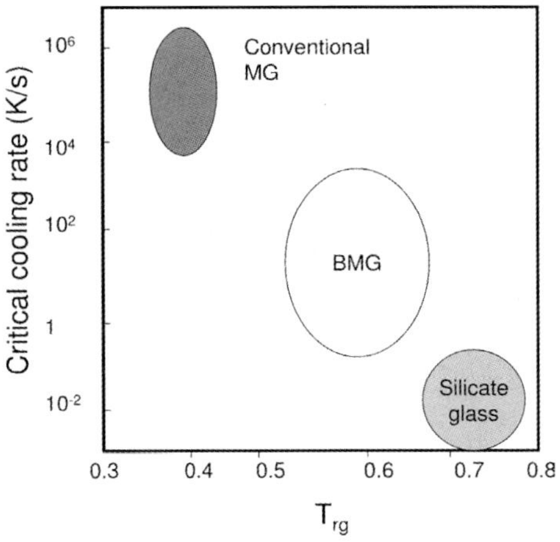

Fig. 5.16 Critical cooling rate for different classes of glass as a function of reduced glass transition temperature $T_{rg} = T_g/T_m$. (Reprinted from Wang et al., 2004, with permission from Elsevier.)

5.6.5 Devitrification

Once the glass is formed, the discussion turns to its devitrification upon a heat treatment. In fact, there have been measurements to determine the rate of growth of crystalline phases in an annealed glass. Once the temperature increases crystal growth may happen whenever the relative nucleation and growth curves overlap or whenever impurities serve as nuclei. Growth rate increases steeply to a maximum and then decreases to zero at liquidus temperature. Such a curve is shown for a silico-soda-lime glass in Fig. 5.17 when the growth of crystals of composition Na_2O–3 CaO–6SiO_2 called devitrite is observed. This phenomenon is to be avoided when a glass article is heat treated unless a glass–ceramic material is to be formed from a parent glass as detailed in the next section. Addition of a few percent of oxides like

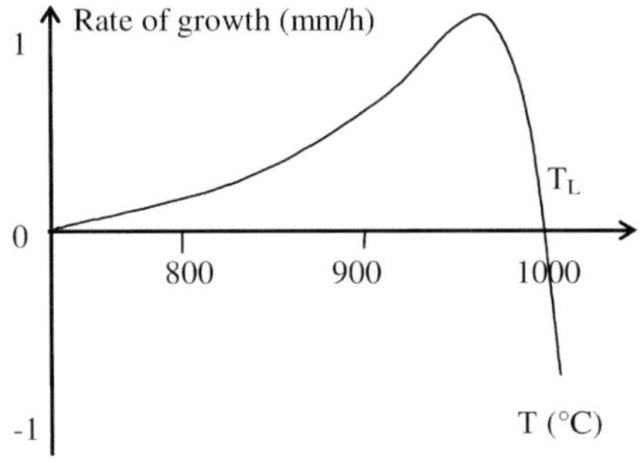

Fig. 5.17 Rate of growth of devitrite in a soda-lime-silica glass (according to Jones, 1956).

MgO or Al_2O_3 allows for reducing devitrification risks in ordinary oxide glasses (window glass). Also, care must be taken regarding preferential sites for nucleation like surfaces, dust, bubbles or inhomogeneities.

5.6.6 Factors That Favour Glass Formation

In view of a number of complex issues it is important to summarize factors that favour glass formation:

i. Viscosity is the controlling parameter, and a high-viscosity regime is searched for at liquidus temperature and around the TTT diagram nose. This means low transport and hence small rearrangement rates leading to low rates of nucleation and crystallization.
ii. Avoid heterogeneous nucleation.
iii. Consider systems with a large crystal–liquid interfacial energy.
iv. Consider systems with a large concentration change between the liquid and the crystal phases.

5.7 Vitroceramic Fabrication

5.7.1 Introduction

Glass–ceramics are fine-grained polycrystalline materials formed when glasses of suitable compositions are heat treated and thus undergo controlled crystallization. The parent glasses have specific compositions comprising precursors as detailed below. Indeed, ordinary window glass is too stable and difficult to crystallize, whereas other compositions crystallize in an uncontrolled manner resulting in undesirable microstructures. Usually, a glass–ceramic is not fully crystalline; typically the microstructure is 50–95 vol.% crystalline with the balance volume being the residual glass. One or more crystalline phases may form during the heat treatment and as their composition is normally different from the parent glass, it follows that the composition of the residual glass is also different from that of the parent glass. The mechanical properties of glass–ceramics are superior to those of the parent glass.

Vitroceramics are fabricated for fire resistance (cooking tops and containers; refer to Eurokera activity, Saint Gobain group and Corning ware) since the combination of crystalline and vitreous states allows for a lower thermal expansion and a higher strength. For astronomy, Zerodur (Schott) has allowed the fabrication of an 8.2 m telescope based in Chile with thermal expansion as low as $10^{-7}\,°C^{-1}$. Surprisingly, only recently have such materials been produced. It must be said that while glasses tend to devitrify, the control of such a phenomenon is very difficult and was reported only by 1950. In fact, it requires well-controlled thermal treatments in view of nucleation and growing curves (Fig. 5.18) and the use of nucleation agents.

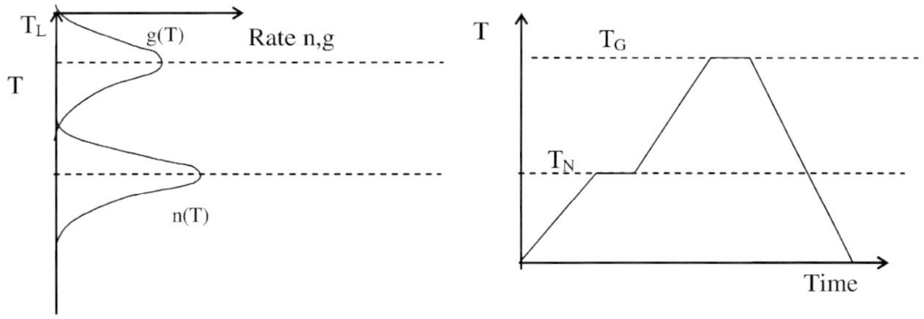

Fig. 5.18 Two-stage heat treatment for a system showing negligible overlap of nucleation and growing curves (according to Rawlings et al., 2006).

5.7.2
Conventional Method (two stages)

The conventional method for producing a glass–ceramic is to devitrify the parent glass by a two-stage heat treatment (Fig. 5.18). The first stage is a low-temperature (T_N) heat treatment that gives a high nucleation rate thus forming a high density of nuclei throughout the interior of the glass (Rawlings et al., 2006). A high density of nuclei is important since it leads to a desirable microstructure consisting of a large number of small crystals. The second stage is a higher temperature (T_G) heat treatment to produce growth of the nuclei. The achievement of such a microstructure is helped by nucleation agents that can be metallic (Au, Cu, Ag, Pt, Pd) or non-metallic (TiO_2, P_2O_5 and fluorides). These aggregate and subsequently induce controlled ceramic nucleation. Designed crystals embedded into the glass matrix are then formed.

5.7.3
Modified Conventional Method (single stage)

The reason for the two-stage heat treatment of the glass is a consequence of the limited overlap between nucleation and growth rate curves. If, on the contrary, there is extensive overlap of the rate curves then nucleation and growth can take place during a single-stage heat treatment at temperature T_{NG} (Fig. 5.19). The rate curves and particularly the nucleation curve are sensitive to composition, and hence optimizing the glass composition (choice of nucleating agents) is in some cases possible to obtain the necessary overlap.

5.7.4
Laser-Induced Method

For some years now, glass to ceramic transformation could be confined using a focused laser beam. This allows for patterning a glass object in three dimensions

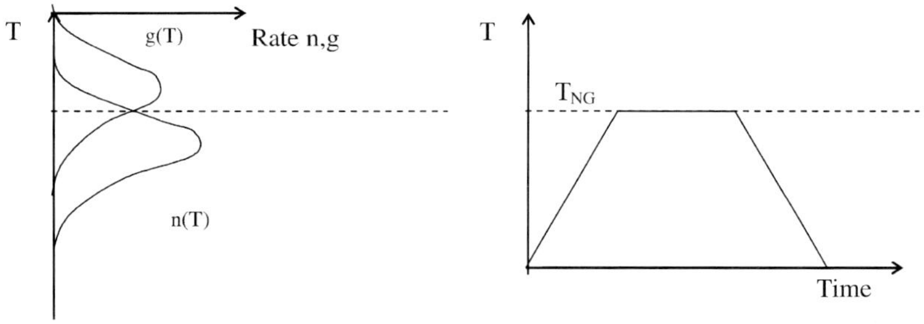

Fig. 5.19 Single-stage heat treatment for a system showing overlap of nucleation and growing curves (according to Rawlings et al., 2006).

(Lehmann, 2005). The nature of the local transformation (cavity, crystallite, composition change) is however still under debate.

5.8
Glass Surface

The first approach to study a glass surface is to start from the bulk structure and move towards the surface. The presence of the surface allows for the interaction of glass with the atmosphere. Surface structure is changed by the physical and chemical interactions. Koike and Tomozawa (2006) report that surface structural relaxation takes place more rapidly than bulk structural relaxation in particular in the presence of water diffusion (see also Chapter 6). Gas molecules penetrate the glass network through its excess volume and may react with the glass. We know that the bulk glass structure is not unique but depends on modifier concentration and network formers. We shall consider here mainly silica and suppose it is modified by alkali (Na^+) knowing that variations are expected for more complex compositions.

5.8.1
Surface Reaction

When silica cools down its surface reacts with water molecules to form silanol bonds (Fig. 5.20). The reaction can be written as

$$Si{-}O{-}Si + H_2O \rightarrow Si{-}OH + HO{-}Si \tag{5.25}$$

The reaction energy is elevated and around 80 kJ mol^{-1}. The water reaction leads to a high density of silanol groups that can reach 4.6 sites per nm^2 (Bather and Gray, 1976). According to the proximity to other silanol groups, hydrogen bonding will be possible. Also, the silanol groups may react with water without dissociation of the water molecules. This reaction is referred to as water physisorption. The glass surface may

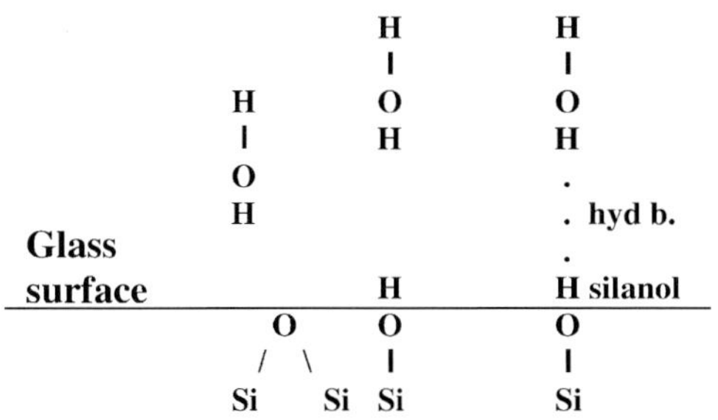

Fig. 5.20 Representation of the hydration of a silica surface.

hence be covered by several monolayers of water depending on the water partial pressure in the atmosphere. Not only water but also organic pollutants may adsorb. It should be noted that although an annealing treatment at 100 °C allows the desorption of water molecules, a treatment at a much higher temperature is necessary to form again siloxane groups (reversing reaction 5.25).

5.8.2
Molecular Diffusion

Molecules diffuse through the glass network. This phenomenon follows diffusion laws (Appendix 13). The smaller the molecule size the more important the diffusion. Diffusion of rare gases (He, Ne, Ar) and molecules (H_2, O_2, H_2O) has been documented[3] for different compositions of glasses (Doremus, 1994) and allows for tailoring glass permeability. Importantly, some of these molecules once in the glass may react and change glass properties and behaviour (viscosity, electrical conductivity). The role of the water reaction is discussed in more detail below.

5.8.3
Glass Network Interaction with Water

In fact, water interaction is not confined to the surface (Section 5.8.1) because of water diffusion through the glass excess volume (Appendix 13). Moreover, when we consider a glass with modifiers, non-bridging oxygen (NBO) atoms appear so that they open the network while mobile ions (alkaline ions) will diffuse. The deleterious role of alkali can be limited by introducing alumina and/or B_2O_3 in the glass melt. The elevated resistance of natural glasses over millions of years is to be attributed to their aluminosilicate-rich compositions while borosilicate glasses are used for

[3] Brown and Kobayashi (1998) entitled one of their papers 'Is your glass full of water?' Water concentration in commercial silicate glasses is in the range 0.01–0.1 wt%.

pharmaceutical packaging (Chapters 3 and 4). In sodium-rich glass, Na^+/H_3O^+ interdiffusion promotes water interaction within the network and can be followed eventually by corrosion of the network depending on the alkalinity of the solution (pH) at the glass surface[4].

5.8.3.1 Water Reaction

Water is present in the glass excess volume. Let us consider first silica. Water reacts with the silicon–oxygen bonds as already discussed for the surface as follows:

$$\text{Si—O—Si} + H_2O \rightarrow \text{Si—OH} + \text{HO—Si} \tag{5.26}$$

This reaction is of utmost importance in glass mechanics since it is enhanced at crack tips and is responsible for the fatigue of glass (Chapter 9). The reaction is enhanced as water concentration is increased.

5.8.3.2 Ion Exchange

Let us consider an alkali-modified silica glass. Because of mobile modifiers, the glass will undergo ion exchange, alkali being leached. Let us consider for instance sodium. Water diffusion (in fact H_3O^+) into the glass network is as follows:

$$\text{Si—O}^-\text{Na}^+ + H_3O^+ \rightarrow \text{Si—OH} + H_2O + Na^+ \tag{5.27}$$

The superficial glass loses sodium ions (glass leaching) while it incorporates water (it may also affect other mobile modifiers). This happens for external glazing or when in contact with water. The formation of a silica gel at a glass surface can be described by Fick's law (Appendix 13), the typical thickness of the gel layer being about 0.1 μm.

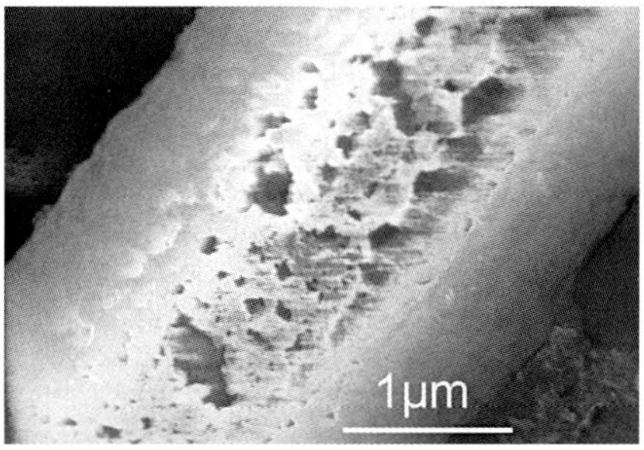

Fig. 5.21 *In vitro* degradation of a glass fibre. (Courtesy Hermine Dika, Pr Bertrand Rihn, Université Henri Poincaré, Nancy 1, France.)

4) $pH = \log[H_3O^+] = \log K_e - \log[OH^-]$,
$= pK_e - \log[OH^-]$, $pK_e = 14$ where the square brackets indicate the

Tab. 5.3 Surface energy (glass–water vapour) and wetting angle according to the type of surface (after Chartier, 1997).

	Surface energy, ζ (mJ m^{-2})	Water contact angle (°)
Native glass surface	400–1000	–
'Clean' glass surface	70	~0
Polluted glass surface	20–40	20–60

Such a layer can be detected by X-ray reflectometry (Table 5.1) because glass subsurface presents different properties as compared to the bulk. Also the hydrated glass surface exhibits lower resistance to contact loading (Chapter 8, Appendix 10) and this is beneficial when abrading the glass edges.

5.8.3.3 Glass Corrosion

Na^+ diffusion towards a glass surface increases the alkalinity (pH) of the aqueous solution in contact with the glass. When pH reaches 9 then siloxane bonds may react with hydroxide ions as follows:

$$\mathrm{Si{-}O{-}Si} + \mathrm{OH^-} \rightarrow \mathrm{Si{-}OH} + {}^{-}\mathrm{O{-}Si} \tag{5.28}$$

Such a pH value is achieved when the glass remains in the same humid and confined environment as for stored glass panels and glass wool. As matter of fact, it may happen that glass products are damaged by corrosion. This is limited when storing glass panels using acidic paper in between. This phenomenon is well known to glass manufacturers and can be tailored using Na_2O-, K_2O-, $CaO-$rich glasses to enhance it or using Al_2O_3-, B_2O_3-, ZrO_2-, TiO_2-rich glasses to limit it (these oxides act as barriers). The development of bioglass (biodegradable fibres in particular) is a further illustration of tailored glass chemical resistance. These have been motivated by the asbestos scare (Fig 5.21). Nowadays, environmentally safe fibres can be produced so that their half-life is less than 40 days in lungs. For the confinement of complex and dangerous wastes from the biosphere, glasses resisting water corrosion are needed (see also Chapter 3). Aluminoborosilicates are being developed to achieve glass loss of less 0.1 % over 100 centuries.

5.8.4 Surface Properties

It should be noted that all reactions described in Section 5.8 change dramatically surface properties, in particular the surface energy (Table 5.3). The complex glass surface can finally be schematized as: glass bulk/hydrated glass/water/pollution. Functionalizing the surface of glass supposes it is first cleaned up; it can also be treated (Chapter 12).

6 Glass Rheology

6.1 Viscosity

6.1.1 Viscosity and Process

One of the most important properties from a glass process point of view is the viscosity, which changes dramatically as the temperature changes. Viscosity determines the melting conditions, the temperature regimes for forming, annealing and tempering, and the maximum temperature to avoid devitrification. Moreover, glass quality is dependent on the way it is homogenized and this is also related to the viscosity regimes. Viscosity measures the resistance of a viscous liquid to shear deformation. The higher the viscosity, the higher the resistance to flow. Consider a liquid between two planes (surface S) and apply a tangential force F to the upper plane (Fig. 6.1). For a solid (infinite viscosity) no flow will occur. On the contrary, a viscous liquid will flow. A velocity gradient will appear in the liquid. If we assume that there is no relative displacement between the plate and the liquid, then velocity varies from zero at the bottom plate to the velocity of the upper plane at distance a (Fig. 6.1). Velocity gradient is then determined as $\nabla v = (v - 0)/a$ and increases when the force F or shear stress ($\tau = F/S$) applied to the upper plate increases. The relationship between the velocity gradient and the shear stress is written

$$\tau = F/S = \eta \nabla v \tag{6.1a}$$

where η is the viscosity in Pa s.

An industrial process can be considered according to its velocity gradient (Table 6.1) and one can observe huge differences between these. Equation (6.1a) means that the viscosity regime (temperature, as discussed below) has to be carefully controlled when forming the glass according to the different processes. Also, it is of utmost importance to prepare the glass under uniform temperature. This operation is referred as fining (Chapter 10). Viscosity controls the quality and the homogenization of the glass. During this step, the elimination of bubbles also depends on the viscosity (Chapter 10, Appendix 4).

Glass; Mechanics and Technology. Eric Le Bourhis

ISBN: 978-3-527-31549-9

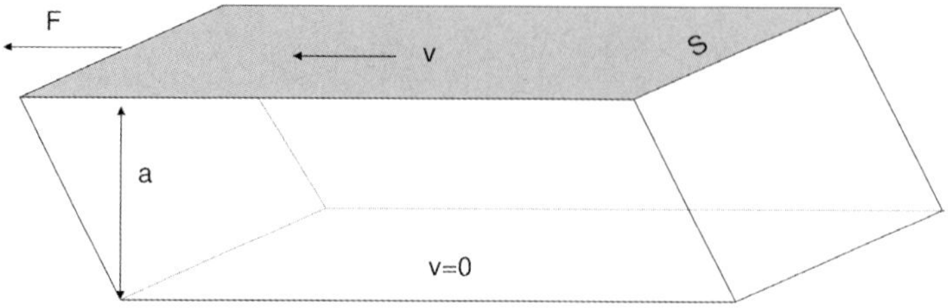

Fig. 6.1 Shear flow in a viscous liquid.

Tab. 6.1 Processes and associated ranges of velocity gradient.

Process	Velocity gradient ∇v (s^{-1})
Moulding	0.1–10
Extrusion	10^2–10^3
Wire drawing	10^4–10^6

The related stresses to be applied by the tools will depend strongly on viscosity regime. From a mechanical point of view (Appendix 9) it is of interest to rewrite Eq. (6.1a) in terms of shear strain. Such an equation shows the time dependence of shear stress τ as a function shear strain γ.[1)]

$$\tau = \eta \nabla v = \eta \bar{\gamma} \tag{6.1b}$$

This behaviour is completely different from that of an elastic solid that shows no time dependence (Appendix 9). In fact in the elastic regime shear stress is proportional to shear strain, $\tau = G\gamma$, where G is the shear modulus. Therefore, while an elastic solid shows a response to stress independent of time, a viscous liquid shows a response that depends not only on stress amplitude but also on time. Hence, the control of the tools for a process involving viscous liquids requires the control of viscosity regime, stress amplitude and loading time. Glassmakers have set several important temperatures that define ranges of viscosity that are important from a process point of view. These temperatures are listed in Table 6.2.

In fact, viscosity depends strongly on the temperature as shown in Fig. 6.2 and process temperatures are set relative to the defined reference temperatures. This will be discussed in detail in Section 6.1.3. Composition also may change the viscosity

1) An ideal Newtonian flow is assumed here with a viscosity independent of strain rate. The plot of $\tau(\gamma)$ is linear. This hypothesis is true at elevated temperature and under low stresses. Instead, non-Newtonian flow (viscosity depending on strain rate) is reported at intermediate temperature closer to T_g and for high strain rates (Simmons et al., 1988) yielding more difficulty in the control of tools. A drastic decrease of viscosity is observed with increasing strain rate. So called rheofluidification may be advantageous from a process point of view if one considers the required stress. For a non-Newtonian flow, the stress-strain relationship is $\tau = \eta(\gamma)\gamma$ with a viscosity depending on shear strain rate γ. It can be approximated by $\tau = \eta\gamma^m$, where m is the strain-rate sensitivity exponent ($m < 1$), similar to that used to describe metal plasticity and it is pointed out that non-Newtonian regime allows for prevention of 'necking' (Jones, 1956). The apparent viscosity is $\tau/\gamma = \eta\gamma^{m-1}$ (Appendix 10).

Tab. 6.2 Reference temperatures and viscosity regime[8].
Practical melting temperature is not the melting temperature associated with the crystal–liquid phase transition (Chapter 5) but the working temperature.

Reference temperature	Viscosity (Pa s)
Practical melting temperature	1–10
Working point	10^3
Littleton softening point	$10^{6.6}$
Annealing point	$\sim 10^{12}$
Glass transition	$\sim 10^{12}$–$10^{12.5}$
Strain point	$\sim 10^{13.5}$

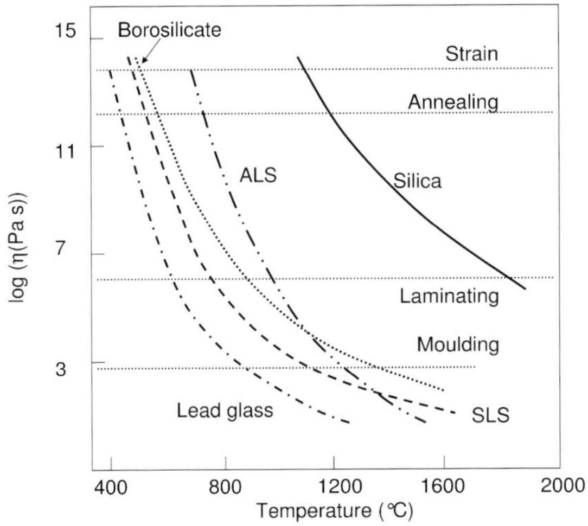

Fig. 6.2 Viscosity variations with temperature and composition (SLS, soda-lime-silica glass; ALS, aluminosilicate glass; according to Zarzycki, 1982).

drastically. Modifiers (Chapter 5) are introduced into the melt to reduce the working temperatures and hence increase industrial yields saving costs in processing. In fact, the same viscosity regime (horizontal line in Fig. 6.2) is obtained for increasing temperatures as glass is changed from lead glass, soda-lime-silica (SLS) glass, borosilicate glass and finally pure silica. Modifiers lead to the formation of non-bridging oxygens and allow viscous flow to occur at much lower temperatures in SLS glasses than in silica (Chapter 5). The ALS glass curve crosses the other with a higher transition temperature than SLS (see also Section 6.7).

2) For comparison we list the viscosity (in mPa s) of common materials: hydrogen, 0.0086; gas, 0.01–0.02; *water*, 1; kerosene, 10; oil, 100; honey, yogurt, 10^3–10^6; glue, paint, resin, 10^2–10^7; asphalt, 10^5–10^8.

To optimize production tools and processing, it is of utmost importance to fabricate a glass with homogeneous temperature and composition or in other words homogeneous viscosity. In fact, the temperature of the glass delivered from the melting furnace and its composition have to be regulated with extreme care. From the control or R&D perspectives, it is necessary to determine carefully and precisely the viscosity as a function of both temperature and composition. This is not straightforward since viscosity varies in a wide range and instruments have been developed accordingly. Interestingly, such measurements revealed that in limited domains empirical relations exist between the compositions and the properties of glass (Priven, 2004). The optimization of the composition for processing can then be made using these empirical laws to define a range of compositions likely to yield a glass with the required properties (see also Chapter 2). Nonetheless, one also has to take into account the complex shape of the produced article as well as its complex thermomechanical loading. This can be achieved again for R&D purposes using numerical modelling to consider the properties of the material and its loading at each point of the specimen (Appendix 11).

6.1.2
Viscosity Measurement

Different instruments have been designed according to viscosity ranges (Table 6.3), and the overall viscosity curve is obtained using complementary techniques. We shall describe some of these – rotating cylinder and fibre elongation viscometers – allowing for low ($10–10^3$ Pa s) and high ($10^6–10^{13}$ Pa s) viscosity measurement respectively. At intermediate temperatures, crystallization may interfere with the measurements and caution should be taken with the resulting data. Viscosity experiments can be divided in several ways: whether they are strain-rate controlled (resultant stress being measured, e.g. concentric cylinders) or stress controlled (resultant strain rate being measured, e.g. fibre elongation), whether they involve pure shear (e.g. concentric cylinders) or a combination of shear and volume stress (beam bending, fibre elongation). Usually the viscometer is calibrated with a liquid of known viscosity; absolute measurements can be made with care.

Tab. 6.3 Measurement methods for viscosity.

Range	Method	Viscosity value, η (Pa s)
Melting	Falling sphere/bubble rise	$<10^4$
	Rotational methods (concentric cylinders, cone-and-plate)	$1–10^5$
Softening and annealing	Parallel plate	$10^5–10^9$
	Penetration	$10^9–10^{11}$
	Fibre elongation	$10^9–10^{14}$
	Beam bending (see also Section 6.3)	$10^7–10^{12}$

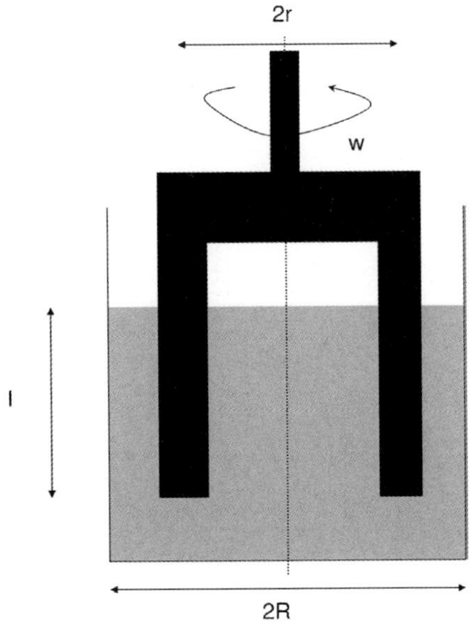

Fig. 6.3 Rotation viscometer principle.

6.1.2.1 **Rotation Viscometer**

This apparatus (also referred as the Couette method) is designed for low-viscosity measurements and comprises concentric cylinders, the strain rate being delivered to the liquid by rotating either the inner or the outer cylinder. The inner cylinder of radius r (spindle either full or hollow) is immersed in the viscous liquid under study to a depth l contained in a cylindrical container of radius R (Fig. 6.3). The spindle is set at a constant spinning velocity w applying a constant torque M. A simple relationship links the torque M needed to maintain a constant spinning speed and the angular velocity that allows the viscosity η to be determined as

$$\eta = C\frac{M}{w} \tag{6.2}$$

with C being a factor depending on the geometrical characteristics of the apparatus and contact height l between the liquid and the cylinder:

$$C = \frac{1}{4\pi l}\left(\frac{1}{r^2} - \frac{1}{R^2}\right) \tag{6.3}$$

These instruments are designed for the low-viscosity regime or elevated temperatures and may require the spindle and cylinder be made of Pt for the high-temperatures regime. The geometry of a cone-and-plate viscometer allows the preparation of starting materials by pre-machining the final form. This avoids the need to pour or melt the material into the cylinder and to immerse the spindle into the liquid at elevated temperatures. This method can also be used to extend the studied range of viscosity.

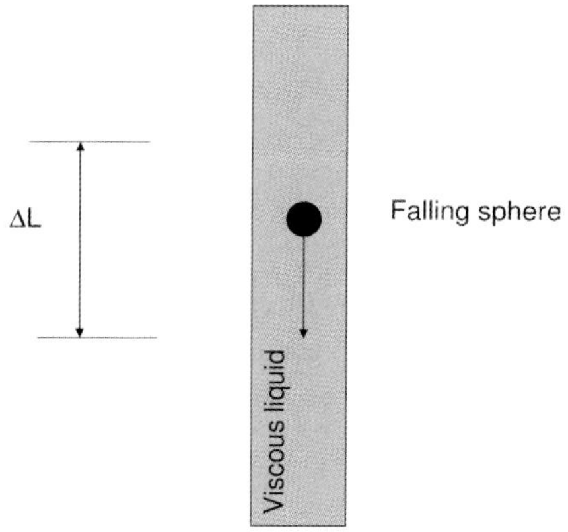

Fig. 6.4 Falling sphere viscometer.

6.1.2.2 **Falling Sphere Viscometer**

This viscometer allows determination of the viscosity of the liquid from the velocity regime of a falling sphere. When a spherical ball (radius r) falls in a viscous liquid under gravity it tends to approach a constant velocity v_l after a transient regime (like a parachutist). This is an important hypothesis that has to be checked (Appendix 4). The movement of such a ball can be obtained from classical mechanics assuming that the sphere friction can be described by Stokes' law (Appendix 4). The velocity of the ball is written

$$v_l = \frac{2gr^2(\rho_s - \rho)}{9\eta} \tag{6.4}$$

Note that densities of both the ball ρ_s and of the liquid ρ are determined prior to experiments (this is not obvious at elevated temperatures). The experiments are carried out as follows. The ball descending from the top of the cylinder containing the liquid reaches a constant velocity that is determined by measuring the time Δt necessary to fall a distance Δl (Fig. 6.4):

$$v_l = \frac{\Delta l}{\Delta t} \tag{6.5}$$

Thereafter, using Eqs (6.4) and (6.5) and knowing the density of the ball and of the liquid allows the determination of the viscosity of the viscous liquid. Falling sphere viscosimetry can be used for high-pressure measurement, as those carried out in a superpress. Pressure indeed appears to change viscosity. Free volume theory refinements (see Section 6.1.3.4) incorporate pressure dependence. Viscosities of silicates and aluminosilicates are reported to decrease with increasing pressure.

6.1.2.3 **Fibre Elongation Viscometer**

This method allows for high-viscosity measurement since fibre extension under gravity (or under an additional applied load for higher viscosity regime) is measured (Fig. 6.5). Glass fibres with a diameter of 0.1–0.3 mm and a length of 10–18 mm are

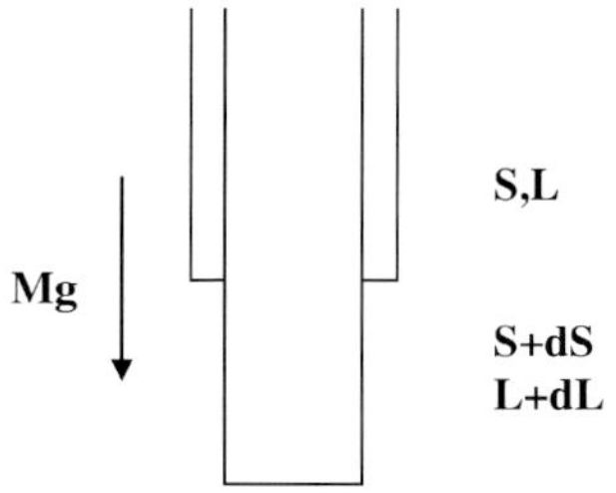

Fig. 6.5 Fibre elongation viscometer principle.

commonly used. This can be carried out in a furnace (or in a vertically mounted dilatometer) when viscosity at elevated temperature is investigated. Under gravity (or applied force), the fibre extends from L to $L + dL$ during time dt. Since viscous flow is volume conservative and calling S the section of the fibre, we have

$$V = SL = (S + dS)(L + dL) \tag{6.6}$$

Relative expansion is written

$$dL/L = -dS/S \tag{6.7}$$

Changing the elastic solution of an incompressible solid $\sigma = E\varepsilon = 3G\varepsilon$ using a time operator allows the viscosity to be determined as (elastic–viscous correspondence detailed in Section 6.3):

$$\sigma = \frac{F}{S} = 3\eta\bar{\varepsilon} = 3\eta\frac{1}{L}\frac{dL}{dt} \tag{6.8}$$

with the elongational viscosity being three times higher than the shear viscosity. Note that surface tension also opposes the applied stress (opposing surface increase while fibre elongates). This is neglected here; one way to determine surface tension is to consider surface tension influence on fibre elongation.

Equation (6.8) is rewritten as

$$\eta = \frac{F}{3}\frac{L}{S(dL/dt)} \tag{6.9}$$

with (we assume no additional applied force)

$$F = Mg \tag{6.10}$$

where M is the fibre mass and g the gravitational acceleration.

The measurement is carried out as follows. The mass of the fibre (only the part that contributes to deformation under gravity is considered) and its section are determined, then its elongation with time is measured using an optical microscope (or using a dilatometer sensor) allowing for dL/dt to be calculated and hence the viscosity to be determined. When measurements are to be carried out at elevated temperatures, the fibre is set into a furnace that allows for the observations of the fibre. For the viscosity regime over 10^{14} Pa s, the method becomes difficult since there is a risk of fracture of the glass specimen under elevated applied loads and deformations are small.

Moreover, delayed elasticity (Section 6.4.4) and structural relaxation (Section 6.5.4) interfere with the viscous deformation. Then only an apparent viscosity is extracted.

6.1.3 Viscosity Variation with Temperature

6.1.3.1 Introduction

Viscosity varies over a very large domain (in fact we will use logarithms) and increases continuously and steeply when freezing a liquid into a glass, as shown in Fig. 6.6 together with the different process ranges of viscosity. At the melting temperature ($\eta < 10$ Pa s) the melt is fluid enough for fining and homogenizing. Blowing and moulding are carried out at viscosities between 10^3 and 10^6 Pa s. At working point (10^3 Pa s) molten glass can be formed and manipulated. Viscosity is low enough for some shear processing (pressing, blowing) but high enough to retain shape after shear stress is removed. At the Littleton softening point, the glass still flows: a rod about 24 cm long and 0.7 mm in diameter elongates 1 mm min^{-1} under its own weight. Glass transition corresponds to a viscosity around 10^{12}–$10^{12.5}$ Pa s. For such a level of viscosity, the supercooled liquid has not enough time to rearrange and then freezes in the glassy state. At the annealing temperature (Table 6.2), viscosity is about 10^{12} Pa s and allows relaxation of residual stresses within 15 min (see relaxation time in Section 6.4). Annealing treatment is necessary to remove tensile residual stresses

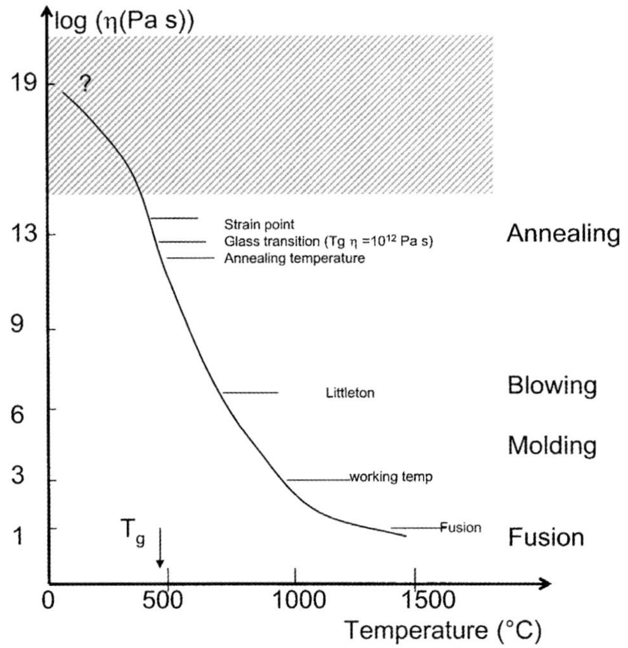

Fig. 6.6 Viscosity as a function of temperature below and over transition temperature (according to Zarzycki, 1982).

that would yield spontaneous fracture and difficulties in scoring glass (Chapters 7, 8 and 10). At the strain point ($10^{13.5}$ Pa s) viscosity is large enough so that the glass can be cooled rapidly down to room temperature without any risk of introducing residual stresses. Since viscosity varies so steeply with temperature while process operations have to be carried out at defined ranges, it is important to control this parameter with precision unless variation in the production is unacceptable. As detailed in Chapter 10, glass production passes through melting and fining operations. The objectives of these steps are to produce a homogeneous melt in terms of composition and temperature, that is, in terms of viscosity.

Figure 6.7 shows log η as a function of the reciprocal temperature for different glasses. Linear variations are obtained over the entire range of temperatures for some glasses (strong glasses) and over reduced ranges for others (fragile glasses). Linear variation is representative of a thermally activated phenomenon following Arrhenius behaviour, as observed for many transport properties:

$$\eta(T) = \eta_o \exp\frac{G}{RT} \tag{6.11}$$

where ΔG and η_o are temperature-independent coefficients called the activation energy (for viscous flow) and the pre-exponential factor respectively. An Arrhenius temperature dependence of viscosity is quite easy to use and allows for rapid prediction of the viscosity at a given temperature.

A strong glass shows such an Arrhenius temperature dependence over the majority of its melt-forming region while a fragile glass shows instead a pronounced 'curvature'.

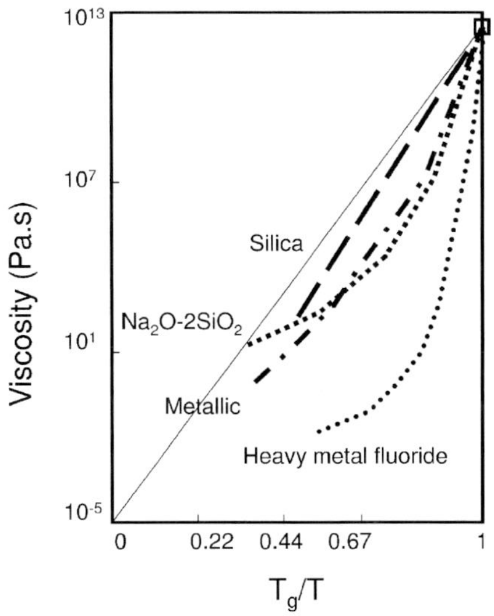

Fig. 6.7 Variation of the logarithm of viscosity as a function of reciprocal temperature for strong and fragile liquids (according to Kurkjian and Prindle, 1998).

In the latter case, the actual behaviour of the supercooled liquid ($T > T_g$) can be fitted empirically by the Vogel–Fulcher–Tammann (VFT; see Section 6.1.3.3) formula that captures relatively well the non-Arrhenius temperature dependence over the melt-forming region. The reason for observing either strong or fragile behaviour has been widely debated. It is believed that fragile liquids lack intermediate-range order; some theoretical backgrounds are developed in Sections 6.1.3.2 and 6.1.3.4.

We may still use an Arrhenius temperature dependence when considering limited domains of temperatures and we adopted this procedure in Chapter 5. Instead, when variations over a large range of temperatures are to be predicted for fragile glass, non-Arrhenius temperature dependence is to be taken into account in order to get suitable predictions of glass behaviour.

At temperatures below T_g (Fig. 6.6) the viscosity curve is still unknown and debated. In fact the question mark is shown on Fig. 6.6 to indicate the domain where viscosity is uncertain. Above 10^{15} Pa s viscosity values should be considered with great care. As plotted in Fig. 6.6, the curve suggests that the rate of increase of viscosity with decreasing temperature becomes greatly reduced. There is a qualitative explanation for such an expected behaviour since the glass below T_g is frozen in a liquid-like structure the configuration of which corresponds roughly to that in equilibrium near the glass transition temperature (see also fictive temperature in Section 6.5.3). Hence glass structure is not at equilibrium. Further consideration of the related viscosity at these temperatures is necessary from the experimental point of view. Indeed, the apparent viscosity commonly measured using penetration viscosimetry is non-Newtonian at these low temperatures and therefore strain rate dependent (Appendix 10). Moreover, the measured deformation is not purely viscous because of delayed elasticity, as stated above and detailed in Section 6.4.4. Hence, it is not possible to measure the equilibrium viscosity at these temperatures. This fact may account for the huge discrepancy in viscosity values published in the literature since apparent viscosity depends on stress and strain rate amplitude, or on experimental conditions. The finding is that the apparent viscosity decreases under these conditions.[2)]

[2)] There exists a very broad range of values in the literature. For instance, Frischat (1987) reports a 10^{20} Pa s value while Macosco (1994) reports 10^{40} Pa s. These values would correspond to relaxation time (refer to precise definition below) from 10^9 s (30 years) to 10^{29} s. These discrepancies yielded sometimes controversy in the observed geometry of stained glasses in cathedrals. In fact, it has been observed that cathedral windows are wavy and thicker at the bottom. Would glass have flowed when standing in place for some hundreds of years? There have been speculations on this subject. In fact, when measuring viscosity at low temperature elevated stresses are to be applied to the specimens. Under such conditions the apparent viscosity is expected to decrease with the amplitude of stress and strain rate (Appendix 10). Discrepancies are then expected between authors as the Newtonian viscosity is not determined. On the other hand, it would be problematic extrapolating high-temperature data (in particular in the case of a fragile melt). It is not possible to predict glass flow under moderate stress. On the other hand, undeformed glass vases (>2000 years old) have been discovered, while natural glasses maintain their structure. In fact, medieval windows could have been blown and subsequently flattened manually and non-uniformly (Chapter 3); then architects would have put the thicker part at the bottom (see also Chapter 12).

6.1.3.2 **Fragility**

Transition temperature is of great importance as well as the properties changes in the melt-forming region. Observation of a viscosity curve on a log $\eta - 1/T$ diagram indicates that indeed the behaviour depends dramatically on the type of glass, with greater curvature being observed for fragile glasses. A strong glass has then a larger viscosity than a fragile glass on approaching T_g. It is quite helpful to employ a unique temperature scale for all glasses using the transition temperature to temperature ratio as the *x*-axis (Fig. 6.7). On such a graph all glass responses meet at a common point corresponding to the glass transition at a viscosity of 10^{12}–$10^{12.5}$ Pa s and T_g/T ratio equal to 1. In such a graph, the glass behaviour is nothing but unique and really composition dependent. The question is why we observe such different behaviours (Doremus, 1994, 2003; Angels, 1995; Kieffer, 2002; Novikov and Sokolov, 2004; Tanaka, 2005). In fact, while the nature of the transformation by which a supercooled liquid 'freezes' to a glass has been the subject of intense investigations and substantial progress has been made in understanding this phenomenon over the past two decades, yet many key questions remain. In particular, the factors that control the temperature-dependent relaxation and viscous properties of the liquid phase as the glass transition is approached, that is, whether the glass-forming liquid is 'fragile' or 'strong', remain unclear (Angels, 1995).[3] The fragility of liquids characterizes the deviation from Arrhenius behaviour or the curvature from a straight evolution (expected from Eyring theory; see Section 6.1.3.4; Fig. 6.2 or Fig. 6.7 replotted with respect to transition temperature). Silica shows strong or almost Arrhenius behaviour since a linear variation is observed for all temperatures (Fig. 6.7). In contrast metal fluoride glasses show important deviations from such an Arrhenius behaviour. Fragility can be numerically defined as the apparent activation energy of shear viscosity η at the glass transition temperature T_g, normalized to T_g (Angels, 1995):

$$m = \left.\frac{\partial \log\eta}{\partial(T_g/T)}\right|_{T=T_g} \qquad (6.12)$$

Hence, *m* characterizes the steepness of the slope of the log η dependence on T_g/T near T_g. A stronger deviation (larger *m*) from Arrhenius behaviour corresponds to a more fragile system. As mentioned above, Fig. 6.7 shows that fragility depends on the material. Silica demonstrates Arrhenius behaviour over the logarithm of viscosity range from 1 to ~13, whereas fragile glasses demonstrate an almost continuously changing slope, or one that can be approximated by two straight lines (one at elevated temperature and another at the transition). Strong behaviour is controlled by continuous Si–O–Si bonding as in silica while the break-up of this bonding controls

[3] The belief that a glass is a supercooled liquid in which atomic motions have slowed to the extent that characteristic relaxation time exceeds the observation period (thousand years for stained medieval windows) is debated and the possibility that glass corresponds to a different physical state from the liquid is considered. This suggests that glass would require an independent set of parameters and this is not done yet in engineering or scientific calculations (Kieffer, 2002).

network modified glasses. Hence, alkali silicates are less resistant to thermal degradation and could flow within alkali-rich regions (Chapter 5). Empirical formulas have been proposed to correct for the deviation from non-Arrhenius temperature dependence (Section 6.1.3.3). The most well known is the VFT empirical law, which is extensively employed in research and industry.

6.1.3.3 **VFT Empirical Formula**

A supercooled fragile liquid shows a viscosity that can be approximated by the VFT (Vogel–Fulcher–Tammann) formula

$$\eta(T) = \eta_o \exp\frac{B}{T - T_o} \tag{6.13}$$

where B and η_o are temperature-independent coefficients. This law can be considered as a correction of the Arrhenius law adding a T_o term. In fact, taking $T_o = 0$ K or assuming $T \gg T_o$ transforms the VFT formula into an Arrhenius temperature dependence. The VFT approximation is still easy to use and allows for close prediction of the actual viscosity. It is an empirical law that allows one to capture the curvature of the experimental $\ln(\eta) - 1/T$ curves. As will be discussed in detail below, different approaches have been proposed to model the actual behaviour represented by the VFT empirical law; these are complex and still matters of debate. In accordance with the VFT equation above, the viscosity of a glass will be infinite at T_o or below T_g (no mobility). This is unlikely to be correct (Section 6.1.3.1, Appendix 10). For a silico-soda-lime glass for instance, the VFT parameters are determined to be $B = 4320$ K, $T_o = 542$ K and $\ln \eta_o = -2.77$ (Guillemet and Gy, 1996).

6.1.3.4 **Microscopic Approach**

Eyring theory for molecular mobility supposes that the movement of each molecule (or structural unity) is thermally activated (Fig. 6.8). Given a barrier energy Δg, the jump frequency is written

$$\nu_E \exp\left(-\frac{\Delta g}{kT}\right) = \nu_E \exp\left(-\frac{\Delta G}{RT}\right) \tag{6.14}$$

where ν_E is the atomic transition frequency in the liquid (order of magnitude $10^{13}\ s^{-1}$; Einstein frequency).

The applied force or shear stress τ modifies the barriers as shown in Fig. 6.8c (Douglas, 1958), flow being biased so that the probability of a forward jump (supported by the shear stress) is higher than for the backward jump (against stress). The jump frequency along the direction of flow ν^+ and in the opposite direction ν^- is written

$$\nu^+ = \nu_E \exp\left(-\frac{\Delta g - \tau\Omega/2}{kT}\right) \tag{6.15}$$

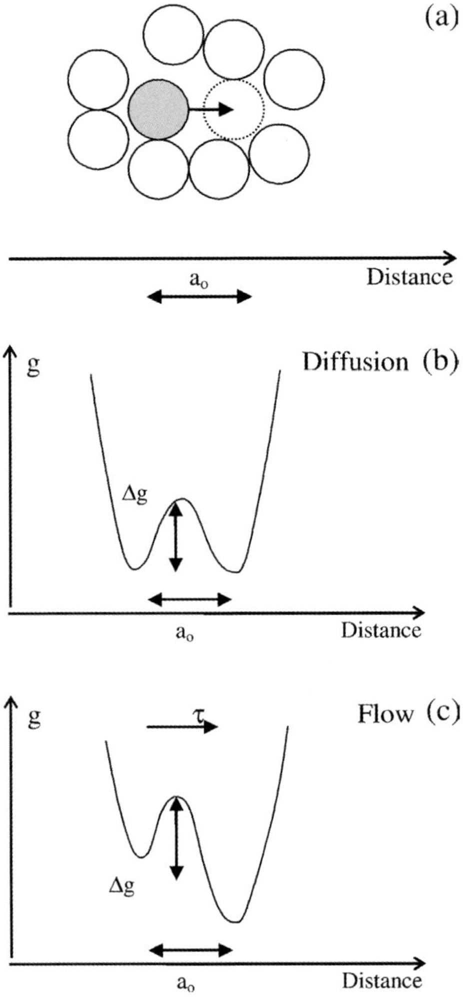

Fig. 6.8 Representation of an individual atomic jump: the basic step for macroscopic diffusion and flow.

$$\nu^{-} = \nu_{\mathrm{E}}\exp\left(-\frac{\Delta g + \tau\Omega/2}{kT}\right) \tag{6.16}$$

where Ω is the atomic volume. The effective jump frequency is written

$$\begin{aligned} \nu = \nu^{+} - \nu^{-} &= \nu_{\mathrm{E}}\exp\left(-\frac{\Delta g}{kT}\right)\left[\exp\left(\frac{\tau\Omega/2}{kT}\right) - \exp\left(-\frac{\tau\Omega/2}{kT}\right)\right] \\ &= 2\nu_{\mathrm{E}}\exp\left(-\frac{\Delta g}{kT}\right)\sinh\left(\frac{\tau\Omega}{2kT}\right) \end{aligned} \tag{6.17}$$

When all the atoms in a specimen make one jump a_o (assuming $a_o = 2r$, where r is the atom radius) in the direction of shear, macroscopic shear strain $\gamma = 1$. Shear strain rate is then written

$$\bar{\gamma} = 2\nu_E \exp\left(-\frac{\Delta g}{kT}\right)\sinh\left(\frac{\tau\Omega}{2kT}\right) \tag{6.18}$$

and for a low applied stress, the strain rate is written (since $\sinh(x) \to x$ for $x \to 0$)

$$\bar{\gamma} = \frac{\tau\Omega}{kT}\nu_E \exp\left(-\frac{\Delta g}{kT}\right) \tag{6.19}$$

We note that Newtonian flow ($\tau \prec \bar{\gamma}$) is obtained under these conditions and that the viscosity is written

$$\eta = \frac{kT}{\Omega\nu_E}\exp\left(\frac{\Delta g}{kT}\right) \tag{6.20}$$

$$\eta = \eta_o \exp\left(\frac{\Delta g}{kT}\right) = \eta_o \exp\left(\frac{\Delta G}{RT}\right) \tag{6.21}$$

with $\eta_o = kT/\Omega\nu_E$. Temperature dependence of η_o is often neglected, in the present framework it is determined to be linearly dependent on T, and hence η_o varies usually far less than an exponential term (see also Appendix 13).

On cooling the liquid, flow becomes impossible because of unavailable free volume for atomic jump. The main theories accounting for this phenomenon are the free-volume model (Cohen and Turnbull, 1959; Turnbull and Cohen, 1961) and the configurational entropy model (Adams and Gibbs, 1965).

Free-volume models follow the idea that on cooling the supercooled liquid, free volume progressively reduces,[4] and hence mobility becomes impossible. The mobility of an atom (or more generally a structural unit) is supposed to be possible only when a free volume which can contain the atom exists next to it. We shall consider in the following the average free volume per atom. Let us call v^* the minimum volume of a cavity that can contain the atom or the mobile structural unit. Diffusion of the atom will depend on the free volume size v_f per atom. Hence either

$$\text{(i) } v_f < v^* \tag{6.22}$$

and there is no diffusion of the atom, or

$$\text{(ii) } v_f > v^* \tag{6.23}$$

and there is diffusion of the atom.

[4] Density fluctuation can be revealed using SAXS techniques (Levelut et al., 2002) and is observed to depend on temperature and thermal history.

The probability of finding an atom with a free volume next to it of size between v and $v + dv$ is given by (Cohen and Turnbull, 1959)

$$\frac{a}{v_f}\exp\left(-a\frac{v}{v_f}\right)dv \tag{6.24}$$

where a is a geometrical factor between $\frac{1}{2}$ and 1.

The probability that the atom is on a potential jump site is then

$$\int_{v^*}^{\infty}\frac{a}{v_f}\exp\left(-a\frac{v}{v_f}\right)dv = \exp\left(-a\frac{v^*}{v_f}\right) \tag{6.25}$$

To obtain the shear strain rate one has to consider that only potential sites for jump contribute and hence

$$\bar{\gamma} = 2v_E\exp\left(-a\frac{v^*}{v_f}\right)\sinh\left(\frac{\tau\Omega}{2kT}\right) \tag{6.26}$$

Again, for low applied stresses, we still obtain Newtonian flow ($\tau \prec \bar{\gamma}$):

$$\bar{\gamma} = \frac{\tau\Omega}{kT}v_E\exp\left(-a\frac{v^*}{v_f}\right) \tag{6.27}$$

while viscosity no longer shows Arrhenius behaviour but instead is written in a WLF form (Williams et al., 1955; Ferry, 1970):

$$\eta = \eta_o\exp\left(a\frac{v^*}{v_f}\right) \tag{6.28}$$

Cohen and Turnbull 1959 further suggested that the free volume can be obtained from thermal expansion coefficients of the amorphous and liquid phases (above the glass transition temperature). Calling the difference in coefficients of thermal expansion $\Delta\alpha = \alpha_l - \alpha_s$ between the liquid and glass (see also Section 6.2) they wrote

$$v_f = 0 \text{ for } T < T_g \tag{6.29a}$$

$$v_f = v_m\alpha(T - T_o) \text{ for } T > T_g \tag{6.29b}$$

Also, pressure dependence can also be incorporated. We obtain

$$\eta(T) = \eta_o\exp\left(a\frac{v^*}{v_f}\right) = \eta_o\exp\left(a\frac{v^*}{v_m\Delta\alpha}\frac{1}{T - T_o}\right) = \eta_o\exp\left(\frac{av^*/v_m\Delta\alpha}{T - T_o}\right) \tag{6.30}$$

This expression is similar to the VFT empirical expression found for viscosity. As the temperature is lowered, the $(T - T_o)$ term causes curvature below Arrhenius behaviour (so-called fragility). However, if $T_o = T_g$ then the curvature is too large and $T_o = T_g - \Delta T$ must be used with $\Delta T \sim 20$–$50\,^{\circ}\mathrm{C}$. Note this model predicts too large a viscosity curvature near T_g and ignores configurational arrangements. Macedo and Litovitz 1965 argued that the free volume model neglects the activation energy required for making a diffusive jump. Thus they developed a hybrid equation that combines the free volume term with the Eyring term (see also Appendix 10):

$$\eta(T) = \eta_o \exp\left[a\frac{v^*}{v_f} + \frac{\Delta g}{kT}\right] \tag{6.31}$$

Then the Arrhenius temperature dependence of the viscosity can occur if either of the following facts are true: (i) the material is a low-expanding one such as fused silica; (ii) free volume is proportional to temperature. In the liquid where

$$a\frac{v^*}{v_f} \gg \frac{\Delta g}{kT}$$

the viscosity reduces to a WLF-type equation further simplified into the VFT formula assuming Eq. (6.28b). The Macedo and Litovitz 1965 approach fixed one of the problems of the free volume model in that it exhibits Arrhenius viscosity for flow under constant free volume conditions (isostructural). Also, some of the free volume model limitations have been addressed by the configurational entropy model where it is postulated that rearrangement of the structure happens when a region containing at least two structural configurational states is available. It is logical to consider the motion as a cooperative process involving a simultaneous rearrangement of a large number of atoms or units. The Adam–Gibbs 1965 model is based on the idea that motion occurs by internal cooperative rearrangement of independent regions of n atoms (Avramov, 2005). When the temperature decreases, the motion of one atom disturbs an increasingly larger number n of its neighbours, or one can say the coherence length of an atom or a molecular motion increases. The Adam and Gibbs 1965 model assumed that the barrier for rearrangement is proportional to n, and determined the temperature dependence of n in terms of the configurational entropy S_c. Finally, viscosity is written

$$\eta(T) = \eta_o \exp\left(\frac{C}{TS_c}\right) \tag{6.32}$$

where C and η_o are temperature-independent coefficients and the configurational entropy S_c is written

$$S_c = \Delta C_p \frac{T - T_o}{T} \tag{6.33}$$

where ΔC_p is the variation of molar specific heat between the liquid and the glass, T_o being the temperature at which configurational entropy becomes null which plays the same role here as that corresponding to the vanishing of free volume. These equations combine into

$$\eta(T) = \eta_o \exp\left(\frac{C}{TS}\right) = \eta_o \exp\left(\frac{C/\Delta C_p}{T - T_o}\right) \tag{6.34}$$

showing the same form as the VFT empirical equation.

6.2
Glass Transition and Its Observation

6.2.1
'Observing' the Glass Transition

As mentioned above, the occurrence of glass depends on the cooling history followed by the viscous liquid. Figure 6.9 shows the volume of a specimen as a function of temperature on cooling (we shall see that heating and cooling curves differ because of structural relaxation; Section 6.5.4). In Figure 6.9, l, g and c correspond to the liquid, glass and crystal respectively.

Let us start from the liquid at elevated temperatures (high-temperature end of the diagram). On decreasing the temperature, the liquid contracts first.

Whenever the cooling rate is too low, then nucleation and growth of crystallites happen as discussed in Chapter 5. Crystals form at the melting or liquidus temperature ($T_L = T_m$) and a discontinuity in the volume V of the specimen is observed.

The cooling rate may be high enough so that the melting temperature is passed without volume discontinuity, the liquid being then supercooled. Next, the supercooled liquid freezes into glass at the transition temperature with less volume

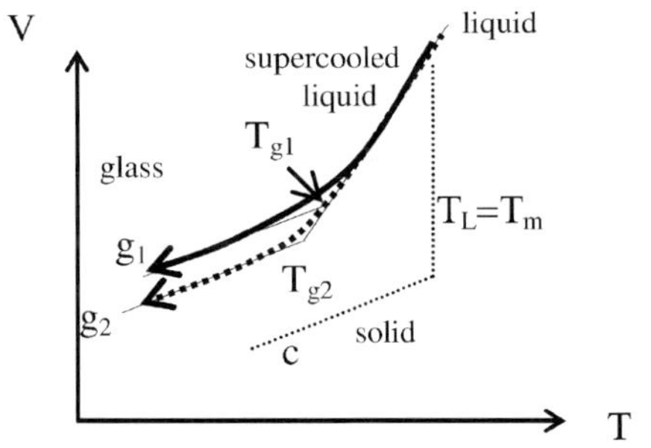

Fig. 6.9 Volume–temperature cooling curves. Note the discontinuity at melting or liquidus temperature ($T_L = T_m$) when crystals form while liquid to glass path is continuous with lower contraction being observed once the glass transition is passed.

Tab. 6.4 Transition temperature ranges.

Glass	Transition temperature (K)
Silica	1500–1600
Soda-lime-silica	800–850
Metallic	550–700
Chalcogenide	300–600

contraction observed. This is called transition instead of transformation since the path is continuous.

It is significant that the viscosity does not show an abrupt change at the glass transition temperature (Section 6.1). This suggests that the viscosity determines the rate at which a property (here the volume) relaxes to some equilibrium value. The presence of the glass transition results from the slowing down of viscous relaxation in the liquid. At a viscosity of 10^{12} Pa s the relaxation time is in the range of minutes to a few hours, corresponding to usual times of experiments. Changes at glass transition allow for transition temperature determination as detailed in Fig. 6.9. Glass structure and hence its density and volume will depend on the cooling rate. When the cooling rate increases freezing happens at a higher temperature. In fact, less time is allowed for atomic rearrangement and a 'higher temperature' configuration is frozen in. Let us consider two glasses (g1, g2) cooled at different rates. On cooling down more slowly, further atom arrangements are made possible and hence a denser structure is obtained. Therefore, the glass volume and transition temperature of glass g2 are lower than those of glass g1. The glass transition temperature is then not unique but instead an interval should be considered. Table 6.4 lists representative values for common glasses. Glass transition temperature can be determined experimentally using dilatometry or calorimetry as detailed in Sections 6.2.2 and 6.2.3.

Now consider a supercooled liquid that can be cooled infinitely slowly. The volume curve followed would be that of the liquid with enough time for a denser structure to organize. The resulting glass (we suppose no crystals appear) would show denser structure (less entropy).[5)] Hence, glasses formed under realistic conditions are expected to be unstable and driven towards a denser and stabilized configuration whenever they are held at temperatures where atomic rearrangements become non-negligible. This structural relaxation is discussed in Section 6.5.4.

Another important feature is that glass will show a lower density than the crystal (see also Chapter 5). As a matter of fact, glass contains excess volume v_e, a part of which called free volume $v_f = v_e - v_o$ appears to play an important role in glass mechanics (see Section 6.1.3.4; Turnbull and Cohen, 1961).

4) Glass entropy value at 0 K is still an open question. Residual entropy of glassy states plays against the third law of thermodynamics that suggests it is zero. As temperature decreases more and more time is required for the liquid to be at equilibrium. The question arises as to what would be the structure of the equilibrium perfectly ordered liquid at zero entropy. If crystalline, it would then suggest that there is a continuous change from the amorphous to the crystalline state.

6.2.2
Dilatometry

A dilatometer allows for the continuous measurement of the length of a given specimen which increases continuously with temperature at a constant heating rate (Fig. 6.10; see also Appendix 3). It is generally in the range 1–50 °C min^{-1}. It has to be controlled carefully since the glass transition depends on the thermal history (see Sections 6.2.1 and 6.5.2). The same remark applies to calorimetry (Section 6.2.3). The expansion of the specimen is monitored by a stylus (or a measuring device) that contacts the specimen under a small force. Note that only the difference between the expansion of the specimen and the cylindrical support is determined (Appendix 3). Therefore, correction of the data has to be made to add the expansion of the instrument itself. Examples of corrected expansion curves are shown in Fig. 6.11.

The curve shows a straight line for silica over the whole temperature range while the other two glasses show first a linear domain (elastic), a transition around 550 and 600 °C to the liquid domain (viscous) with a change in the observed slope and then a maximum of the curve (around 620 and 680 °C). The first transition corresponds to

(a)

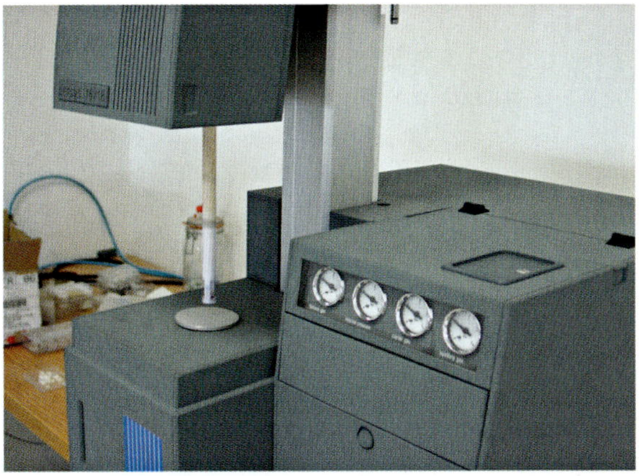

(b)

Fig. 6.10 Dilatometer: (a) general view and (b) sample–stylus contact.

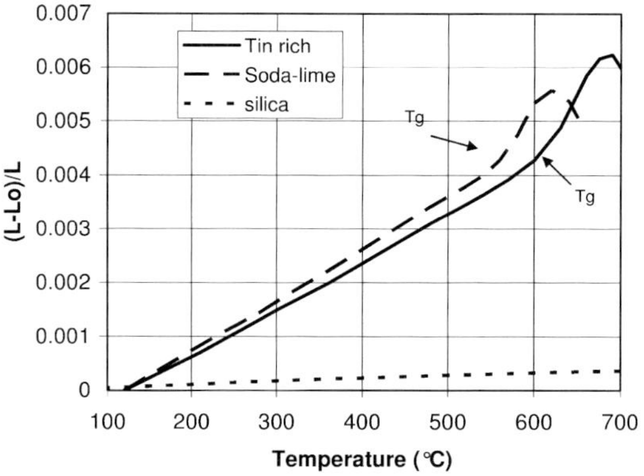

Fig. 6.11 Thermal expansion heating curves for silica, silico-soda-lime and tin-rich silico-soda-lime glasses.

the glass transition while the second one corresponds to the dilatometric softening point (obtained for a viscosity of $\eta = 10^{10}$ Pa s). Glass transition temperature can be determined at the intersecting point of the tangents of the solid and liquid domains. It is about 560 and 620 °C for the SLS and tin-rich glasses shown in Fig. 6.11. In the linear domain the slope of the curve allows one to determine the thermal expansion of the solid glass, that is, the relative expansion per temperature unit:

$$\alpha_s = \frac{1}{L}\frac{dL}{dT} \tag{6.35}$$

where L is the specimen length and T the temperature. Glass thermal expansion is determined from Fig. 6.11 to be about $9 \times 10^{-6}\,°C^{-1}$ (8.9×10^{-6} and $8.3 \times 10^{-6}\,°C^{-1}$ for the SLS glass and the tin-rich glass respectively; Le Bourhis, 1998b). The increase in slope after transition indicates that the thermal expansion of the liquid α_l is larger than that of the glass. From the curves shown in Fig. 6.11, the liquid thermal expansion is estimated to be about three to four times that of the solid. However, care has to be taken with the values obtained from dilatometry for $T > T_g$, since expansion is measured by a contacting stylus which progressively penetrates into the viscous liquid and thermal expansion is underestimated. Further penetration of the stylus explains the end and maximum of the curve. In fact, when viscosity falls to about 10^{10}–10^{11} Pa s, the glass becomes so soft that it no longer behaves as a rigid specimen. It sags if lying horizontally and suspended at only a small number of points, and it may deform under the action of the small force necessary to keep the measuring device (stylus) in contact with the specimen. In practice, whatever the arrangement adopted, the apparent maximum is found at nearly the same temperature, that is, the softening temperature.

An important characteristic of modified glasses is that they show larger thermal expansion than silica. When a specimen is annealed the average distance between atoms changes according to the asymmetry of the potential energy curve or bond. When modifiers are introduced non-bridging oxygen (NBO) atoms are formed

(Chapter 5) leading to bond asymmetry and hence more thermal expansion (Appendix 3). In contrast, borosilicates offer the possibility to form the glass at lower temperatures as in a modified glass while B atoms participate to the formation of the glass network. As a matter of fact no NBO atoms are formed and thermal expansion is reduced. As discussed in Chapter 2, these glasses offer good resistance to thermal shock.

6.2.3 Differential Scanning Calorimetry

A differential scanning calorimeter allows the continuous measurement of the heat flow rate Φ between a given specimen (S) and a reference (R) as a function of temperature (Figs. 6.12 and 6.13). The reference is often the empty crucible, the same as that used for containing the specimen.

The difference in heat flow rate Φ allows for the determination of the specific heat of the specimen. Let us consider a period of time from t_1 to t_2 during which the specimen temperature rises from T_1 to T_2. Heat Q is written

$$Q = \int_{t_1}^{t_2} \Phi(t)\mathrm{d}t = mc_\mathrm{p}(T_2 - T_1) \tag{6.36}$$

where m is the mass of the specimen and c_p its specific heat. When glass transition is passed there is a change in specific heat that is detected by the instrument as schematically represented in Fig. 6.14 (see also Fig. 6.32). Glass transition temperature can be determined at the inflection point. DSC softwares offer three different determinations of the transition as shown schematically in Fig. 6.14. T_g^ip is determined at the point of inflection, T_g^on at the onset of the transition and $T_\mathrm{g}^\mathrm{end}$ at the end of

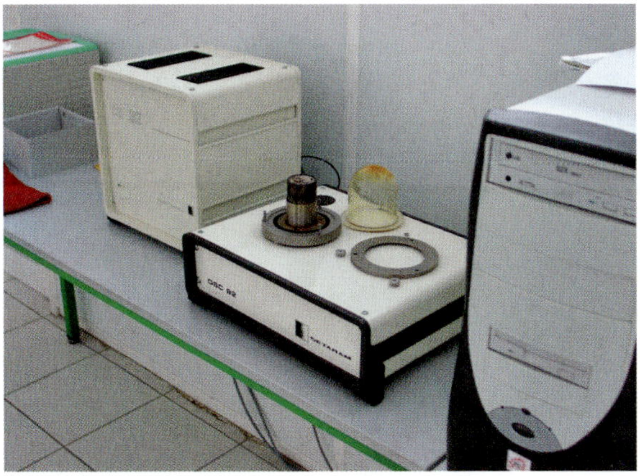

Fig. 6.12 Differential scanning calorimeter set-up.

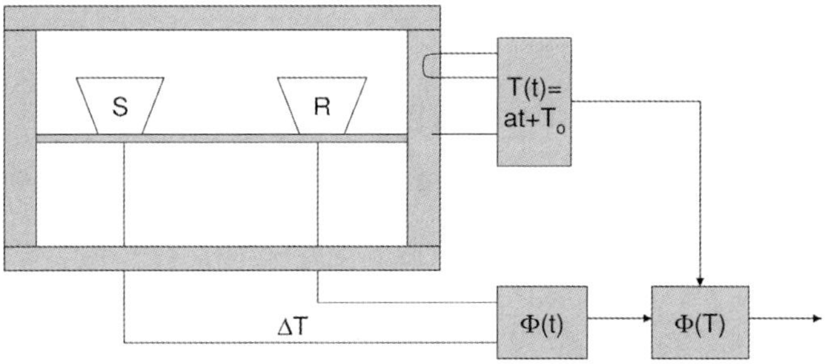

Fig. 6.13 Differential scanning calorimeter principle (R, reference; S, for specimen; according to Höhne et al., 1996).

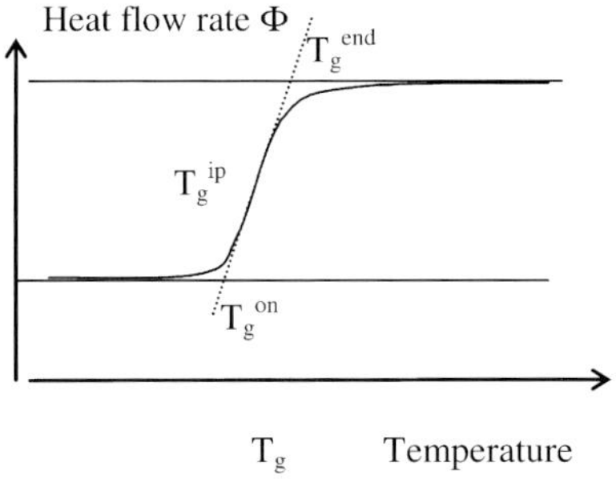

Fig. 6.14 Heat flow rate Φ at glass transition as a function of temperature.

the transition. Notably, experimental curves often show a hump at T_g^{end} because of relaxation phenomena; this is illustrated in Fig. 6.32 and discussed in detail in Section 6.5.4 (Höhne et al., 1996; Heilmaier and Eckert, 2005). This may lead to erroneous transition temperature determination on heating curves. Below and above the glass transition, it is possible to determine the specific heat of a specimen of known weight employing Eq. (6.36) and analysing the heat flow in the respective domains. When temperature is further increased, crystallization may appear (exothermic flow) that further melts at melting temperature (endothermic flow).

6.3
Viscous Response of Glass

An important consequence of the viscous behaviour is that glass shows time-dependent properties at temperatures approaching the glass transition temperature.

In fact, these phenomena appear when the characteristic time of the observations scales with the relaxation time of the glass, $T_R = \eta/G$, where η is the viscosity and G the shear modulus. This relaxation time will be introduced in more detail in Section 6.4.2.

The time dependence appears in the shear stress–shear strain equation introduced above:

$$\tau = \eta\bar{\gamma} \tag{6.37}$$

It should be noted again that, in contrast, an elastic material shows a response independent of time:

$$\tau = G\gamma \tag{6.38}$$

From these equations a phenomenological correspondence has been proposed to pass from an elastic solution to a viscous one in which the elastic constants in the elastic equations are replaced with a time-dependent operator. Following Goodier 1936, the analogy between the steady viscous motion and the elastic deformation enables one to take over from the elastic theory solutions for viscous flow. One has to suppose two isotropic and homogeneous bodies of the same shape, one purely viscous the other one purely elastic and incompressible. The two are loaded in the same way. 'If certain displacements are prescribed as given function in the elastic body, then the corresponding velocities in the viscous body are prescribed as the same functions when the coefficient of viscosity η is substituted for G as long as the deformation makes no significant change in the boundaries' (Goodier, 1936). Applying this simple rule allows the tensile representation of a viscous solid to be determined passing from $\sigma = E\varepsilon = 2(1+\nu)G\varepsilon$ to $\sigma = 2(1+0.5)\eta\bar{\varepsilon} = 3\eta\bar{\varepsilon}$, viscous flow being volume conservative (so-called prefactors 1 and 3 for shear and tensile loading respectively). Elongational viscosity is then obtained to be three times higher than shear viscosity introduced in Section 6.1.2.3.

Let us now consider the forming of glass by gravity (Fig. 6.15). This is used for production of laminated windshields for instance. The elastic solution is given by

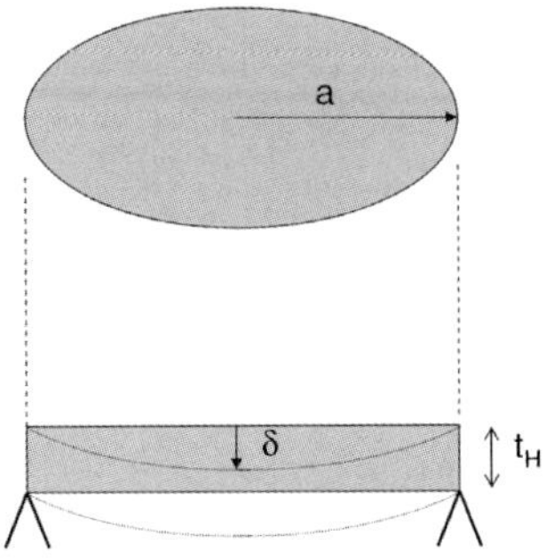

Fig. 6.15 Diagram of the deformation of a plate under gravity.

Timoshenko (1951):

$$\delta = k\frac{\rho g a^4}{Et_{\mathrm{H}}^2} = k\frac{\rho g a^4}{2(1-\nu)Gt_{\mathrm{H}}^2} \tag{6.39}$$

where δ is the elastic displacement of the sheet centre, a and t_{H} the dimension and the thickness of the sheet, ρ and M its density and its mass, E the Young's modulus, g the gravitational acceleration and k a shape factor ($3/4\pi$ for a circular plate). The solution of Eq. (6.39) is valid for $\delta < t_{\mathrm{H}}$. Changing the elastic solution using a time operator allows determination of the viscous solution:

$$\delta = k\frac{\rho g a^4}{3t_{\mathrm{H}}^2}\int\frac{\mathrm{d}t}{\eta} \tag{6.40}$$

Here, we neglect the elastic part of the deformation and still suppose $\delta < t_{\mathrm{H}}$. This limit is quite unrealistic for commonly produced shapes and the nonlinear solution is to be considered. Adding the elastic and viscous solutions (Eqs (6.39) and (6.40)) results in a Maxwell modelling of the plate behaviour as detailed in Section 6.4. Equation (6.40) cannot be integrated simply unless the temperature (the viscosity) is supposed to be constant, which is rarely the case when an industrial process is considered.

From this simple approach it is clear that complex shapes are achieved when the thermal route can be carefully controlled. In practice, two sheets of glass are twinned and subsequently treated together to form laminated windshields. When the sheet thickness and colour are identical the thermomechanical route is the same, assuming no thermal gradient is present through the sheets when heating and cooling rates are low enough. This allows one to obtain parallelism of the formed sheets and hence optimized optical properties. In practice, glass sheets are different (either in terms of thickness or colour) and the processing line is operated so that the top sheet bows onto the bottom sheet (without contact damage; see Chapter 8).

6.4 Viscoelastic Response of Glass

6.4.1 Introduction

Between the elastic and viscous extremities stands the viscoelastic behaviour that is expected around the glass transition. The purely viscous response of a melt is often termed 'liquid' behaviour whereas purely elastic response is termed 'glassy' behaviour. The intermediate transition is not sharp but shows a viscoelastic regime. Therefore, at the transition both the elastic and the viscous behaviour are to be combined. At short times the behaviour will predominantly be elastic while viscous behaviour appears at times long enough to attain steady-state flow. Then viscoelastic behaviour is expected at intermediate times. The simplest approach is to combine in series or parallel these behaviours as detailed below. We will introduce an important parameter, the relaxation time (that combines both properties), which allows prediction of whether for a given period of time the material behaves like an elastic, a viscous or a viscoelastic material.

6.4.2
Maxwell and Kelvin Solids

Further understanding of viscoelasticity requires in a first step the consideration of simple models. We shall consider an elastic solid to be represented by a spring ($\tau = G\gamma$) while a viscous solid will be represented by a dash pot ($\tau = \eta\bar{\gamma}$; Fig. 6.16).

Hence, as said above the viscous description of a solid can be found by replacing the elastic constants in the elastic solution and taking into account the time dependence.

Associating the two solids (elastic and viscous) allows an insight into the viscoelastic behaviour (intermediate between purely elastic or purely viscous). The simplest association of these two elements is in parallel (Kelvin's solid) and in series (Maxwell's solid). Let us consider the respective combinations (Fig. 6.17).

For a Kelvin solid we combine $\tau = \tau_1 + \tau_2$; $\tau_1 = G\gamma$ and $\tau_2 = \eta\bar{\gamma}$ and obtain

$$\tau = G\gamma + \eta\bar{\gamma} \tag{6.41}$$

For a Maxwell solid we have $\gamma = \gamma_1 + \gamma_2$; $\tau = G\gamma_1$ and $\tau = \eta\bar{\gamma}_2$. Then taking a time derivation of the first and second equation and combining $\bar{\gamma} = \bar{\gamma}_1 + \bar{\gamma}_2$; $\bar{\tau} = G\bar{\gamma}_1$ and $\tau = \eta\bar{\gamma}_2$ we get

$$\tau/\eta + \bar{\tau}/G = \bar{\gamma} \tag{6.42}$$

These solids allow creep and relaxation to be modelled with a characteristic relaxation time T_R.

Consider for instance a Maxwell solid submitted to a constant deformation $\gamma(t) = \gamma_o$ so that $\bar{\gamma} = 0$. Shear stress relaxation with time is found solving

$$\tau/\eta + \bar{\tau}/G = \bar{\gamma} = 0 \tag{6.43}$$

Taking

$$T_R = \eta/G \tag{6.44}$$

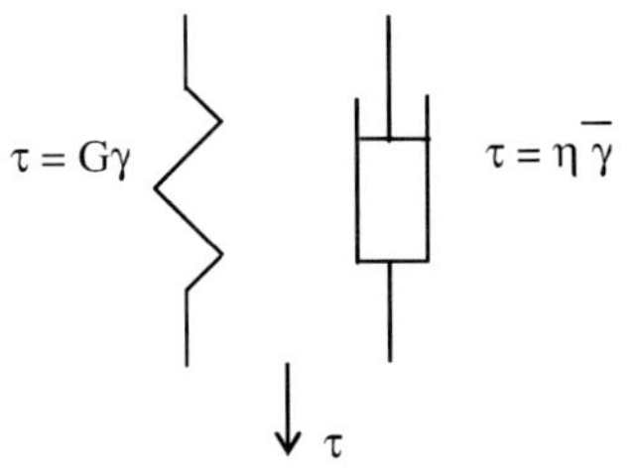

Fig. 6.16 Spring and dash pot.

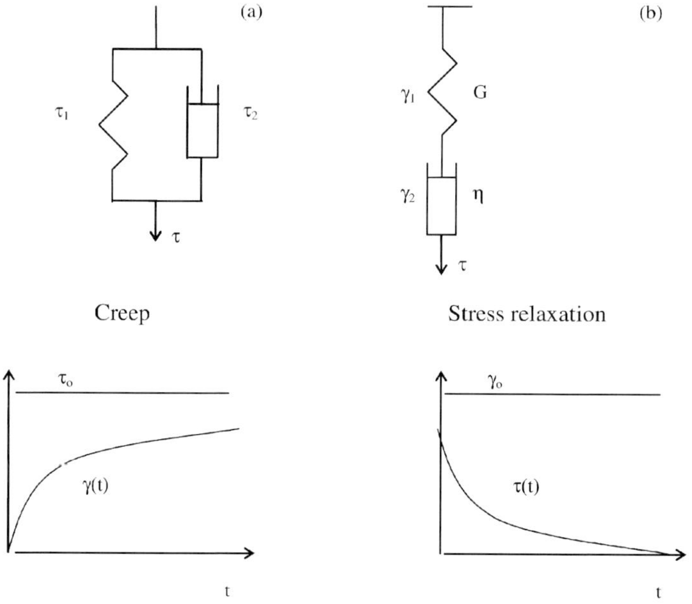

Fig. 6.17 (a) Kelvin and (b) Maxwell solids and related creep and stress relaxation.

we find

$$\tau(t) = \tau_o \exp(-t/T_R) \qquad (6.45)$$

which is plotted in Fig. 6.17b. We can observe that to apply a constant deformation to a Maxwell solid, a vanishing stress has to be applied. A very important parameter has appeared: the relaxation time T_R. For $t \ll T_R$ the solid behaves elastically, while for $t \gg T_R$ viscous flow is prominent.

Similarly considering a Kelvin solid under constant applied stress τ_o we have

$$\tau = \tau_o = G\gamma + \eta\bar{\gamma} \qquad (6.46)$$

The solution to this equation is

$$\gamma(t) = \tau_o/G[1 - \exp(-t/T_R)] \qquad (6.47)$$

Hence, it is found that a Kelvin solid flows under constant stress once time is of the order of T_R (Fig. 6.17a). Some authors refer to Kelvin T_R as retardation time. As discussed below glass delayed elasticity and flow can be captured with a Burger solid that combines in series a Kelvin and a Maxwell solid. A Kelvin solid yields retardation while a Maxwell one yields relaxation. Relaxation time informs on the time scales at which a viscoelastic solid will behave elastically or relax. Let us consider glass transition: the viscosity is 10^{12}–$10^{12.5}$ Pa s while shear modulus of most glasses scale with 10^{10}–10^{11} Pa so that relaxation time is of the order of 100 s. This means that at the glass transition temperature stress relaxation happens after a couple of minutes.

6.4.3
Dynamic Mechanical Analysis

Dynamic mechanical analysis (DMA) consists in applying a periodical stress field to a material. We can anticipate from Section 6.4.2 that the behaviour of a viscoelastic material will depend strongly on the frequency ω of the applied stress.

Consider a shear stress applied periodically to a viscoelastic solid. We take a sinusoidal function having in mind that any periodic function is a Fourier series of sinusoidal functions. Shear stress is written

$$\tau = \tau_o \cos\omega t \qquad (6.48)$$

A generalized relation between stress and strain is obtained using complex notation ($\tau^* = \tau_o\cos\,\omega t + i\tau_o\sin\,\omega t = \tau_o\exp\,i\omega t$) in the form

$$\tau^* = G^*\gamma^* \qquad (6.49)$$

G^* being a complex number of the form $G^* = G' + iG''$. The real part or storage modulus describes the capacity of the material to store energy elastically upon deformation. The imaginary part is called loss modulus since it provides a measure of the energy dissipated by the material. Both moduli show a dependence on frequency. The phase shift δ between stress and strain is given by

$$\tan\delta = G''/G' \qquad (6.50)$$

where δ is the loss angle and $\tan\delta$ the loss factor related to the dissipated mechanical energy ΔW by $\Delta W/W_e = 2\pi\tan\delta$ (where W_e is the stored elastic energy). $Q^{-1} = \Delta W/W_e = 2\pi\tan\delta$ is also called the damping coefficient.

The elastic, viscous and viscoelastic behaviours are shown schematically in Fig. 6.18. In the elastic regime, stress and strain are in phase (no time dependence). In the viscous regime stress and strain are phase shifted by 90°. Viscoelastic behaviour corresponds to an intermediate phase shift.

Let us illustrate this behaviour for the case of a Maxwell solid. The modulus expression for G^* can be found taking again Eq. (6.42) with the complex notation (with $\bar{\tau}^* = i\omega\tau^*$):

$$\tau^*/\eta + \bar{\tau}^*/G = \tau^*/\eta + i\omega\tau^*/G = \tau^*(1/\eta + i\omega/G) = \bar{\gamma}^* = i\omega\gamma^* \qquad (6.51)$$

We extract

$$\tau^* = i\omega/(1/\eta + i\omega/G)\gamma^* \qquad (6.52)$$

and obtain

$$G^* = \frac{i\omega}{1/\eta + i\omega/G} \qquad (6.53)$$

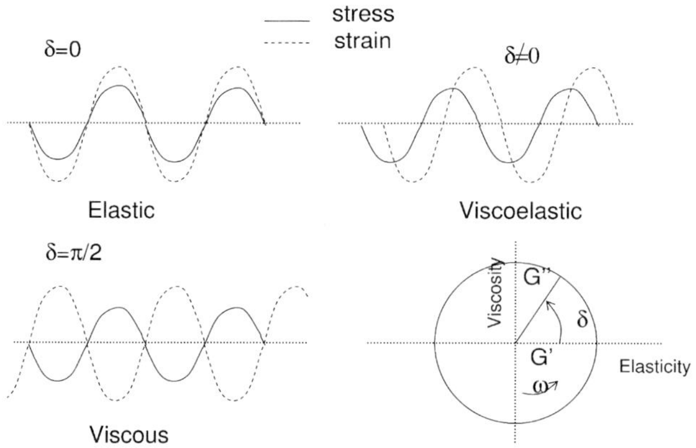

Fig. 6.18 Elastic, viscous and viscoelastic dynamic response.

Storage and loss moduli are

$$G' = \frac{\omega^2 T_R^2}{1 + \omega^2 T_R^2} G \tag{6.54a}$$

$$G'' = \frac{\omega T_R}{1 + \omega^2 T_R^2} G \tag{6.54b}$$

These expressions yield for

$$\omega T_R \to \infty, G' \to G, G'' \to 0, \delta \to 0 \tag{6.55a}$$

that is, a purely elastic regime for large frequencies, while for

$$\omega T_R \to 0, G' \to 0, G'' \to 0, \delta \to \pi/2 \tag{6.55b}$$

that corresponds to a purely viscous behaviour. At intermediate frequencies the material shows a viscoelastic behaviour with an intermediate loss angle ($0 < \delta < \pi/2$).

These trends can be generalized. In fact when forced at a high frequency, a glass or a fluid reacts rigidly. Deformations are small and all energy introduced into the system can be regained upon releasing strain. The storage modulus is large and the loss modulus small (Fig. 6.19). At intermediate frequencies, deformations become large enough for structural elements in the liquid to be permanently removed from their original position. Such displacements dissipate energy. At a particular frequency, a resonance occurs where the loss modulus peaks while storage modulus goes through an inflection (Fig. 6.19; for a Maxwell solid it is expected for $w = 1/T_R$). Towards low frequencies, the loss modulus decreases again. The fluid flows effortlessly, resulting in large irreversible deformations, indicating that the storage modulus becomes smaller.

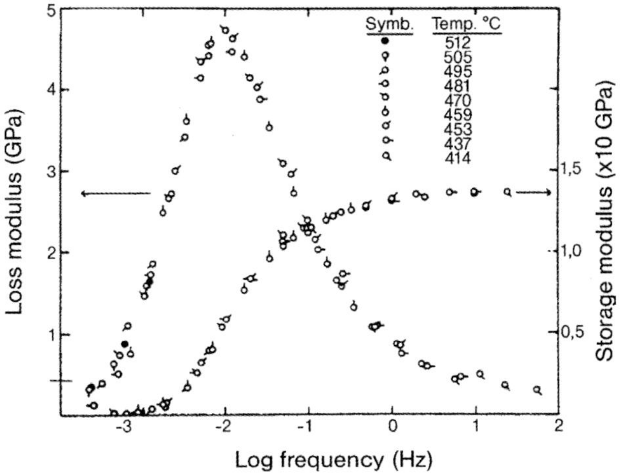

Fig. 6.19 Loss and storage moduli as a function of frequency for $2SiO_2$–Na_2O glass (master curve; see also Section 6.4.5.4). (Reprinted from Mills, 1974 with permission from Elsevier.)

Alternatively, glass behaviour can be investigated as a function of temperature for a given frequency (Fig. 6.20). The graph obtained for a chalcogenide glass shows that the dynamic response (at a given frequency, Fig. 6.20b) varies considerably with temperature. The Young's modulus E curve shows a dramatic decrease of storage modulus while damping (proportional to loss factor) increases dramatically near the glass transition ($T_g = 280\,^\circ\mathrm{C}$). In fact, at low temperatures glass shows an elastic behaviour (rigid with no loss) that progressively changes into viscoelastic behaviour.

6.4.4
Modelling Real Solids

The modelling of glass behaviour requires the association of several Maxwell elements and/or Kelvin elements. The simplest model allowing capture of instantaneous and delayed elasticity as well as viscous flow is the Burger association of a Maxwell (G_1, η_1) element and a Kelvin (G_2, η_2) element in series. Under an applied constant stress, strain is written (Fig 6.21):[6)]

$$\gamma(t) = \tau_o \left[\frac{1}{G_1} + \frac{1}{G_2}\left(1 - \exp - \frac{t}{T_R^2}\right) + \frac{t}{T_R^1}\right] \tag{6.56}$$

where $T_R^1 = \eta_1/G_1$ and $T_R^2 = \eta_2/G_2$. The solution is plotted in Fig. 6.21.

6) $\tau = \tau_K = \tau_M$, $\gamma = \gamma_K + \gamma_M$; $\tau_o/\eta_1 = \dot{\gamma}_M \rightarrow \gamma_M = \tau_o/\eta_1 t + \tau_o/G_1$; $\tau_o = G_2\gamma_K + \eta_2\dot{\gamma}_K \rightarrow \gamma_K = \tau_o/G_2[1 - \exp(-t/T_R{}^2)]$.

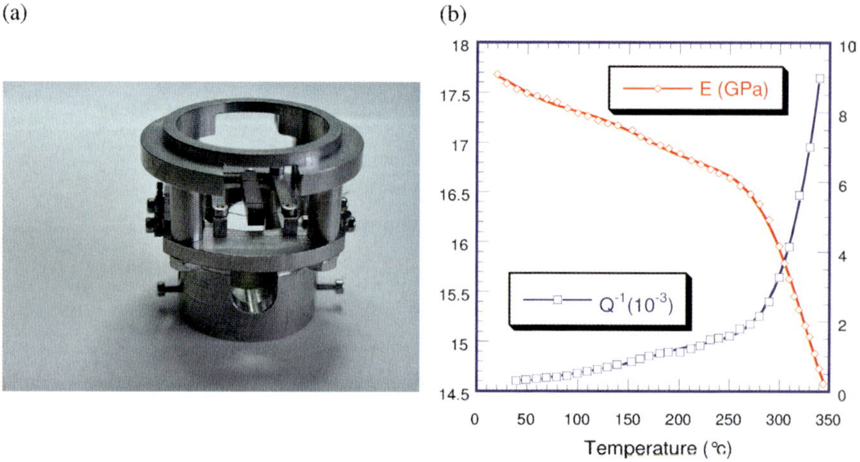

Fig. 6.20 (a) Dynamic resonance method (DRM) head (a parallelepiped beam is observed at the centre of the head). (Courtesy P. Gadaud, ENSMA Poitiers.) (b) Young's modulus and damping coefficient $Q^{-1} = 2\pi \tan\,\delta$ as a function of a temperature for a GeAsSe chalcogenide glass obtained with a DRM device. (From Le Bourhis et al , 2001a with permission from Elsevier.)

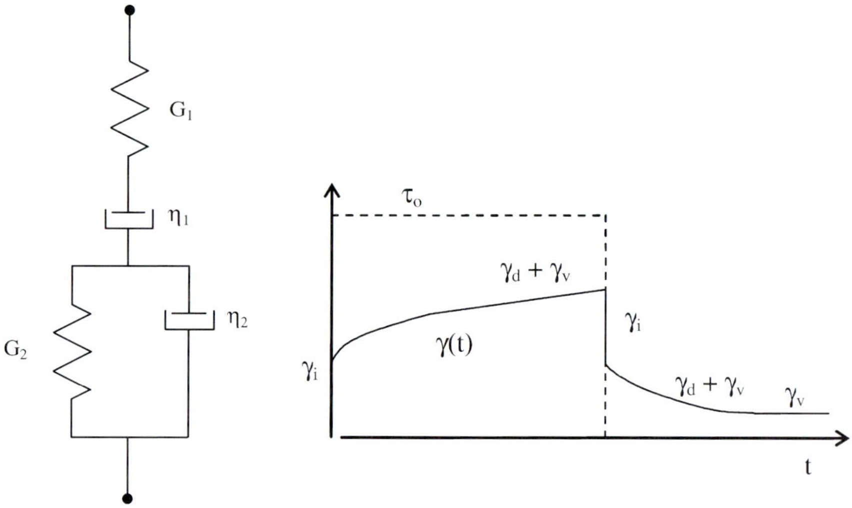

Fig. 6.21 Burger solid and its response to a constant applied stress. On loading under τ_o an instantaneous elastic strain γ_i, a delayed elastic strain γ_d and a viscous strain γ_d appear. On unloading only elastic strain γ_i and delayed elastic strain γ_d recover.

Equation (6.56) produces an instantaneous elastic strain (first term, $\gamma_i = \tau_o/G_1$), a delayed elastic strain (second term, $\gamma_d(t) = (\tau_o/G_2)[1 - \exp(-t/T_R^2)]$) and a viscous strain (third term, $\gamma_v(t) = \tau_o t/T_R^1$). The first two terms recover on unloading, whereas viscous flow is irreversible. Often used in the literature, anelasticity refers to the instantaneous plus delayed recoverable deformation while viscous flow, in contrast, is not recoverable.

Generalized Maxwell and Kelvin models are combinations of several Maxwell elements in parallel or Kelvin elements in series respectively. They were introduced to describe discrete relaxation times. The generalized Maxwell model is written

$$\tau(t) = \gamma_0 G(t) = \gamma_0 \sum_{k=1}^{n} G_k \exp(-t/T_k^R) \tag{6.57}$$

with the relaxation times of the different elements being $T_k^R = \eta_k/G_k$. This distribution of relaxation time may be due to a real spatial distribution of regions having different viscosities (because of temperature and composition heterogeneities). More fundamentally, such a distribution can be associated with the fact that the heights of the potential barriers impeding atomic or molecular motions are not equal. We generally only consider the average value while in fact there might be a broad and continuous distribution of barrier heights and hence relaxation times.

6.4.5 Functional Formulation

6.4.5.1 Creep

Let us consider the stress being applied instantaneously and take a Heaviside function

$$\tau(t) = \tau_o H(t - t_j) \tag{6.58}$$

with

$$H = 0 \text{ for } t - t_j < 0 \tag{6.59a}$$

$$H = 1 \text{ for } t - t_j \geq 0 \tag{6.59b}$$

Deformation is taken as a function of history (it is history dependent):

$$\gamma(t) = J(t - t_j)t_o \tag{6.60}$$

where J is the compliance characteristic of the material. For a Kelvin solid $J(t) = \gamma_o/\sigma_o[1 - \exp(-t/T_R)]$.

Now we consider stress is applied with n successive steps:

$$\tau(t) = \sum_{j=1}^{n} \Delta\tau_j H(t - t_j) \tag{6.61}$$

Assuming linear viscoelasticity, strain becomes

$$\gamma(t) = \sum_{j=1}^{n} \Delta\tau_j J(t - t_j) \qquad (6.62)$$

Equation (6.62) can be generalized to a continuously varying stress as

$$\gamma(t) = \int_0^t \frac{d\tau(t')}{dt'} J(t - t')dt' \qquad (6.63)$$

For the case of successive steps and continuously varying stress between steps then one has to add Eqs (6.62) and (6.63).

With a single creep measurement, one can theoretically calculate the entire creep compliance. Thereafter, any response can be calculated using this compliance and changing the stress history $\tau(t)$. There are experimental difficulties in imposing a true step stress function and in measuring $J(t)$ at long times. In practice, glass is assumed to have simple thermorheological behaviour. The mechanisms are considered to be independent of temperature, and only the kinetics is changed with the temperature. The creep function at a temperature can be determined from experiments at other different temperatures, the overall evolution with time arbitrarily plotted for a reference temperature (Section 6.4.5.4).

6.4.5.2 **Stress Relaxation**

Using the same approach as for creep we consider strain to be

$$\gamma(t) = \gamma_o H(t - t_j) \qquad (6.64)$$

Stress relaxation is written

$$\tau(t) = G(t - t_j)\gamma_o \qquad (6.65)$$

where G is the relaxation modulus of the material. For a Maxwell solid $G(t) = \sigma_o/\gamma_o \exp(-t/T_R)$.

For a continuously varying strain we have

$$\tau(t) = \int_0^t \frac{d\gamma(t')}{dt'} G(t - t')dt' \qquad (6.66)$$

These relations can be generalised for a three-dimensional loading (Appendix 9). The deviator s_{ij} and hydrostatic σ stress tensor being determined using the following similar equations:

$$s_{ij} = \int_0^t \frac{de_{ij}}{dt'} G(t - t')dt' \qquad (6.67)$$

Tab. 6.5 Shear relaxation times of the Prony series according to Rekhson (1986); $\sum_{k=1}^{n} w_{1k} = 1$.

k	1	2	3	4	5	6
T_{1k}^{R} (s)	19	291.9	1843	11 800	49 490	171 700
w_{1k}	0.0438	0.0611	0.0899	0.2516	0.2974	0.2561

$$\boldsymbol{\sigma}(t) = \int_0^t \frac{d\varepsilon}{dt'} K(t - t')dt' \tag{6.68}$$

where e_{ij} is the deviator strain tensor and ε the hydrostatic strain tensor. $G(t)$ and $K(t)$ are the shear and hydrostatic relaxation moduli:

$$G(t) = 2G\psi_1(t) \tag{6.69}$$

$$K(t) = 3K_e - (3K_e - 3K_g)\psi_2(t) \tag{6.70}$$

where ψ_i are the relaxation functions, G and K_g the shear and bulk instantaneous elastic constants and K_e the bulk equilibrium elastic modulus that accounts for both instantaneous and delayed volumetric elasticity.

Kurkjian 1963 and Rekhson 1986 proposed the use of generalized Maxwell models in the form of Prony series (Table 6.5) for the shear relaxation function $\psi_1(t)$:[7)]

$$\psi_1(t) = \sum_{k=1}^{n} w_{1k}\exp(-t/T_{1k}^{R}) \tag{6.71}$$

Moreover the hydrostatic relaxation function relaxes much more slowly and is often supposed to be (Guillemet and Gy, 1996).

$$\psi_2(t) = \psi_1(t/10) \tag{6.72}$$

The Prony series proposed by Rekhson 1986 for the shear relaxation is composed of six different elements with shear relaxation times which are to be considered by their relative contribution (Table 6.5).

7) This function can be written for a continuous spectrum of relaxation time as $\psi_1(t) = \int_0^\infty F(T_R)\exp(-t/T_R)dT_R$ with $\int_0^\infty F(T_R)dT_R = 1$. Taking $p = 1/T_R$ and $N(p) = 1/p_2 F(1/p)$ allows one to show $\psi_1(t)$ to be the Laplace transform of $N(p)$: $\psi_1(t) = \int_0^\infty N(p)\exp(-pt)dp$. This formalism can be employed to rationalize elastic-viscoelastic correspondence discussed in Section 6.4.5.3.

6.4.5.3 **Elastic–Viscoelastic Correspondence**

Let us consider the stress relationship obtained from the functional formulation for a viscoelastic material. We obtain

$$\tau(t) = \int_0^t \frac{d\gamma}{dt'} G(t - t')dt' \tag{6.73}$$

Let us compare this relation to that of an elastic solid $\tau = G\gamma$.

The elastic–viscoelastic correspondence says that the viscoelastic solution can be generated from the incompressible ($\nu = 0.5$) elastic one using a Boltzmann hereditary integral (hereditary for past dependent) with respect to past time from $t' = 0$ to the present time $t' = t$. This is illustrated as

$$G \rightarrow \int_0^t \frac{d}{dt'} G(t - t')dt' \tag{6.74}$$

This procedure is very useful for complex loading conditions (contact loading for instance; see Appendix 8). As a matter of fact, the solutions to these integral equations are completely dependent on the choice of material relaxation modulus $G(t)$.

6.4.5.4 **Superposition Principle (simple thermorheological behaviour)**

We shall assume that glass shows a simple thermorheological behaviour or, in other words, the mechanisms are independent of the temperature and only the kinetic is changed. Once the behaviour is known at a temperature (so-called reference temperature), it can be determined at any temperature using shift factors $\phi(T)$ for the relaxation time (Ferry, 1970), or at different times, relaxation curves are derived from another by a change in time scale (Figs. 6.22 and 6.23). Kurkjian 1963 studied extensively glass stress relaxation at different temperatures. Glass rods were first stabilized at a given temperature and twisted. The overall results were observed to adjust on a single master curve (Fig. 6.23). This can occur only if all relaxation times characterizing the material have the same temperature dependence and the weighting factors are constant. This implies that all the processes involved in the relaxation have the same activation energy and that the same processes are active at all temperatures. Indeed if a new process appears, the weighting factors must change, altering the shape of the curve. Fortunately, real glasses do exhibit thermorheological simplicity to a high level of approximation.

Like Lee et al. 1965 we shall introduce the thermorheological simplicity of glass by using the reduced time defined by

$$\xi = \int_0^t d\xi \tag{6.75a}$$

with

$$d\xi = \phi(T)dt' \tag{6.75b}$$

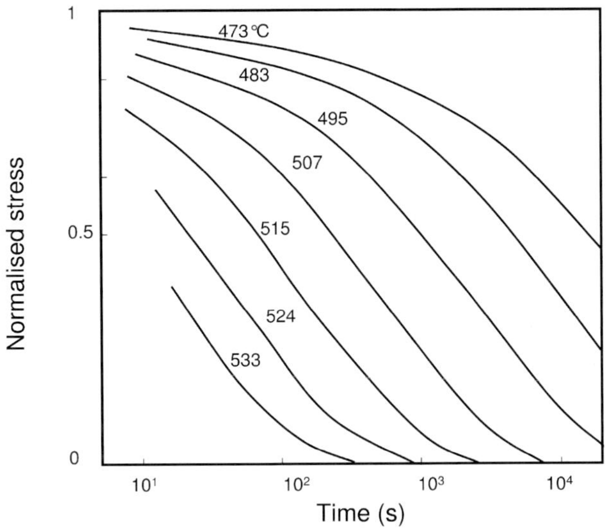

Fig. 6.22 Normalized stress as a function of logarithm of time, at seven experimental temperatures (according to Kurkjian, 1963).

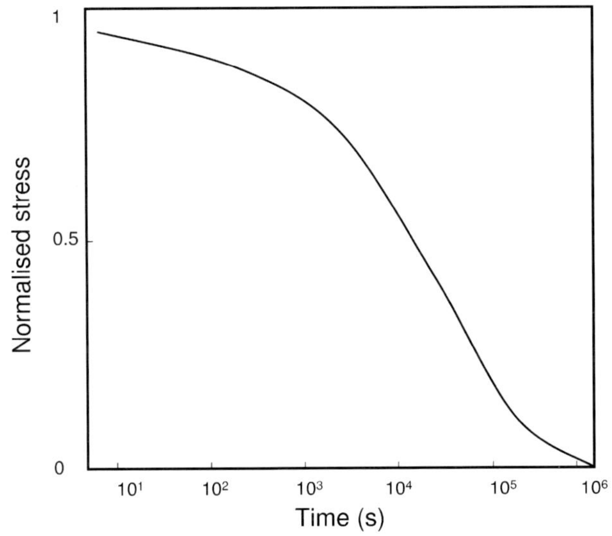

Fig. 6.23 Master curve at 473 °C produced from the data obtained at various experimental temperatures (Fig. 6.22; according to Kurkjian, 1963).

In general temperature varies and the integral cannot be simplified. When the temperature is constant, reduced time is proportional to time and is written $\xi = \phi(T)t$. We assume that at any temperature T, the stress relaxation function is $\psi_T(t) = \psi_R(\xi)$, where ψ_R is the stress relaxation at reference temperature ($\phi(T_R) = 1$).

The shifting procedure can be illustrated in a graph showing the ψ dependence with $\log(t)$. Consider only two different temperatures (reference and studied temperature):

$$\psi_T(\log t) = \psi_R(\log\xi) = \psi_R(\log[\phi(T)t]) = \psi_R(\log t + \log[\phi(T)]) \quad (6.76)$$

Then, the simple thermorheological behaviour says that the stress relaxation function at any temperature is obtained in the log scale by shifting the curve by $\log[\phi(T)]$. A reverse operation consists of determining the behaviour of the glass and shifting the creep functions (so called equivalence between time and temperature) to obtain the master curve evolution with time at a reference temperature (Section 6.4.3; Figure 6.24).

Shift factor can be obtained from viscosity data:

$$\phi(T) = \frac{\eta_R}{\eta} \quad (6.77)$$

where η_R is the viscosity at the reference temperature. Under a continuously varying temperature, reduced time incorporates the varying scale factor (see also Section 6.5).

Assuming a Prony series, the reference temperature is to be fixed by the following equation:

$$T_R = \frac{\eta_R}{G} = \sum_{k=1}^{n} w_{1k} T_{1k}^{R}$$

Then, the relaxation functions can be obtained at any temperature using the shift factor presented above. For instance, for a soda-lime-silica glass, the Prony series is shown in Table 6.5 and is to be taken for a reference temperature of 508 °C with $\eta_R = 1.86 \times 10^{15}$ Pa s (Guillemet and Gy, 1996).

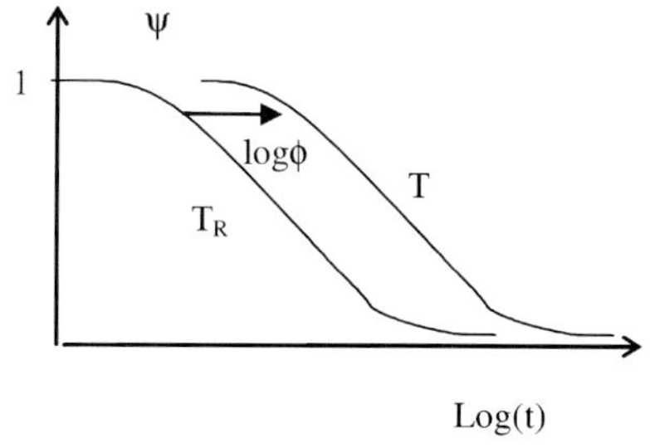

Fig. 6.24 Shear relaxation function as a function of time for two different temperatures.

6.5 Tempering of Glass

6.5.1 Introduction

The thermal history of glass has to be carefully controlled in order to design the level of residual stresses. When the material passes through the viscous–elastic domain submitted to a thermal gradient, compressive residual stresses appear at the surface (tensile in the core). Fast cooling rate will allow tempering of the glass while slow cooling rate and annealing of the glass will allow stabilization of the structure and reduction of residual stresses. A tempering set-up comprises either a ventilator or a compressor injecting air through nozzles that reaches the glass surface (Figs. 6.25 and 6.26). A parabolic stress profile showing compressive and tensile zones is represented and is typical of thermal tempering. Figure 6.27 shows a tempered backlight showing signatures of the process (nozzle air marks) since glass photo-elasticity allows for thermal residual stress heterogeneities to be observed (see also Chapter 11).

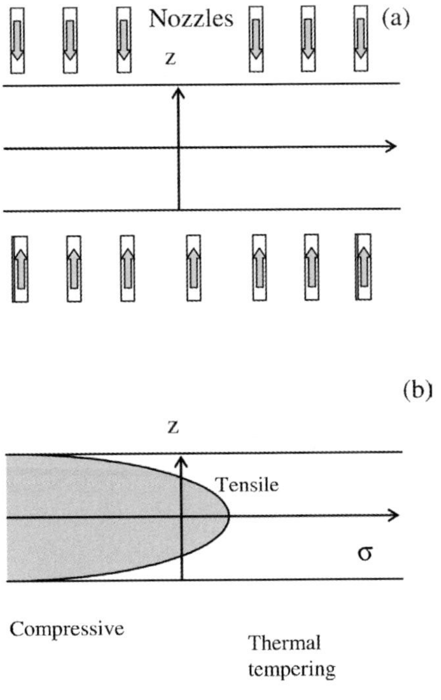

Fig. 6.25 (a) Schematic representation of tempering with nozzles injecting cooling air (grey arrows) and (b) residual stress profile in a thermally tempered glass sheet.

Fig. 6.26 Bending and tempering unit. (Courtesy Tamglass Ltd, http://www.tamglass.com/.)

6.5.2
Freezing Theory

Glass tempering prediction is a complex problem since

i. thermal transfer involves unknowns,
ii. thermal and structure histories are complex, and
iii. triaxial stress build up is to be determined.

Fig. 6.27 Nozzle photoelastic signatures (due to tempering) are observed here on a backlight (they correspond to residual stress heterogeneities and are better observed at dawn; see also Chapter 11).

In a first approach, Adams and Williamson 1920 proposed a thermoelastic model to determine the stresses in a glass sheet cooled at a constant rate c_R. The outer layers of the glass sheet will become rigid first, then as the inside cools, it tries to shrink, but is restrained by the rigid outer layers. Then, a simple representation of the tempering is to separate the sheet into three parts (the two surfaces and the core) and allow viscous flow in the core until it freezes (Fig. 6.28). A thermal gradient appears throughout the sheet. On passing the glass transition with cooling down to room temperature, the difference in displacements between the surface and the core cannot be accommodated any longer and residual stresses build in.

If the glass is cooled at a constant rate, the temperature profile is parabolic and temperature variation through the sheet (between the surface and the centre) is written

$$\Delta T = t_H^2 c_R c_P / 8k \tag{6.78}$$

where t_H, c_R, c_P and k are the thickness of the sheet, the cooling rate, the specific heat and thermal conductivity of the glass respectively.

The biaxial stresses are

$$\sigma(z) = \frac{E\alpha}{1-\nu}[\overline{T} - T(z)] \tag{6.79}$$

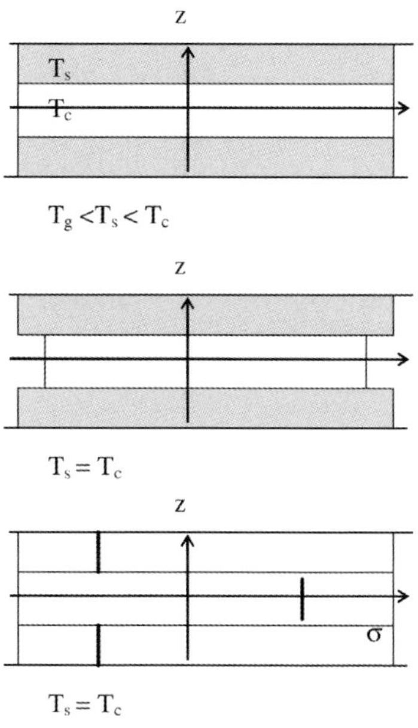

Fig. 6.28 Schematic representation of residual stress generation in a thermally tempered glass sheet. Viscous flow is possible over T_g. Thermal gradient builds in. On passing the glass transition and obtaining uniform temperature, stress builds in.

where $\overline{T}$ is the mean temperature, $T(z)$ the temperature at position z, E the Young's modulus, ν the Poisson ratio ($E/(1-\nu)$ is often called the biaxial modulus) and α the thermal expansion coefficient of the glass. When the temperature distribution is parabolic, maximum tensile stress appears at the centre of the glass sheet:

$$\sigma_c = \alpha_s E/3(1-\nu)\Delta T = A t_H^2 c_R \tag{6.80}$$

This very simple relation shows that a thicker sheet yields larger thermal gradient and hence larger residual stresses. The effect of using larger cooling rate is also straightforward leading to more tempering. It can be anticipated that tempering of thin glasses is difficult since it requires technology for large cooling rates. One must also consider that such tempering may mark the glass as strong air flows have to be injected through the nozzles which may deform the glass surface and damage the optical quality.

The three-part model can to be extended to a continuous model allowing for a parabolic profile of stresses to be obtained (Aben and Guillemet, 1993; Figure 6.25). Equibiaxial stress field is written

$$\sigma_x = \sigma_y = 3\sigma_c\left[\frac{1}{3} - \left(2\frac{z}{t_H}\right)^2\right] \tag{6.81}$$

where σ_c is the tensile residual stress at the sheet centre (at $z=0$, core stress). At the surface, residual stress is $\sigma_s = -2\sigma_c$. Stress is null for $z/t_H = 0.3$ or at a distance of 1/5t_H from the surface.

6.5.3
Stress Relaxation

The first approach described above supposes that stresses are frozen at T_g while stress relaxation happens at characteristic times that depend on the temperature because of the viscous–elastic behaviour of glass. Supposing glass has a simple thermorheological behaviour (Section 6.4.5.4), we consider reduced time and have

$$\xi = \int_0^t d\xi \tag{6.82}$$

with

$$d\xi = \frac{\eta_R}{\eta}\,dt' \tag{6.83}$$

Stresses build in and partially relax because of the viscoelasticity of glass. Stresses are written assuming an equibiaxial stress field (subscript B, this is to be reconsidered close to an edge or a hole, see Chapter 11):

$$\sigma(z,t) = \int_0^t G_{BR}(\xi-\xi')\frac{\partial}{\partial t'}[\varepsilon(t') - \alpha_s T(z,t')]dt' \tag{6.84}$$

where G_{BR} is the relaxation modulus for biaxial stress which is deduced from the shear and hydrostatic relaxation moduli, $\varepsilon(t')$ is the strain at time t' which is the same for all values of z (the deviation of the induced thermal strains from the uniform actual value generates the stresses in the different layers of glass) and is defined by the equation of mechanical equilibrium, α_s is the thermal expansion coefficient for glass and $T(z, t')$ is the temperature at position z and time t'. The viscoelastic approach was revealed to yield a non-negligible discrepancy between modelling and experiments (Gardon, 1978; Narayanaswamy, 1978), because the rapid cooling of glass causes also its structure to deviate considerably from the stabilized (equilibrium) state. Hence realistic models have to consider the glass structural relaxation in order to be more predictive (Section 6.5.4). As a matter of fact, R&D numerical tools incorporate stress and structure relaxations.

6.5.4
Structural Relaxation

Not only stress relaxation has to be considered when modelling glass tempering but also the structural relaxation, since the glass structure is known to be out of equilibrium. This issue was addressed first by Acloque (1951). At a given temperature glass will tend to approach the liquid configuration. This indeed happens without applying any stress to the material. When a supercooled liquid freezes it departs from the liquid curve (see for instance Section 6.2.1). Then, the liquid freezes in a less dense configuration than the supercooled liquid would present at the same temperature. When time is allowed for the glass to rearrange it will progressively tend to the denser liquid configuration and hence glass structure is expected to change with time. At room temperature the time scale is beyond the human time scale and hence structural relaxation of glass is not detected. The situation is different when the glass is set at a higher temperature. Some examples are shown in Fig. 6.29 where two pieces of glass were set at the same temperature while previously being stabilized at two different temperatures (Moynihan, 1996). The glass that was stabilized at a temperature higher than that imposed relaxes more rapidly than the one that was stabilized at a lower temperature. The first glass presents a more open structure as it was stabilized at a higher temperature and shows a greater average volume, and hence a reduced resistance to configurational changes. In contrast, the glass stabilized at a lower temperature shows a denser structure and hence more resistance to structural changes. When tempering the glass, this phenomenon has to be taken into account as well.

In the transformation range, the glass no longer has enough time to reach its equilibrium state at each temperature because of its high viscosity. Glass structure will depend on tempering conditions (cooling rates; Gardon, 1978; Narayanaswamy, 1978; Scherer, 1982, 1996; Guillemet, 1990; Aben and Guillemet, 1993).

Markowsky and Soules 1984 proposed a way to predict the structure of glass as a function of its thermal history using the concept of fictive temperature first introduced by Tool 1946. Contrary to a liquid whose structure is known from its actual temperature, glass structure can be obtained knowing both the temperature T

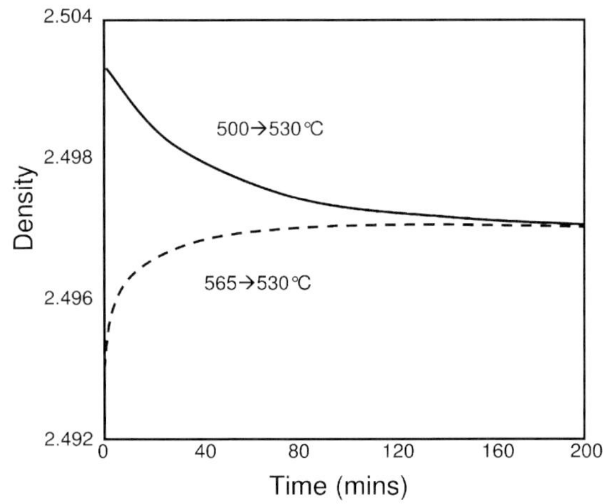

Fig. 6.29 Structural relaxation of two pieces of glass stabilized at two different temperatures (according to Moynihan, 1996).

and the fictive temperature T_f, that is, its instantaneous temperature and its prior history. When glass approaches transition, equilibrium state cannot be reached because of the elevated viscosity. By definition, the fictive temperature T_f of a glass corresponds to the temperature of the liquid which is in the same structural state as the glass under study. In a stabilized glass, $T_f = T$ while in a non-stabilized glass $T_f \neq T$. In practice, the fictive temperature corresponding to the state A of a glass is represented in Fig. 6.30 by point B and can be obtained by taking the line parallel to the expansion curve of the solid and determining its intercept with the expansion

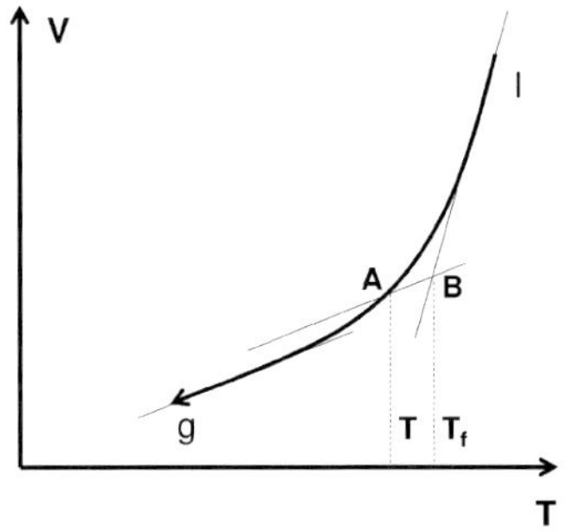

Fig. 6.30 Volume–temperature cooling curves showing fictive temperature geometrical determination (l and g, liquid and glass domains; A, the point at which fictive determination is made; B, the intercept of the expansion curve of the liquid and the line parallel to glass expansion curve passing through A). The temperature at B is by definition the fictive temperature at point A.

curve of the liquid. A glassy state can be viewed as a liquid-like state which has been frozen in. The liquid-like structure is then known from the fictive temperature of the liquid-like structure. When a supercooled liquid is cooled more rapidly, the glass freezes at a higher temperature, the liquid-like structure corresponds to a higher temperature and hence the fictive temperature is higher. Let us consider again Fig. 6.29. The glass that was stabilized at a higher temperature shows a higher fictive temperature, and when forced at a lower temperature it relaxes to a liquid-like structure presenting a lower fictive temperature. Then, the second glass fictive temperature will instead increase progressively. Fictive temperature evolution depends on thermal history and this is the reason for the difference between the two glass behaviours. The glass presenting a higher fictive temperature that has a more open structure relaxes more rapidly. Markowsky and Soules 1984 proposed an algorithm to determine the fictive temperature evolution. Structural evolution being a nonlinear phenomenon, an iterative computational scheme was proposed.

Both stress and structural relaxation phenomena are to be taken into account in the modelling of glass tempering, structural relaxation accounting for approximately 25 % of the total residual temper stresses (Narayanaswamy, 1978; Rekhson, 1986). Shear stress relaxation happens at characteristic time scaling with T_R while structure is accepted to relax about ten times slower ($\sim KT_R \sim 9T_R$; Guillemet, 1990; Gy et al., 1994; Guillemet and Gy, 1996). A Prony series is then adopted again with characteristic relaxation times shifted by approximately one order of magnitude. Alternatively, a simpler KWW function often fits well the relaxation of glasses (Debast and Gilard, 1963; Scherer, 1996).

Let us consider an increment in time dt; from a given step numbered $n-1$ to the following step numbered n, we have

$$d\xi = \frac{\eta_R}{\eta} dt \tag{6.85}$$

ξ being referred to as the reduced time (Aben and Guillemet, 1993) and subscript R indicates the reference temperature. Here, the viscosity function has to capture structural relaxations (fictive temperature). This is obtained from the Adam and Gibbs theory of cooperative relaxations (Section 6.1.3.4) and is written

$$\eta(T, T_f) = \eta_o \exp \frac{B}{T[1 - (T_o/T_f)]} \tag{6.86}$$

When glass is at an elevated temperature its structure is at equilibrium and is that of the melt so that $T_f = T$ and we get a VFT expression for the viscosity. Close to glass transition, glass freezes progressively, T_f departs from T and viscosity is affected. After each increment of time the fictive temperature (structure) can be calculated following the Markowsky and Soules (1986) algorithm that suggests the use of 'partial fictive temperatures' T_f^k and a Prony series:

$$T_f^{k,n} = T^n - (T^n - T_f^{k,n-1}) \exp(-d\xi/K\tau_k) \tag{6.87a}$$

$$T_{\mathrm{f}} = \sum_{k=1}^{6} w_k T_{\mathrm{f}}^{k} \tag{6.87b}$$

$$T_{\mathrm{f}}(z, 0) = T_{\mathrm{i}} \tag{6.87c}$$

where T_{i} is the initial temperature.

Figure 6.31 shows the fictive temperature (structure) evolution following this algorithm on cooling and then re-heating a glass at the same rate and then on cooling the same glass at a higher cooling rate. It can be seen that the two curves obtained with the same rate (on cooling and heating) coincide only well below T_{g} and well above it. In fact, hysteresis results from structural relaxation and can be detected experimentally using DSC as shown in Fig. 6.32 and care has to be taken in analysing experimental results (Höhne et al., 1996). In Section 6.2.1, we actually focus on one single route (cooling path). The actual behaviour can be commented on in view of the discussion regarding Fig. 6.29. In fact, on heating the glass the starting point is a liquid-like structure at a lower fictive temperature than the supercooled liquid following the cooling curve. This means that the starting point for the heating curve corresponds to a denser structure and rearrangement will be slower. Then the heating curve will always be below the cooling curve (for the same cooling and heating rate). Note also that on cooling a glass more rapidly, a larger fictive temperature is obtained, or a less dense glass (Fig. 6.31; silica shows an anomaly with a reverse trend).

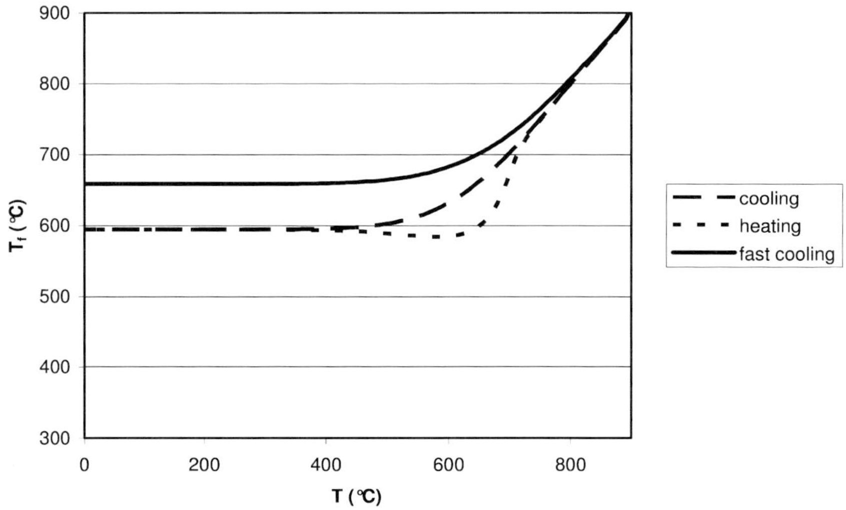

Fig. 6.31 Structure (fictive temperature, T_{f}) as a function of temperature on cooling and heating a glass piece. For comparison fast cooling route is shown and yields a glass with a higher fictive temperature (less dense glass).

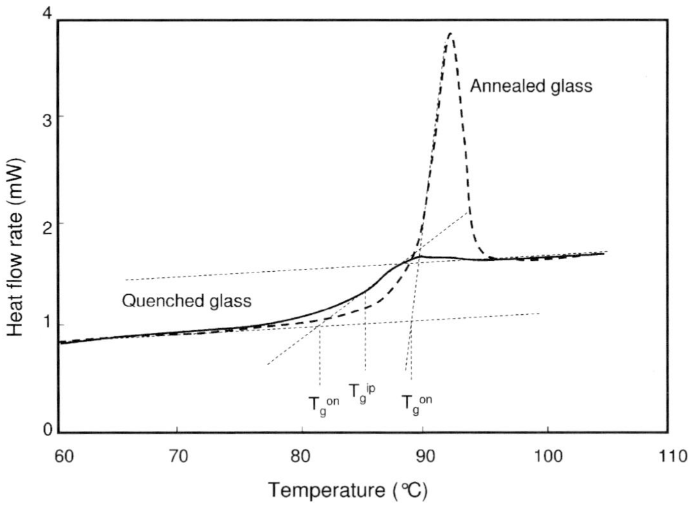

Fig. 6.32 DSC heating curves of annealed and quenched glasses (according to Höhne et al., 1996). Note relaxation peaks, the higher being observed for annealed glass resulting from the non-equilibrium nature of glass. DSC analysis should be considered with care since it would appear that the conventional glass transition temperatures of quenched glass is lower than that of annealed glass while it is the contrary in fact.

The calculation of residual stresses requires consideration of the thermal expansion mismatch between the liquid and the glass when the actual structural state is to be determined as described above. Assuming equibiaxial stress field (subscript B), stress is given by a formula incorporating structural relaxation through the fictive temperature parameter:

$$\sigma(z,t) = \int_0^t G_{BR}(\xi - \xi') \frac{\partial}{\partial t'} [\varepsilon(t') - \alpha_s T(z,t') - (\alpha_l - \alpha_s) T_f(z,t')] dt' \qquad (6.88)$$

where G_{BR} is the relaxation modulus for biaxial stress, $\varepsilon(t')$ is the strain at time t' which is the same for all values of z and is defined by the equation of mechanical equilibrium, α_s and α_l are the thermal expansion coefficients for glass and liquid respectively and $T(z, t')$ is the temperature at position z and time t' (we ignore edges here). Figure 6.33 shows the amplitude of core tensile residual stresses that build in under different experimental conditions. It shows a very good agreement between the calculated and experimental values allowing for great precision in the predictions. Figure 6.33 shows that, in fact, to generate residual stresses the initial temperature of the glazing needs to be over the transition temperature. This is not unexpected, since cold glass is elastic and therefore cannot be tempered. The potential for temper is fully developed only for quenching from temperatures well above the transition

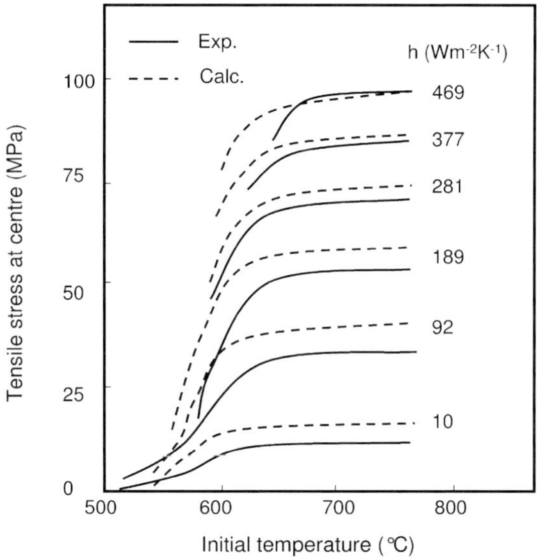

Fig. 6.33 Dependence of residual mid-plane tensile stress ($\sigma_c = \sigma(0, \infty)$) on the initial temperature T_i of a glass and on the heat transfer h (W m^{-2} K^{-1}); thickness 6.1 mm, natural convection corresponds approximately to the lowest curve (according to Narayanaswamy, 1978). They actually used the viscosity function $\eta(T, T_f) = \eta_o \exp \{B[(x/T) + (1-x)/T_f]\}$ with $x \sim 0.5$ instead of Eq. (6.86).

temperature. Also higher heat transfer (air flow) induces higher plateau-level temper residual stresses and increases the temperature at which the plateau is reached. When quenching the glass, it is important not to mark it by nozzles air impact. In R&D, it is important to consider also the complex geometry and thermal history of the glazing (holes, edges) and containers. This requires numerical methods and adjustment to observed results. These complex geometries can be considered using finite element analysis (FEA; Appendix 11). It is beyond the scope of this book to describe these methods in detail. It has been realized that before designing a new tool to produce for instance backlights, different series of numerical experiments will allow one to approach the optimized tool. This tool once fabricated shows great potentiality to produce a glazing prototype close to the demanded one. This of course allows for economical gains since many experimental tests are saved.

6.6 Transient Stresses

The simple gradient model shown in Fig. 6.28 can also be used to model thermal shock (see Chapter 2). When the glass sheet is annealed at $T < T_g$ the material remains elastic (i.e. no viscous flow happens). On cooling, transient compressive and tensile stresses build up and vanish when uniform temperature is reached. Stress is written

$$\sigma = \alpha_s E/(1-\nu)\Delta T \qquad (6.89)$$

where ΔT is the temperature variation through the sheet (between its surface and its centre), α_s the thermal expansion coefficient of the glass, E the Young modulus and ν the Poisson ratio. These transient tensile stresses at the surface are dangerous for the glazing since fracture may originate from a superficial flaw when it is under tension (see Chapter 7). As a matter of fact, thermal shock with $\Delta T < 0$ (on cooling down) is the most dangerous, as illustrated in Fig. 6.34, since tensile stresses develop at the surface where flaws are known to exist (see Chapter 8). Much greater temperature differences can be withstood in sudden heating than in sudden cooling because only when cooling is the surface placed under tension. Also, thermal endurance can be greatly increased by a toughening process which places the surface under compression. Transient stresses are proportional to Young's modulus and thermal expansion coefficient. In Chapter 2, we discussed extensively the way to select materials in view of relevant parameters that are the thermal expansion, the elastic modulus and the ultimate stress at fracture. We showed the potential of silica-rich glasses and borosilicates for applications requiring a high resistance to thermal shocks.

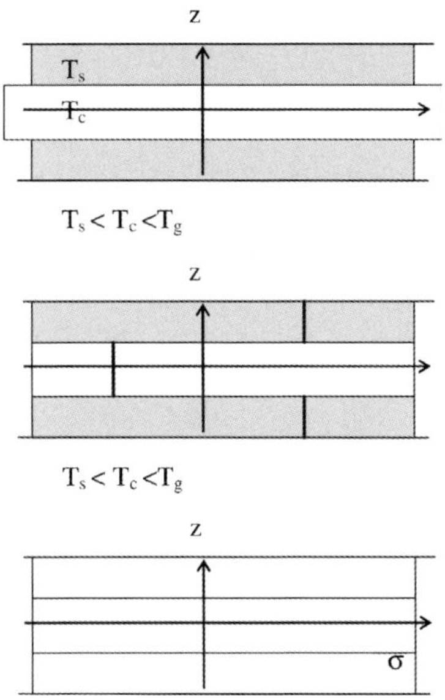

Fig. 6.34 Schematic representation of the transient stress generation in a glass sheet under thermal shock ($\Delta T < 0$, $T < T_g$) showing transient tensile stress at the surface and transient compressive stress in the centre. When temperature is uniform, transient stresses vanish.

Transient stresses should be very carefully considered also during the tempering process. Their partial relaxation leads to permanent stresses as discussed above. Under fast cooling there is a risk of the glass breaking under transient stresses. As discussed in Chapters 8 and 10, manufacturers control carefully production lines in order to avoid any abrasion of the glass surface (from the rollers, while packing) and this is surveyed up to the tempering operations. These procedures permit a gain in production yield as transient stress-induced breakages are avoided. The determination of transient stresses when the transition temperature is reached is more complex than when $T < T_g$. As reviewed above, for such temperatures stress and structural relaxation of the glass have to be taken into account.

6.7 Chemical Tempering of Glass

6.7.1 Introduction

Chemical tempering is applied for dedicated applications like aeroplane and high-speed train windshields, requiring high safety levels, and for thin glazing for which conventional thermal tempering would mark the surface or would not allow for sufficient strengthening. Although production yields are much less important, chemical tempering has to be used to achieve very high level of stresses required in the range of 500–600 MPa. The treated depth of glass (diffusion length) is reduced to hundreds of micrometres. The glass to be treated is immersed in a salt bath containing cations the radii of which are larger than those to be exchanged in the glass (Fig. 6.35). Ion exchange induces compression at the glazing surface. Generally, the

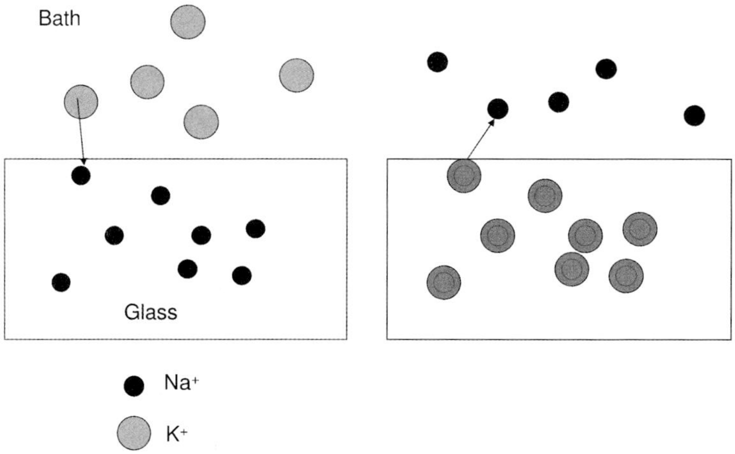

Fig. 6.35 Chemical tempering principle.

cations are alkali metal ions since they diffuse in glass much more rapidly than any other ions (Appendix 13). For instance, a glass containing sodium is strengthened by immersion into a potassium salt bath (generally KNO_3). To enhance diffusion elevated temperatures (400–500 °C) are used for the treatment. Unfortunately, these temperatures, while being below the transition temperature of the glass, are high enough for stress relaxation to take place and this phenomenon balances diffusion effects in common glasses. Therefore, aluminosilicate (ALS) glasses were developed to replace soda-lime-silica (SLS) glasses because alkali metal diffusion is more rapid in ALS glasses while viscous relaxation at treatment temperature is slower. Both effects allow for achieving higher stress levels. Incidentally, as the viscosity of ALS glasses is larger than that of SLS ones at the same temperature (Fig. 6.2), ALS glass fabrication is more difficult and expensive than that of SLS glass.

6.7.2 Ion Exchange and Stress Build-Up

The concentration of introduced ions can be calculated using Fick's diffusion law. As detailed in Appendix 13 since surface concentration c_s remains approximately constant, concentration $c(z, t)$ at depth is written

$$c(z,t) = c_s \left[1 - \operatorname{erf}\left(\frac{z'}{2\sqrt{Dt}}\right)\right] \tag{6.90}$$

where $z' = t_H/2 - z$ is the distance to the surface, c_s the concentration at the surface, D the coefficient of interdiffusion between the two kinds of atoms and t the immersion time.

We shall assume in a first step that no stress relaxation takes place at the bath temperature. The biaxial stresses are

$$\sigma(z) = \frac{EB}{1-\nu}[\langle c \rangle - c(z,t)] \tag{6.91}$$

where $\langle c \rangle$ is the mean concentration of the exchanged ions, $c(z, t)$ the concentration at position z and time t, E the Young modulus, ν the Poisson ratio, B the ion exchange-induced dilation coefficient and t the total exchange time. B is analogous to the thermal expansion coefficient in that the concentration c replaces the temperature. Thus B is the linear strain produced per unit exchange of ions (Tyagi and Varshneya, 1998):

$$B = \frac{1}{3V}\frac{dV}{dc} \tag{6.92}$$

A typical residual stress profile is shown in Fig. 6.36. As compared to thermally tempered glazing only a thin layer is under compressive stresses (Fig. 6.25).

Chemical tempering allows for treating more complex glass shapes than thermal tempering. However, care has to be taken so that the treatment penetrates the surface over flaw extension (~100 μm; see also Section 7.7.2).

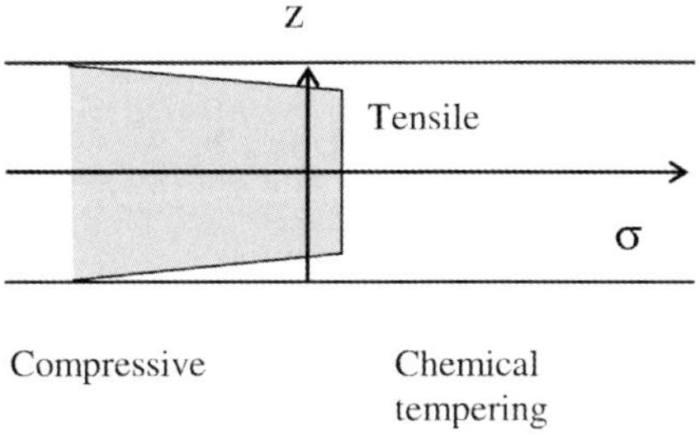

Fig. 6.36 Residual stress profile in a chemically tempered glass sheet.

6.7.3
Stress Relaxation

As mentioned above, stress relaxation may take place at treatment temperatures. This phenomenon is now considered. One has to make further assumptions: (i) diffusion does not modify the viscosity of the superficial layers and (ii) relaxation happens at the treatment temperature and we neglect relaxation while cooling rapidly the specimens. Assumption (i) is quite realistic while assumption (ii) allows for using time instead of reduced time (we neglect structural relaxation). Then, stress relaxation is written (Shen and Green, 2004)

$$\sigma(z,t) = \int_0^t G_{BR}(t-t')B\frac{\partial[\langle c\rangle - c(z,t')]}{\partial t'}\,dt' \tag{6.93}$$

where G_{BR} is the biaxial relaxation function.

6.7.4
Engineered Stress Profile Glasses

Recently, a new approach for chemical strengthening has been proposed by Green et al. 1999. The key feature is to carefully design the residual stress profile in such a way that the maximum compression is moved away from the external surface while stress gradient is carefully controlled in the surface region (Fig. 6.37). These glasses have been termed engineered stress profile (ESP) glasses. Normally, ion-exchanged glasses are processed to maximize the magnitude of the surface compression and the extent over which it acts (within economic constraints). In ESP glasses, it is not the overall depth of the surface compression that is important but rather the form of the stress profile in the near-surface region.

In fact, cracks are 'trapped' within the surface region (Fig. 6.38). These glasses show multiple cracking as a warning of the final failure. The processing may not necessitate deep compressive layers. This is economically attractive as it may lead to

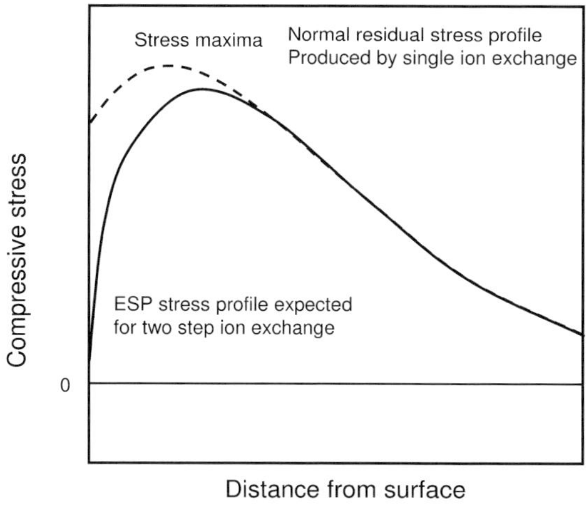

Fig. 6.37 Stress profile in ESP glazing obtained after a two-step ion exchange. (From Abrams et al., 2003, with permission from Elsevier.) For comparison a single ion exchange profile is shown; note even for single ion exchange there is surface stress relaxation. This is avoided at low temperature with smaller treated depth (Guillemet and Gy, 1996).

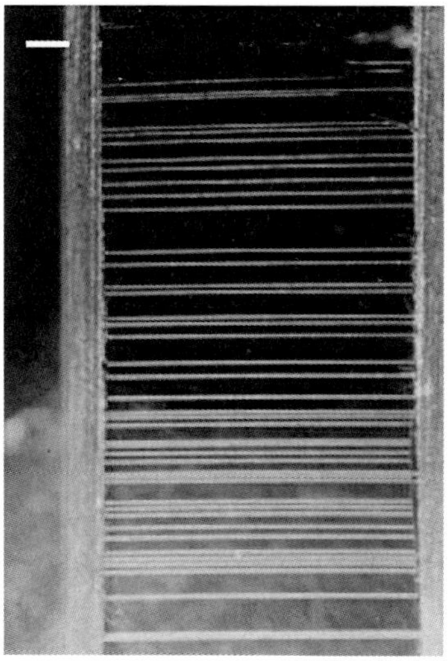

Fig. 6.38 Multiple surface cracks in a two-step ion-exchanged glass tested under four-point bending. Loading axis is perpendicular to crack direction (HF acid etching was used to widen cracks for better visibility; the white bar is 1 mm long). (From Abrams et al., 2003, with permission from Elsevier.)

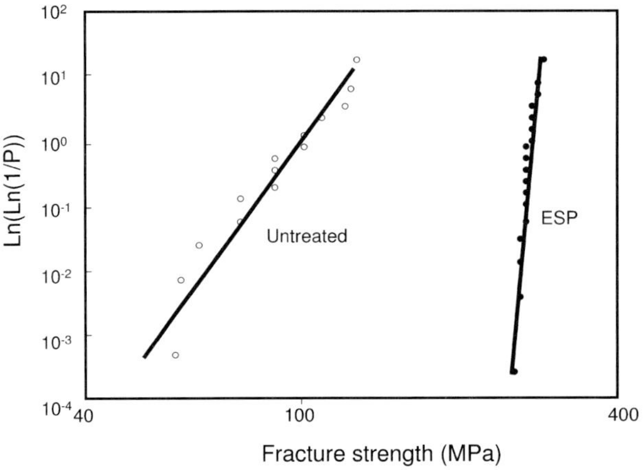

Fig. 6.39 Weibull distribution of ESP glass compared to untreated glass. (From Abrams et al., 2003, with permission from Elsevier.)

reduced processing times and, thereby, give rise to energy savings. The treatment is in two steps. In the first step, molten KNO_3 is used as a salt bath (at a temperature of 450 °C) to treat the aluminosilicate glasses. Here again soda-lime-silicate glass is difficult to ion exchange, as exchange depths are limited by slow diffusion rates and stress relaxation. Therefore ALS glasses are preferred. A second exchange uses a mixed KNO_3/$NaNO_3$ salt bath at some lower temperature (400 °C). The effect of the second exchange is to make the stress gradient steeper near the surface and control the position of maximum compression.

Another important issue for these glasses is the low strength variability while improving the average strength, as illustrated in Fig. 6.39. Strength variability will be discussed in more detail in the next chapter. While annealed silicate glasses were tested in the Weibull modulus range of 5–10, ESP Weibull modulus may exceed 60. Both strength and variability are important issues as regards glass application. To the author's knowledge, these ESP glazings have not met commercial success.

7 Mechanical Strength of Glass

Glass failure at temperatures below the glass transition is brittle and statistical. Fracture happens at much lower stresses when the specimen is loaded under tension than when loaded under compression (see also Appendix 10) and is of utmost importance for glass applications.

7.1 Theoretical Strength

Orowan 1949 first proposed a rupture criterion in terms of elastic strength, surface energy (two new surfaces being generated upon fracture) and atomic structure. Fracture was expected to happen under tensile stress σ_{th} determined as (Appendix 5)

$$\sigma_{th} = \sqrt{\frac{E\zeta}{r_o}} \tag{7.1}$$

where E is Young's modulus and ζ the surface energy (ζ is usually identified with the free surface energy of the cleavage surface, although this is still under debate; Mattoni et al., 2005). Equation (7.1) shows that stress at fracture increases together with surface energy and Young's modulus. Taking for a soda-lime-silica glass $E = 70$ GPa, $\zeta = 0.6$ J m^{-2} (Table 7.1) and $r_o = 0.15$ nm (Si–O distance; see Section 5.2) one finds a theoretical strength σ_{th} as high as 16 GPa. Unfortunately, this value is well above experimental and usual values (50 MPa for flat glass). Such a huge discrepancy was attributed to the presence of flaws at the glass surface as discussed in detail below (see also Appendix 5). In fact, it was observed that strength was much dependent on the glass object. For instance, glass fibres show a much higher strength than flat glass. It became more and more obvious that flaws (or defects) were responsible for the deteriorated strength of glass as compared to the estimated strength.

Glass; Mechanics and Technology. Eric Le Bourhis

ISBN: 978-3-527-31549-9

Tab. 7.1 Toughness, Young's modulus and surface energy[2] values for different glasses (Aben and Guillemet, 1993; Barton and Guillemet, 2005).

Glass	K_c (MPa m$^{1/2}$)	E (GPa)	ζ (J m^{-2})
Silica	0.74–0.81	73	0.65
Soda-lime-silica	0.72–0.82	70–74	0.4–1
Borosilicate	0.75–0.82	64–89	0.63
Aluminosilicate	0.85–0.96	83–91	0.63
Lead–silicate	0.62–0.73	58–65	0.44

7.2
Tensile Resistance of Glass

The simplest way to test a glass in tension is in principle to pull a rod or fibre, although one must note that in practice there are several difficulties in gripping the specimens. Three- and four-point bending is often used as well as biaxial flexure and brasilian or diametral compression tests (Zhou et al., 2006). The latter allows generation of a tensile zone into the specimen (Le Bourhis, 1996). Except for fine and long fibres (Appendix 6) where gripping problems can be solved, one should make the central and tested zone smaller in diameter on using uniaxial tensile tests so that failure happens from the fibres and not at the grips. Glass is a so-called brittle material at temperature well below T_g and deforms elastically until fracture (Fig. 7.1). In the elastic domain (domain 1) the uniaxial tensile loading curve $\sigma(\varepsilon)$ of a glass specimen is then linear.[1]

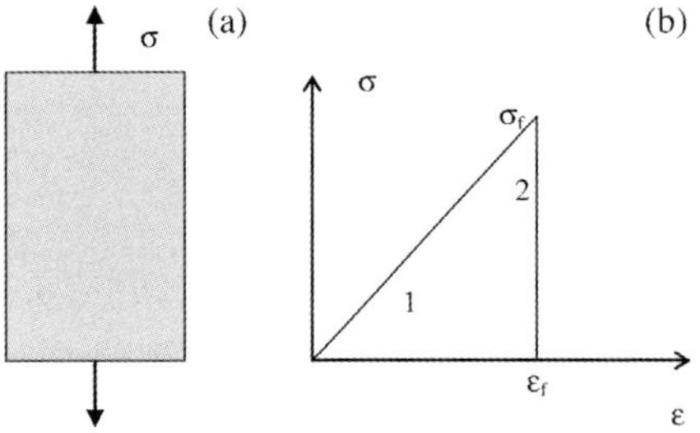

Fig. 7.1 (a) Tensile stressed specimen; (b) σ–ε tensile curve. Domain 1, slope is Young's modulus E. Domain 2 corresponds to fracture. σ_f and $\varepsilon_f = \sigma_f/E$ are the stress and strain at rupture respectively.

1) Deviation from Hooke's law is observed under elevated strains (~10%). Glass becomes either stiffer (silica, germania) or less stiff (SLS; Hillig, 1962; Krause et al., 1979). Such elevated strains can be commonly obtained in coated fibres (Shiue et al., 1996).

2) Native surface (see Chapter 5). Considering that $G_c = K_c^2/E$ is the energy release on fracture, one finds with $E = 70$GPa and $K = 0.7$ MPa m$^{1/2}$, $G_c = 7$ J m^{-2}, that is, one order of magnitude larger than the actual value. This discrepancy is attributed to the rearrangement (relaxation) of the surface after breakage. Another reason might be the dissipation of energy in the plastic zone confined at the crack tip (see Section 7.4).

Stress σ is written

$$\sigma = E\varepsilon$$

where ε is the elastic strain and E the elastic modulus (Young's modulus). A larger value of E means a stiffer material, that is, a larger resistance to elastic deformations. While the material deforms along the loading axis, also a side deformation is induced the amplitude of which depends on the Poisson ratio (Appendix 9). Elastic modulus depends on glass composition and reflects both the interatomic bonding strength and connectivity of the network (packing density). The moduli of glasses are observed to increase together with the glass transition temperature (Fig. 7.2a). Metallic glasses

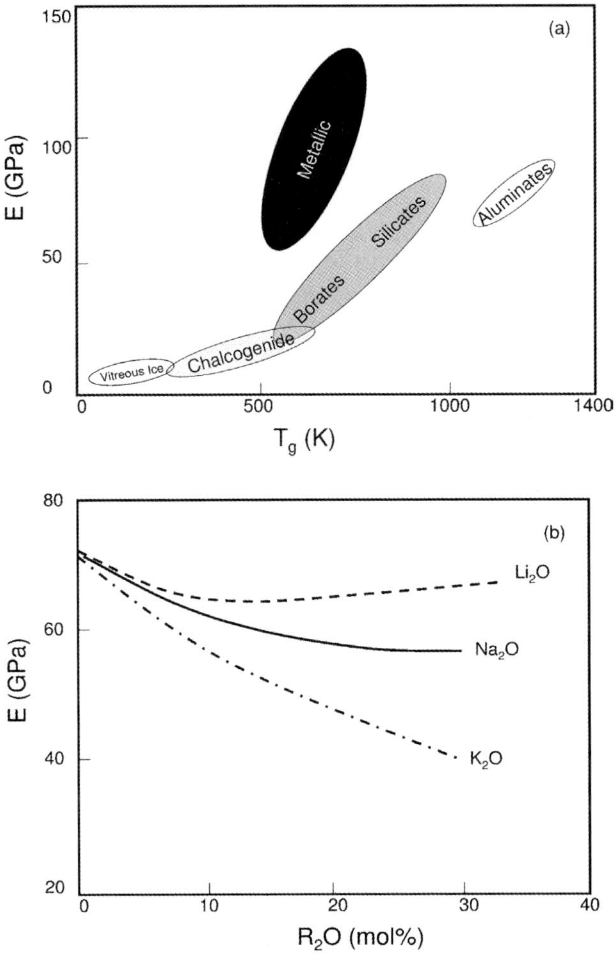

Fig. 7.2 (a) Young's modulus E (at 293 K except for amorphous ice at 77 K) as a function of transition temperature (according to Rouxel, 2006); (b) Young's modulus E of silicate glass as a function of modifier concentration (according to Scholze, 1991).

while showing transition temperatures below those of oxide glasses show much higher Young's modulus because of their higher packing density ($C_g \sim 0.7$ as compared to ~ 0.5 for common glasses; Chapter 5). When modifiers are introduced into the oxide glasses network, these atoms locate in the excess volumes while they weaken the network inducing NBO. These two phenomena play opposite roles. The resulting trends in silicate glass are shown in Fig. 7.2b demonstrating that the latter phenomenon is predominant since moduli of silicate glasses are observed below that of silica whatever the introduced alkali. The softening of the glass is more pronounced for a modifier having less Dietzel field intensity or for a modifier that is expected to show weaker interaction with oxygen (see Chapter 5) as seen in Fig. 7.2b. In contrast to a modifier, B_2O_3 oxide participates in the glass network and allows for maintaining and or even increasing Young's modulus (Livshits et al., 1982).

The elastic domain is terminated by catastrophic failure, the area under the linear curve corresponding to the elastic energy. Metallic glasses allow large elastic energy storage since they present elevated strengths. Failure is a dramatic event that has to be prevented. As discussed in Chapters 6, thermal and chemical tempering have been developed to strengthen glass articles. This issue is discussed in Section 7.7. In practice, the stress field is much more complex than in the specimen studied here under uniform and uniaxial tensile stress. Fracture will happen in tensile stressed zones. Hence, it is of utmost importance to understand fracture in order to improve strength and predict the domain where glass articles can be used safely. As we discuss now, fracture depends on intrinsic properties but more importantly on extrinsic ones (process-induced and damage-induced flaws) that make the problem very complex. Having all these parameters in mind, finite element analysis (FEA) can be carried out to predict the behaviour of glazing under realistic and complex loading conditions (Appendix 11).

Failure originates from a defect in the glass generally at the surface of the specimen. We reject here bubbles as an origin since these can be avoided nowadays in industrial processes thanks to fining or control operations (Chapters 10 and 11). As stated above, when we assume there is no defect in the glass and fracture originates from individual bonds, then tensile strength values are predicted to be much higher than those observed (Section 7.1 and Appendix 5). In fact, fracture starts at a critical defect or flaw and further propagates. The flaw at the origin of fracture can be easily detected in a flat glass among the broken pieces because it is surrounded by a smooth mirror zone (Fig. 7.3) the extension of which allows empirical determination of the level of stress at failure (Aben and Guillemet, 1993; Carré, 1996). This phenomenon has been of much help in detecting the origin of spontaneous fracture and to develop a treatment against NiS inclusions instability (Chapter 9; Barry et al., 1998). Stress at rupture is determined from $\sigma_r = M/\sqrt{r}$, with M the mirror constant and r the mirror radius. M is about $2\,\mathrm{MPa\,m^{1/2}}$ for a silico-soda-lime glass. This equation is of considerable practical importance since it allows estimation of fracture stress without having to measure it directly. This is very useful in complex shaped glass articles. In Fig. 7.3 the superficial origin can be easily located at the glass edge surface (on the top surface shown in the photograph). The crack expands first at a low speed creating flat or mirror-like surfaces. At higher speeds, rougher, less reflective surfaces are formed, a 'mist' area being observed around the mirror zone. Finally very rough and irregular

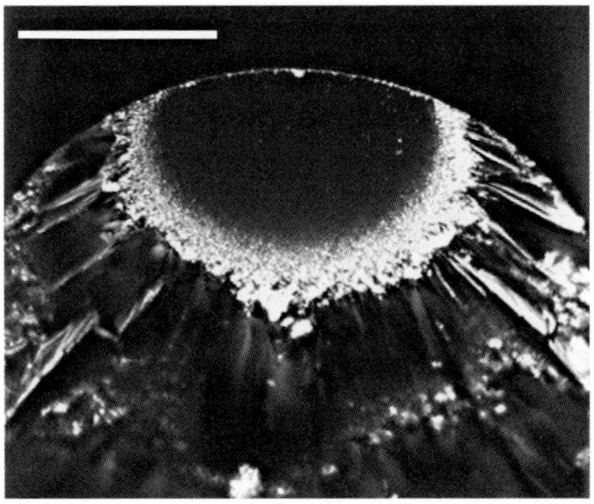

Fig. 7.3 Origin of fracture observed at the centre of the mirror zone (white bar is 1 mm long).

faceted surfaces form the so-called 'hackle' area. Further expansion of the crack yields crack branching that is a multiplication of the number of cracks. Surface origin of fracture is further confirmed by etching the glass surface (Pavelchek and Doremus, 1974). It was also realized for a long time that a fire-finished surface presented a higher strength than a polished one although this effect was not permanent, because of further surface damage after a period of time. Experiments carried out with hydrofluoric acid (HF) reveal that strength does not increase much when etch depth is less than 1–2 μm (Doremus, 1994). As the etch process is continued, some samples show great strength presumably because they are etched free of original flaws and fracture starts from shallow ones. The way superficial flaws are generated in a flat glass under a contact is extensively discussed in Chapter 8. Flaw generation may happen in the production line and in use. The nature of the flaws in glass fibres is more controversial. It is observed that fibre strength approaches the ultimate strength and hence flaws are to be searched for at the microstructural scale. They might be microscopic bubbles which are difficult to eliminate. Fining period (Chapter 10) would be too short as compared to the long period of time required to eliminate the smallest bubbles that are the slowest in rising to the melt surface (Appendix 4). Melting or fining operations are continued until visible bubbles are detected but invisible bubbles might still remain. Moreover, the glassy state presents a liquid-like structure, with a characteristic fictive temperature (Chapter 6) close to the glass transition temperature, or a structure frozen at the glass transition. The glass presents, as detailed in Chapters 5 and 6, an open structure giving rise to microscopic flaws. Moreover, glass fibre is quenched at much larger cooling rates than a flat glass because of a higher surface to volume ratio. Fibre fictive temperature is hence higher than that of a flat glass and presents a more open structure.

Let us consider a specimen containing a flaw of size $2a$ loaded under a uniform tensile stress σ (Fig. 7.4). It will fail when stress concentration (stress intensity factor, SIF)

$$K = \sigma Y \sqrt{a} \tag{7.2}$$

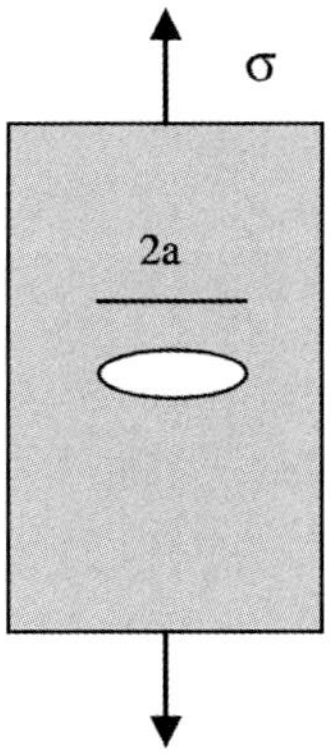

Fig. 7.4 Schematic representation of a specimen under tensile loading containing a flaw with extension $2a$.

Tab. 7.2 ***Y*** for most used crack geometry.

Y	**Crack geometry**
$\sqrt{(\pi)}$	Straight crack in infinite specimen
$1.12\sqrt{(\pi)}$	Edge crack in semi-infinite specimen
$2/\sqrt{(\pi)}$	Penny crack in infinite specimen

reaches a critical value called fracture toughness K_c that is related to Young's modulus and surface energy by $K_c = \sqrt{2E\zeta}$.[3] Y is a geometrical factor depending on the flaw geometry that can be found in Sih (1972) and Schzalke (1981). We take usually $Y = \sqrt{\pi}$ valid for a straight crack in an infinite specimen (Table 7.2).

Equation 7.2 is written at failure as

$$K_c = \sigma_f Y \sqrt{a} \tag{7.3a}$$

or

$$\sigma_f = \frac{K_c}{Y\sqrt{a}} \tag{7.3b}$$

The criterion of Eq. (7.3a) corresponds to an energy balance between the energy required to form two surface energy $2\zeta a$ and that elastically released $Ya^2\sigma^2/2E$ as detailed below. Equation (7.3b) shows that fracture stress amplitude increases with the toughness and decreases with the flaw size.

We assume that there are no residual stresses (glass being fully annealed). Otherwise, these contribute negative stress intensity factor when compressive

[3] This relation can be rewritten in terms of critical mechanical energy release rate G_c: $K_c = \sqrt{G_c E}$ (Lawn, 1993). G_c is the energy required to create two new fracture surfaces and for ideally brittle materials is 2ζ, where ζ is the surface energy per unit area. Plane stress conditions (for a thin plate; see also Appendix 9). Under plane strain conditions (for a thick plate)

$$K_c = \sqrt{\frac{G_c E}{1 - \nu^2}}$$

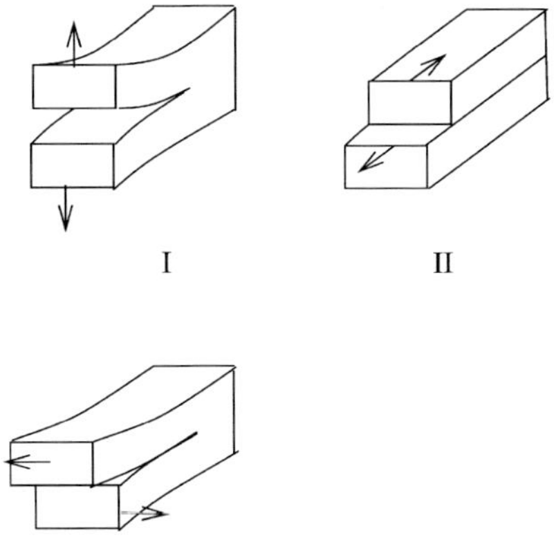

Fig. 7.5 Schematic representation of the three modes of fracture, namely modes I, II and III.

($K_r < 0$; see Section 7.7). Furthermore, we considered mode I corresponding to the opening of the flaw that is most commonly observed as compared to modes II and III corresponding respectively to sliding and 'tearing' (Fig. 7.5). From here onward, we only consider mode I and note the fracture toughness K_c.

Table 7.1 reports the toughness value at room temperature for different oxide glasses. These values are within a small range 0.7–0.8 MPa m$^{1/2}$ (see also Chapter 2 for comparison with other materials). In contrast, metallic glasses show high toughness ~50 MPa m$^{1/2}$ and can be deformed at low temperature ($<T_g$) even as bulk (Pampillo and Chen, 1974; Spaepen, 1977; Argon, 1979; see Appendix 10). Thereafter, the mechanical resistance of oxide glass specimens depends mostly upon the defects (flaws). In fact, there is a dramatic difference between glass fibre resistance and that of a flat glass since the respective process and use induce flaws at different scales. Glass manufacturers are well aware of this issue and sizing is applied to protect the fibres from abrasion (Chapter 10). Surface damages deleteriously affect the performance as shown schematically by an arrow towards lower strengths in Fig. 7.6. The fact that very high strength can be obtained from fibres plays much in favour of reinforcing polymers with glass fibres (Chapter 2).

7.3 Stress Concentration and Griffith Energy Balance

7.3.1 Stress Concentration

The first understanding of why fracture starts from flaws is due to Griffith (1920, 1924) who used the study of Inglis (1913) on the concentration of stresses at an elliptical flaw (Fig. 7.7). Let us assume that a flaw of semi-axes a and b ($a > b$) is

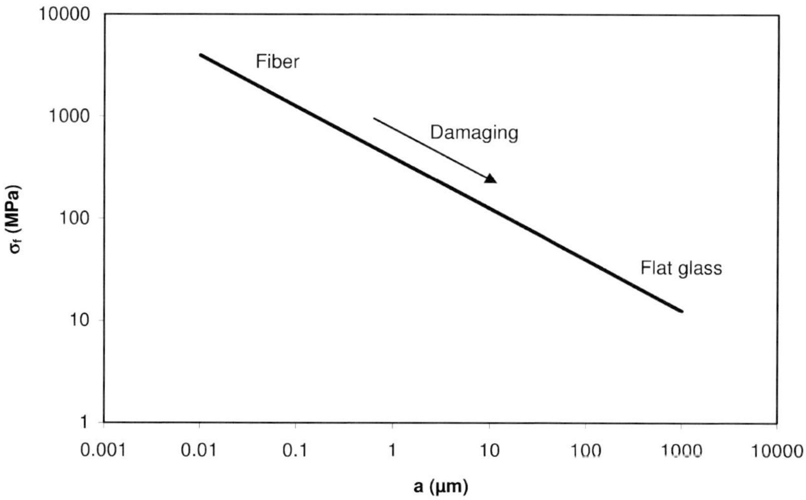

Fig. 7.6 Strength σ_f as a function of critical flaw dimension a ($K_c = 0.7\,\text{MPa}\,\text{m}^{1/2}$, $Y = \sqrt{\pi}$). Damaging the glass surface yields reduced performances (see also Chapter 8).

present in a beam submitted to uniform tensile stress σ. The flaw concentrates stresses at the points of minimum curvature, and the maximum stress σ' is given by

$$\sigma' = \sigma(1 + 2a/b) \tag{7.4}$$

If one considers a shape factor a/b of about 4.5, Eq. (7.4) tells us that stress concentration reaches $\sigma'/\sigma = 10$. Hence, the stress σ' at the minimum curvature is one order of magnitude larger than the uniform applied stress σ. This simple calculation shows that stress concentration may reach very high values when the flaw

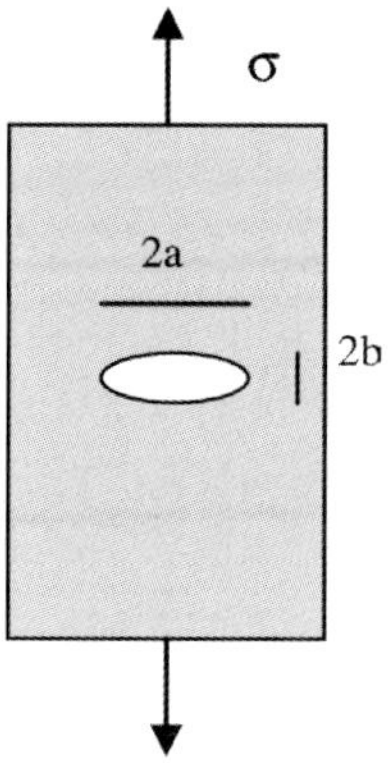

Fig. 7.7 Elliptical flaw of semi-axes a and b in a beam under uniform tensile stress σ.

tip becomes sharper. Fracture will then extend at the flaw tip and further propagate until specimen fracture.

7.3.2 Energy Balance

Further understanding of fracture is accomplished considering the total energy of the fractured specimen U. U is the sum of the mechanical U_m and surface U_s energies:

$$U = U_m + U_s \tag{7.5}$$

When fracture propagates a distance da the energy change is

$$dU = dU_m + dU_s = -Gda + dU_s \tag{7.6}$$

where G is the mechanical energy release and dU_s the variation in energy induced by creation of new surfaces.

Equilibrium is written

$$dU/da = 0 = -G_c + dU_s/da \tag{7.7}$$

where G_c is the critical mechanical energy release.

Considering, as Inglis (1913) did, an infinitely narrow elliptical flaw ($b \rightarrow 0$), we have per unit width along the crack front (in fact two fronts)

$$U_m = -\frac{\pi a^2 \sigma^2}{E} \tag{7.8}$$

$$G = -\frac{dU_m}{2da} = \frac{d(\pi a^2 \sigma^2 / E)}{2da} \tag{7.9}$$

while

$$U_s = 4a\zeta \tag{7.10}$$

Total energy is

$$U = -\frac{\pi a^2 \sigma^2}{E} + 4a\zeta \tag{7.11}$$

Both terms (U_m and U_s) are plotted in Fig. 7.8. U_s varies linearly and dominates for small crack size values. In contrast, for large crack sizes, U_m dominates and the total energy passes through a maximum (dU/da = 0) that allows determination of equilibrium for

$$\sigma_f = \sqrt{\frac{2E\zeta}{\pi a}} = \sqrt{\frac{EG_c}{\pi a}} \tag{7.12}$$

Over this value catastrophic failure happens.

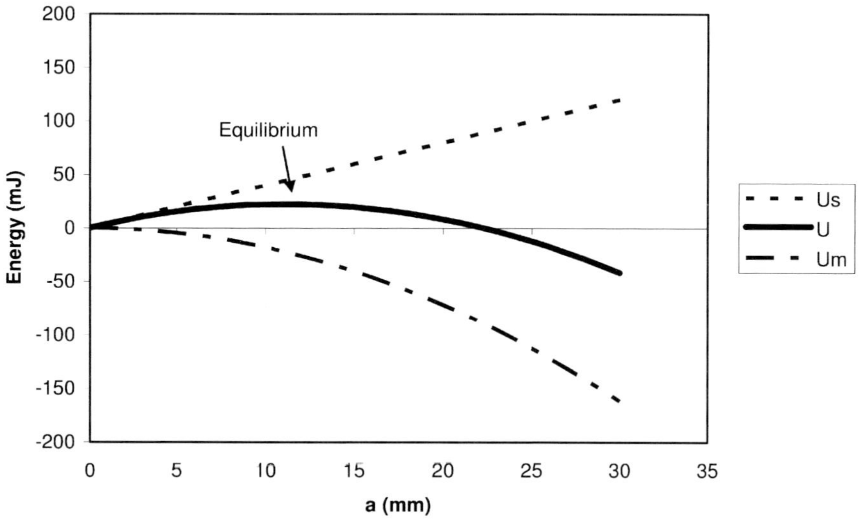

Fig. 7.8 Energy of an infinitely narrow elliptical flaw under uniform tensile stress σ. Crack equilibrium corresponds to U maximum ($\zeta = 1\,\mathrm{J\,m^{-2}}$, $E = 70\,\mathrm{GPa}$, $\sigma = 2\,\mathrm{MPa}$).

7.4
Linear Elasticity Crack Tip Stress Field

Considering mode I, the stress field at the crack tip proximity is determined to be (Irwin, 1958; Paris and Sih, 1965; Fig. 7.9)

$$\begin{pmatrix} \sigma_x \\ \sigma_y \\ \tau_{xy} \end{pmatrix} = \frac{K}{\sqrt{2\pi r}} \begin{pmatrix} \cos(\theta/2)[1 - \sin(\theta/2)\sin(3\theta/2)] \\ \cos(\theta/2)[1 + \sin(\theta/2)\sin(3\theta/2)] \\ \sin(\theta/2)\cos(\theta/2)\cos(3\theta/2)] \end{pmatrix} \tag{7.13}$$

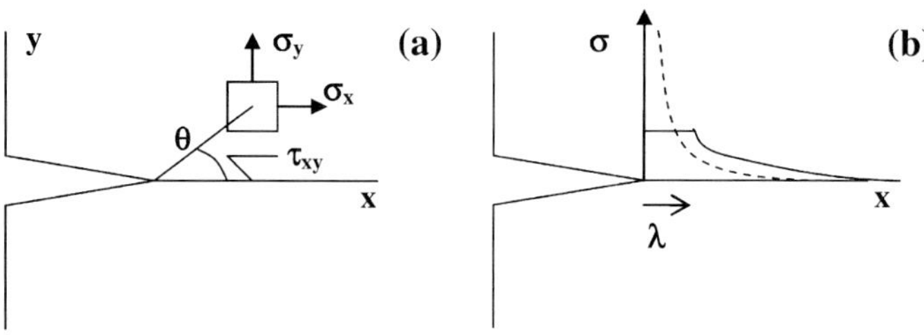

Fig. 7.9 (a) x, y axis and stress components; (b) schematic representation of stress concentration at crack tip (dashed line) and influence of a plastic zone (continuous line).

This reduces to

$$\sigma_{ij} = K(2\pi r)^{-1/2} f_{ij}(\theta) \qquad (7.14)$$

showing stress concentration at the proximity of the crack tip with dependence of the form $r^{-1/2}$.

It should be noted that a mathematical singularity is obtained at $r=0$ because the tip extremity curvature radius was taken as 0. We have not considered that stress concentration may induce a confined plastic deformation that shifts the singularity (Fig. 7.9b).

Using the Dugdale (1960) model one may estimate the size λ of the plastic zone at the crack tip to be

$$\lambda = \frac{\pi}{8}\left(\frac{K_c}{Y}\right)^2 \qquad (7.15)$$

where Y is the yield stress. For a soda-lime-silica glass ($K_c = 0.7\,\text{MPa m}^{1/2}$, $Y \sim H/3 \sim 2\,\text{GPa}$; Appendix 8) one finds a plastic zone at the tip extremity of size $\lambda \sim 50$ nm. This very small value is to be compared to that obtained for a metal: $\lambda \sim 1$ mm. This result emphasizes the actual brittle behaviour of glass since, contrary to a metal, only a few bonds at the crack tip support the elevated stress concentration.

The presence of a plastic zone at the crack tip has been until recently quite speculative given the relevant length scale. Using atomic force microscopy (AFM), Creuzet et al. (1993) could measure the deformation of a glass specimen at the emergence of a crack (or the topography of the surface at crack emergence). They showed that the experimental results in the vicinity of crack emergence were not in agreement with linear elasticity. More recently Célarié et al. (2003) suggested that 'Glass breaks like metal, but at a nanometer scale', although further experimental evidences are demanded (Guin and Wiederhorn, 2004). Furthermore, Mattoni et al. (2005) emphasize that for very a short crack it is difficult to discriminate between bond breaking and incipient plasticity yielding departure from classic Griffith theory.

7.5 SIF under Non-uniform Stress

To this point we supposed the specimen was uniformly loaded. Instead let us consider a straight crack ($0 < x < a$) in an infinite specimen under a varying stress $\sigma(x)$ normal to the crack surface. SIF is given by

$$K = \frac{2\sqrt{a}}{\sqrt{\pi}} \int_0^a \frac{\sigma(x)}{(a^2 - x^2)^{1/2}}\,\mathrm{d}x \qquad (7.16a)$$

The term $1/(a^2 - x^2)^{1/2}$ can be considered as a weight factor on stress $\sigma(x)$ and is called the Green's function. It maximizes at the crack tip vicinity. For a uniformly loaded specimen one finds again (taking for the integration $x/a = \sin\delta$)

$$K = \frac{2\sqrt{a}}{\sqrt{\pi}}\sigma\int_0^a \frac{\mathrm{d}x}{(a^2 - x^2)^{1/2}} = \sigma\sqrt{\pi}\sqrt{a} \tag{7.16b}$$

For an edge crack in a semi-infinite specimen, the integration is to be corrected with $Y = \sqrt{\pi}$ replaced by $1.12\sqrt{\pi}$ (Table 7.2):

$$K = \frac{2Y\sqrt{a}}{\pi}\int_0^a \frac{\sigma(x)}{(a^2 - x^2)^{1/2}}\mathrm{d}x \tag{7.16c}$$

7.6 Toughness Measurement

Toughness measurements require that specimens are prepared with care introducing a sharp crack of known size a. As detailed below, glass is characterized by a distribution of flaws that yields statistical fracture, the introduced crack being the critical one from where fracture initiates. Different loading modes are used and give their name to the related methods. When increasing the applied load F, stress concentration increases and yields crack growth. Stress concentration reaches toughness value at rupture. Stress concentration K as a function of applied load is given for the three main loading geometries, a being the crack length, and b, L and e being geometrical factors (Fig. 7.10).

7.6.1 Compact Tension

Under compact tension (CT) stress concentration is

$$K = \frac{F\sqrt{a}}{be}\left[29.6 - 185.5\frac{a}{b} + 655.7\left(\frac{a}{b}\right)^2 - 1.017\left(\frac{a}{b}\right)^3 + 638.6\left(\frac{a}{b}\right)^4\right] \tag{7.17a}$$

7.6.2 Notch Beam Test

For the notch beam test (NBT or SENB) stress concentration is

$$K = \frac{3FL\sqrt{a}}{2b^2e}\left[1.96 - 2.75\frac{a}{b} + 13.66\left(\frac{a}{b}\right)^2 - 23.98\left(\frac{a}{b}\right)^3 + 25.22\left(\frac{a}{b}\right)^4\right] \tag{7.17b}$$

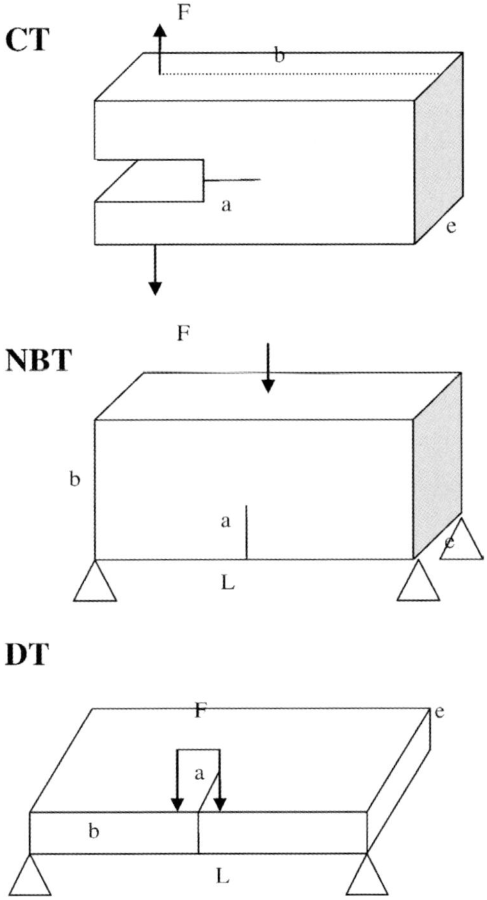

Fig. 7.10 Specimens for toughness measurement and loading conditions.

7.6.3 Double Torsion

Under double torsion (DT) stress concentration shows no dependence on introduced flaw size a and is

$$K = Fb\left[\frac{3(1+\nu)}{Le^4}\right]^{1/2} \qquad (7.17c)$$

All expressions show a linear variation of the stress intensity factor K with the applied force F. Interestingly on loading the specimens, the crack propagation can be measured in a domain where $K < K_c$, that is, before catastrophic failure. This domain is referred to as the subcritical domain. Subcritical crack growth has important implications on the ageing of glass, that is, its behaviour with time. This issue is addressed extensively in Chapter 9. It should also be noted that an indentation method may be employed to determine toughness when small quantities of material are available. This issue is discussed in more detail in Chapter 8.

7.7 Influence of Residual Stress on Strength and Fragmentation

As mentioned above, compressive residual stresses contribute negative SIF ($K_r < 0$). Stress concentration superimposes; hence we have

$$K = \sigma Y \sqrt{a} + K_r \tag{7.18a}$$

Also, fracture happens when the stress intensity factor K reaches the critical value K_c:

$$K_c = \sigma_f Y \sqrt{a} + K_r \tag{7.18b}$$

Equation (7.18b) is sometimes rewritten in the form $T = \sigma_f Y \sqrt{a} = K_c - K_r$ with T the apparent toughness including the residual stresses as a part of the crack resistance of the material. For compressive residual stresses ($K_r < 0$) there is a beneficial effect on T.

Equation (7.18b) can be written (for compressive stresses we take the absolute value)

$$\sigma_f = \frac{K_c - K_r}{Y\sqrt{a}} = \frac{K_c + |K_r|}{Y\sqrt{a}} \tag{7.19}$$

Hence, for compressive residual stresses, fracture happens under greater applied stresses than for the residual stress-free state. In fact, tempering is used commonly to strengthen the glass surface. As discussed in Chapter 6, compressive stresses build at the surface and are equilibrated by tensile stresses in the core of the glass. Therefore, breakage happens either (i) when tensile applied stresses allow for Eq. (7.19) to reach the critical value σ_f at a critical flaw present most probably at the surface (see also Section 7.8 and Chapter 8) or (ii) when an impact-generated flaw reaches the tensile zone (Fig. 7.11). The latter situation corresponds to complex loading and complex

Fig. 7.11 High-speed photographic sequence showing the catastrophic failure of a thermally tempered soda-lime-silica glass block caused by the impact of a 2 mm diameter tungsten carbide sphere at a velocity of 150 m s^{-1}. The fringes are isochromatics, the arrow points to the black fringe, which marks the region of zero stress (see Chapter 11). The catastrophic failure begins in frame 4. Note the change in shape of the stress pattern as well as the bifurcation of the cracks in frames 6 to 12. Surface compressive stresses are 200 MPa, inter-frame time is 1 μs. (Reprinted from Chaudhri and Liangyi, 1986, with permission from Macmillan Publishers Ltd.)

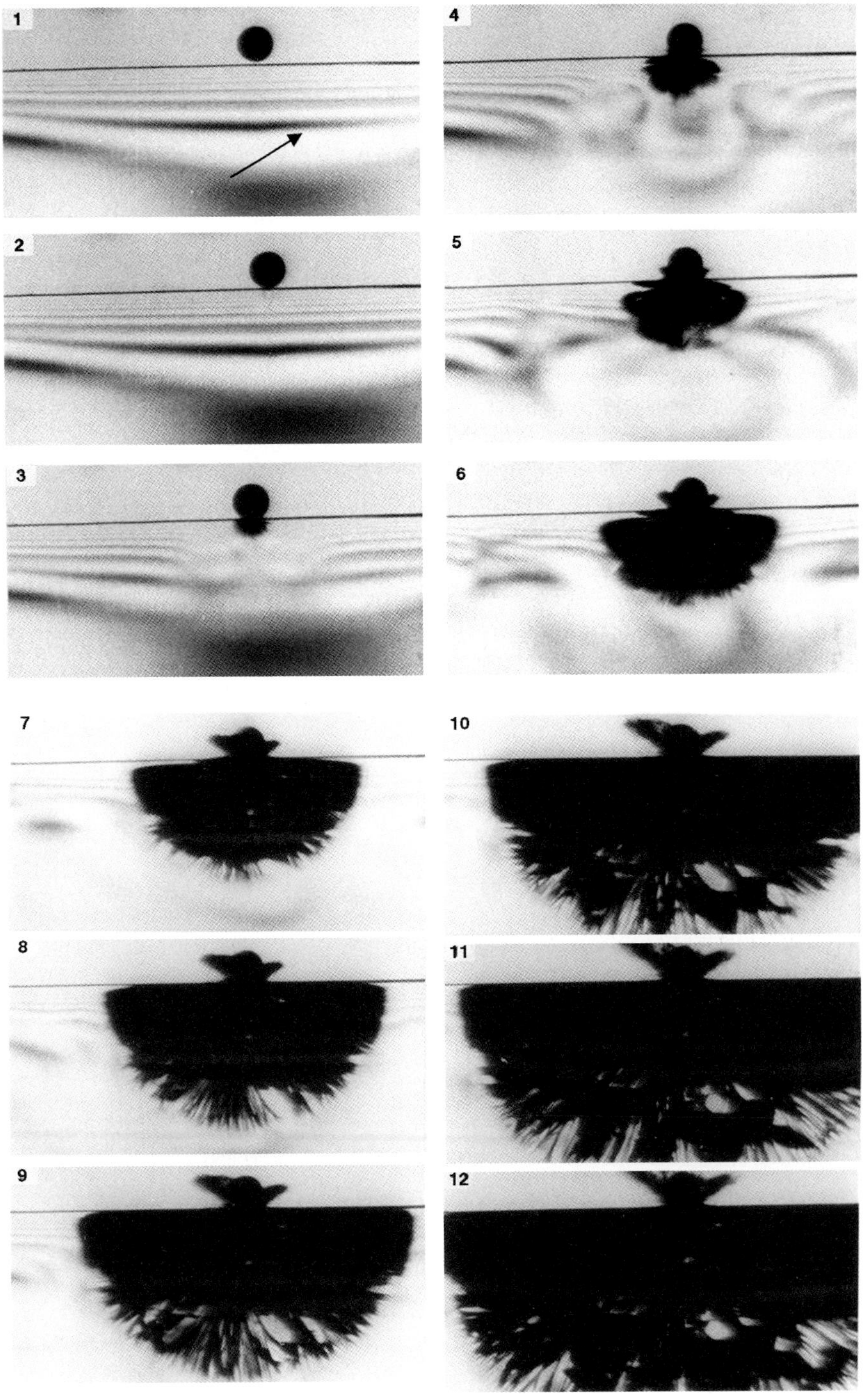
1
2
3
4
5
6
7
8
9
10
11
12

fracture path as revealed by high-speed photography (Fig. 7.11). This is discussed in Section 7.7.4.

Let us consider first situation (i) with a critical flaw present at glass surface either in a thermally or chemically tempered glazing, or under slowly and rapidly varying residual stresses respectively (Figs. 6.25 and 6.36). We suppose the glazing is stressed uniformly and illustrate then the beneficial role played by strengthening since applied stress at rupture increases accordingly.

7.7.1 Influence of Residual Thermal Stress on Strength

Thermal residual stress varies slowly with depth. We suppose it can be considered constant over crack length. We suppose cracks are present at the glass surface and take the residual stress equal to the surface compressive stress; hence $K_r = \sigma_s Y\sqrt{a}$ ($\sigma_s < 0$). We may consider crack length is negligible as compared to glass thickness $a \ll t_H$ so that $Y = 1.12\sqrt{\pi}$ (valid for an edge crack in a semi-infinite solid). Stress at rupture is

$$\sigma_f = \frac{K_c}{Y\sqrt{a}} - \sigma_s = \frac{K_c}{Y\sqrt{a}} + |\sigma_s| \tag{7.20}$$

Therefore, strengthening at the glass surface is directly related to the amplitude of the residual surface stresses. This means that thermally tempered glazing strength is shifted by the thermal residual surface stress amplitude.

7.7.2 Influence of Residual Chemical Stress on Strength

Chemical residual stresses vary much more rapidly with depth than thermal residual stresses. Let us suppose the compressive stress varies linearly and is null at a depth h (so-called exchanged depth; Chapter 6). Then, residual stress variation with depth z' ($= t_H/2 - z$) is written $\sigma_r = (1 - z'/h)\sigma_s$.

We have (Section 7.5)

$$\begin{aligned} K_r &= \frac{2Y\sqrt{a}}{\pi}\int_0^a \frac{\sigma_r(z')}{(a^2 - z'^2)^{1/2}}\,dz' = \frac{2Y\sqrt{a}}{\pi}\int_0^a \frac{\sigma_s[1-(z'/h)]}{(a^2 - z'^2)^{1/2}}\,dz' \\ &= Y\sqrt{a}\sigma_s\left(1 - \frac{2a}{\pi h}\right) \end{aligned} \tag{7.21}$$

Finally stress at rupture is

$$\sigma_f = \frac{K_c}{Y\sqrt{a}} - \sigma_s\left(1 - \frac{2a}{\pi h}\right) = \frac{K_c}{Y\sqrt{a}} + |\sigma_s| \times \left(1 - \frac{2a}{\pi h}\right) \tag{7.22}$$

Again, we find that stress at rupture increases with the residual stress amplitude. In the case of chemical tempering, not only the stress level is to be considered in evaluating the beneficial role of residual stresses, but also the treated layer thickness through the a/h ratio that plays an important role. This results from the fact that the superficial flaw size is not negligible as compared to the exchange depths reached under commercially available conditions. In fact, let us consider a high exchanged depth for which $a/h \rightarrow 0$. Then stress at rupture can be approximated by $\sigma_f \approx (K_c/Y\sqrt{a}) + |\sigma_s|$ and we obtain a result similar to that obtained for thermal tempering.

7.7.3 Influence of Residual Stress on Fragmentation

On breakage, release of tempering-induced elastic energy happens. When crack speed approaches half the Raleigh wave speed, it is possible for the crack to split into multiple branches (Fig. 7.11). Crack multiplication allows for dissipating more energy since more surfaces are generated. This phenomenon happens for a critical value of SIF about four times K_c (around $3\,\mathrm{MPa\,m}^{1/2}$ for a silico-soda-lime glass). Akeyoshi et al. (1967) showed that the density of the fragments generated on breakage is to be related to the level of residual stresses (degree of temper) (Fig. 7.12). It has been argued that the elastic energy stored in the tensile zone is the parameter controlling the glass breaking (Bouyne and Gaume, 2001).

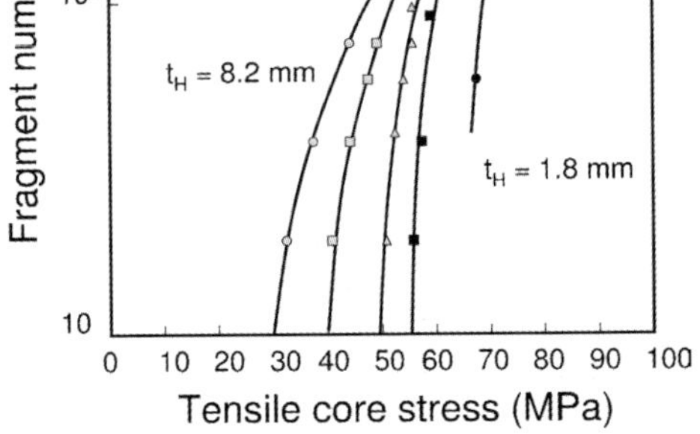

Fig. 7.12 Fragment densities (in a $5 \times 5\,\mathrm{cm}^2$ area) as a function of residual tensile stress (according to Akeyoshi et al., 1967).

The number of fragments increases with residual stress level and the thickness of the glass sheet, that is, the amplitude of elastic energy stored during the tempering process. Akeyoshi et al. (1967) carried out extensive experimental tests to determine the fragment densities in broken samples of different thickness and tempering degree. They produced a map showing the fragment density as a function of tensile core residual stress amplitude for commercially available flat glass thicknesses. This map is still in use and allows for determination and control of residual stresses in flat glass. A glass article is taken from the batch and fractured. The fragment density is then determined and residual stress amplitude extracted from the Akeysohi et al. chart. This control is demanded to fit the R43 international standard for tempered automotive glazing. Not only must tempering produce strength, but also on breakage the residual stresses are to be designed so that harmless fragments are generated. The R43 standard fixes the number of glass fragments in all $5 \times 5\ \mathrm{cm}^2$ areas to be over 40 and below 400. Such a control of the residual stresses is destructive and non-destructive methods are preferred (Chapter 11). Notably, the Akeyoshi et al. (1967) approach for thermally tempered glasses can be generalized to chemically tempered glasses (Bouyne and Gaume, 2001).

The consequence of the above results is that toughened glass cannot be cut or otherwise modified in shape as this would disturb the system of pre-stresses, and with fracture beginning at any point in the glass, the whole sheet would shatter immediately into small (although harmless) fragments. Hence, glass tempering is carried out offline after cutting, drilling and edge grinding operations.

7.7.4 Impact-Induced Fracture

There has been little experimental work dedicated to the way an impact crack generates and further enters the tensile region causing fracture. Figure 7.11 is hence of much interest as it shows a high-speed photographic sequence in cross-section of a thermally tempered soda-lime-silica glass impacted by a hard sphere thrown at $150\ \mathrm{m\,s^{-1}}$ towards its surface. The dark fringe corresponds to zero stress (these are photoelastic images; see also Chapter 11). As the sphere impacts the block, stress field changes as revealed by the photoelastic survey of the zone (frames 1–3). Interestingly between frames 4 and 5 the cracks enter the region of tensile stress and this leads to the initiation of catastrophic failure of the sample. The cracked area is observed as a dark and growing surface as time increases. The velocity of the cracks depends on the direction they propagate. The velocity of the cracks moving towards the surface of the impact is much smaller ($200–300\ \mathrm{m\,s^{-1}}$) than the velocity of the cracks spreading along the loading axis. In frames 7–12 cracks further spread and bifurcate, the images showing chaotic features. In fact, finite element analysis (FEM) happens to be of great help in getting insight into impact failure (Appendix 11) while we further discuss crack initiation under contact loading in Chapter 8.

7.8
Statistical Weibull Analysis

7.8.1
Introduction

The strength of a glass and more generally of a brittle material depends on its toughness K_c that is generally known and the critical flaw size a that depends on the industrial process and use, the ultimate stress being extracted from the following relationship (Section 7.2):

$$K_c = \sigma_f Y \sqrt{a} \tag{7.23}$$

Therefore, the determination of the glass mechanical performance requires the knowledge of flaw population which has to be determined statistically. Weibull (1951, 1952) statistics has been commonly used to characterize the statistical variation of the fracture strength of brittle materials such as ceramics, glasses and solid catalysts. It is based on the 'weakest link theory' which means that the most serious flaw in the material will control the strength, like a chain breaking if its weakest link fails (Fig. 7.13). The most serious flaw is not necessarily the largest one because its severity also relies on its location and orientation. In fact, strength is controlled by the flaw subjected to the highest SIF.

7.8.2
Functional Formulation

We must consider that flaws are distributed into the specimens. We shall suppose that the probability that two independent events occur is the product of the respective probabilities (Kittl and Diaz, 1988). Let us consider a volume V of the specimen and decompose it into two smaller volumes V_1 and V_2:

$$V = V_1 + V_2 \tag{7.24}$$

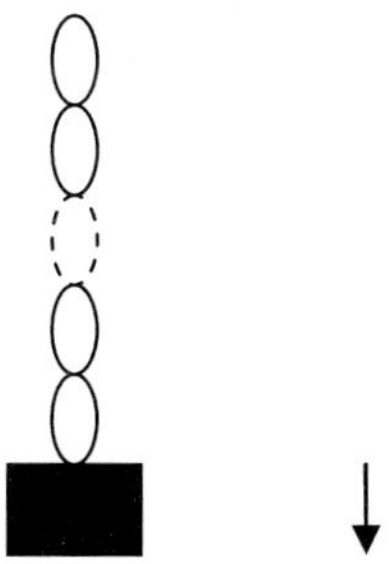

Fig. 7.13 Loaded chain of n links. The chain fails from its weakest link.

$P_s(V, \sigma)$ is the probability that a given volume V resists the stress σ. We have

$$P_s(V,\sigma) = P_s(V_1 + V_2,\sigma) = P_s(V_1,\sigma)P_s(V_2,\sigma) \tag{7.25}$$

Using this relationship for a volume of $V+\Delta V$ allows one to write

$$P_s(V + \Delta V,\sigma) = P_s(V,\sigma)P_s(\Delta V,\sigma) \tag{7.26}$$

while we have also

$$P_s(V + \Delta V,\sigma) = P_s(V,\sigma) + P_s'(V,\sigma)\Delta V \tag{7.27}$$

Thereafter we have

$$P_s'(V,\sigma) - [P_s(V + \Delta V,\sigma) - P_s(V,\sigma)]/\Delta V \tag{7.28}$$

Using Eq. (7.26), Eq. (7.28) becomes

$$P_s'(V,\sigma) = [P_s(V,\sigma)P_s(\Delta V,\sigma) - P_s(V,\sigma)]/\Delta V \tag{7.29}$$

Then

$$P_s'(V,\sigma)/P_s(V,\sigma) = [P_s(\Delta V,\sigma) - 1]/\Delta V \tag{7.30}$$

which for $\Delta V \rightarrow 0$ becomes

$$P_s'(V,\sigma)/P_s(V,\sigma) = P_s'(0,\sigma) = -F(\sigma)/V_o \tag{7.31}$$

where $F(\sigma)$ is the risk function for stress σ and V_o a unit volume.

Equation (7.31) is integrated as follows:

$$\ln P_s(V,\sigma) = -\int_V \frac{F(\sigma)}{V_o}\,\mathrm{d}V \tag{7.32}$$

so that

$$P_s(V,\sigma) = \exp\left(-\int_V \frac{F(\sigma)}{V_o}\,\mathrm{d}V\right) \tag{7.33}$$

Weibull suggested the following risk function:

$$F(\sigma) = \begin{cases} \left(\dfrac{\sigma - \sigma_u}{\sigma_0}\right)^m & \sigma \geq \sigma_u \\ 0 & 0 \leq \sigma \leq \sigma_u \end{cases} \tag{7.34}$$

where σ_u is the stress for which the probability for rupture is null, and is taken as $\sigma_u = 0$ for a brittle material, and m is the Weibull modulus.

Therefore we have

$$F(\sigma) = \left(\frac{\sigma}{\sigma_0}\right)^m \tag{7.35}$$

We have to consider two situations to integrate Eq. (7.33): stress is either uniform (Section 7.8.2.1) or not uniform (Section 7.8.2.2).

7.8.2.1 **Uniform Tensile Stress**

Under uniform tensile stress (for instance in a uniaxially tensile stressed fibre) Eq. (7.33) becomes

$$P_s(V,\sigma) = \exp\left[-\frac{V}{V_o}\left(\frac{\sigma}{\sigma_o}\right)^m\right] \tag{7.36}$$

This is the most commonly used form of Weibull resistance probability. Parameter m is called the Weibull modulus and is indicative of strength scatter; σ_o is indicative of strength level and corresponds to fracture stress with a failure probability of 63.2%. It is closely related to the mean strength σ_m of the distribution by $\sigma_m = \sigma_o \Gamma(1 + 1/m)$, with $\Gamma(x) = \int_0^\infty e^{-t} t^{x-1} dt$ (see also Appendix 8). For a Weibull modulus in the range of 10 typical for ceramics, $\Gamma(1 + 1/m)$ is about 0.9 to 1.

7.8.2.2 **Non-uniform Tensile Stress**

When stress varies in the volume of the material, Eq. (7.33) becomes

$$P_s(V,\sigma) = \exp\left(-\int_V \frac{F(\sigma)}{V_o} dV\right) = \exp\left[-\frac{1}{V_o}\int_V \left(\frac{\sigma}{\sigma_o}\right)^m dV\right] \tag{7.37}$$

Integration of Eq. (7.37) requires knowledge of the stress field.

Let us consider for instance a beam under flexure (Chapter 2; Fig. 7.14). Stress changes from compressive at the top surface to tensile at the bottom surface. The neutral line is where stress amplitude vanishes (Appendix 7). Accordingly, the stress

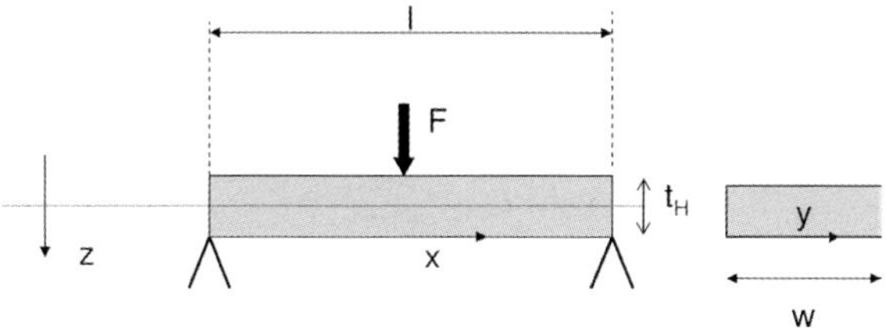

Fig. 7.14 Beam under flexure.

field is written

$$\sigma(x,z)=\sigma_M\begin{pmatrix}\dfrac{z}{t_H/2}\dfrac{l/2-x}{l/2} & \text{for } x\in[0,l/2]\\ \dfrac{z}{t_H/2}\dfrac{l/2+x}{l/2} & \text{for } x\in[-l/2,0]\end{pmatrix} \tag{7.38}$$

where σ_M is the maximum stress (at the beam surface).

Integration on volume V (in fact considering only the so-called tensile zones) yields[4)]

$$P_s(V,\sigma')=\exp\left[-\frac{V}{V_o}\frac{1}{2(m+1)^2}\left(\frac{\sigma_M}{\sigma_o}\right)^m\right] \tag{7.39}$$

Equations (7.36) and (7.39) were obtained using the same risk function. However, one notes that for the same given applied stress (tensile stress σ equal to maximum flexure stress σ_M) the probability that the specimen resists is higher under flexure. This is further discussed below and results from an intrinsic size effect, the maximum tensile stresses being applied to only a small part of the bent specimen.

3) Crack opening happens in tensile zones. Since stress is tensile for $z\in[-t_H/2,0]$ integration is over half of the beam volume. The beam volume V is $V=lwt_H$. We have

$$P_s(V,\sigma)=\exp\left[-\frac{1}{V_o}\int_V\left(\frac{\sigma}{\sigma_o}\right)^m dV\right]$$

$$\int_V\left(\frac{\sigma}{\sigma_o}\right)^m dV=\iiint_V\left(\frac{\sigma}{\sigma_o}\right)^m dxdydz=\int_{-l/2}^{0}\int_{-w/2}^{w/2}\int_{0}^{t_H/2}\left(\frac{\sigma}{\sigma_o}\right)^m dxdydz$$
$$+\int_{0}^{l/2}\int_{-w/2}^{w/2}\int_{0}^{t_H/2}\left(\frac{\sigma}{\sigma_o}\right)^m dxdydz.$$

$$\int_V\left(\frac{\sigma}{\sigma_o}\right)^m dV=2\left[\left(\frac{\sigma_M}{\sigma_o}\frac{1}{t_H/2}\frac{1}{l/2}\right)^m\left[-\frac{(l/2-x)^{m+1}}{m+1}\right]_0^{l/2}w\left[\frac{z^{m+1}}{m+1}\right]_0^{t_H/2}\right]$$
$$=2w\left(\frac{\sigma_M}{\sigma_o}\frac{1}{t_H/2}\frac{1}{l/2}\right)^m\frac{(l/2)^{m+1}(t_H/2)^{m+1}}{(m+1)^2}$$

$$\int_V\left(\frac{\sigma}{\sigma_o}\right)^m dV=2w\left(\frac{\sigma_M}{\sigma_o}\right)^m\frac{(l/2)(t_H/2)}{(m+1)^2}=\frac{V}{2(m+1)^2}\left(\frac{\sigma_M}{\sigma_o}\right)^m$$

then $P_s(V,\sigma)=\exp\left(-\dfrac{V}{V_o}\dfrac{1}{2(m+1)^2}\left(\dfrac{\sigma_M}{\sigma_o}\right)^m\right).$

Tab. 7.3 Weibull modulus for different classes of materials.

Material	Weibull modulus
Glass container, flat glass	3–7
Glass fibre	10–30
Ceramics	~10
Metals	~40

7.8.3
Population of Flaws

Glass specimens present a flaw size distribution. The determination of flaw population passes through an experimental study of a representative series of samples that are to be tested. Weibull statistics is subsequently used to fit the resistance probability $P_S(V)$ as a function of stress (Appendix 6). Weibull statistics is also applied extensively for studying brittle fracture even at the nanoscale (Pugno and Ruoff, 2006) as well as the interfacial rupture (Piellard, 2005). The scatter of flaw size distribution and of the fracture strength is indicated by the so-called Weibull modulus m (Table 7.3). From an engineering point of view, a high Weibull modulus can be as important as a high strength. A higher m value leads to a steeper flaw size distribution and thus to a lower dispersion of the fracture strength (Fig. 7.15). Interestingly a glass edge can be carefully abraded to obtain m as high as 10 (Barton and Guillemet, 2005). The highest Weibull moduli for glasses are obtained from fibres that can be produced almost flaw-free, m approaching 30 and tensile strength being close to the intrinsic one. Modern

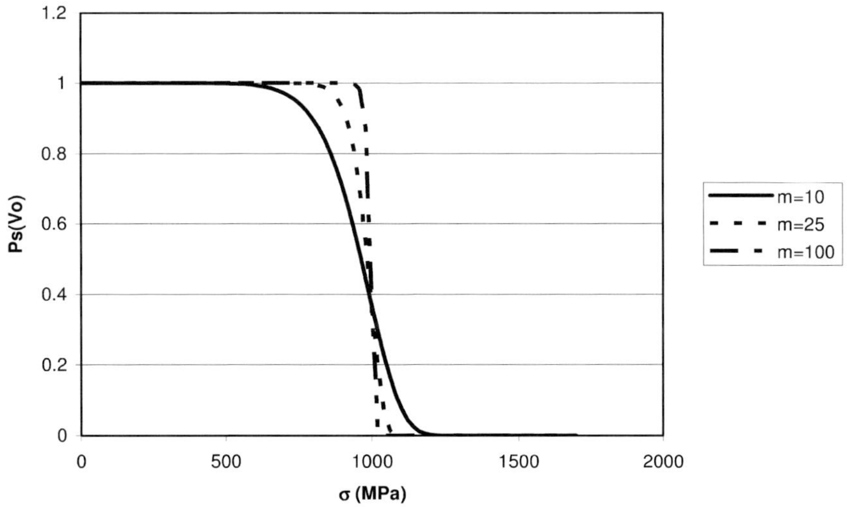

Fig. 7.15 Weibull probability $P_s(V_o)$ to resist stress σ ($\sigma_o = 1$ GPa). Three different Weibull moduli have been considered ($m = 10$, 25 and 100).

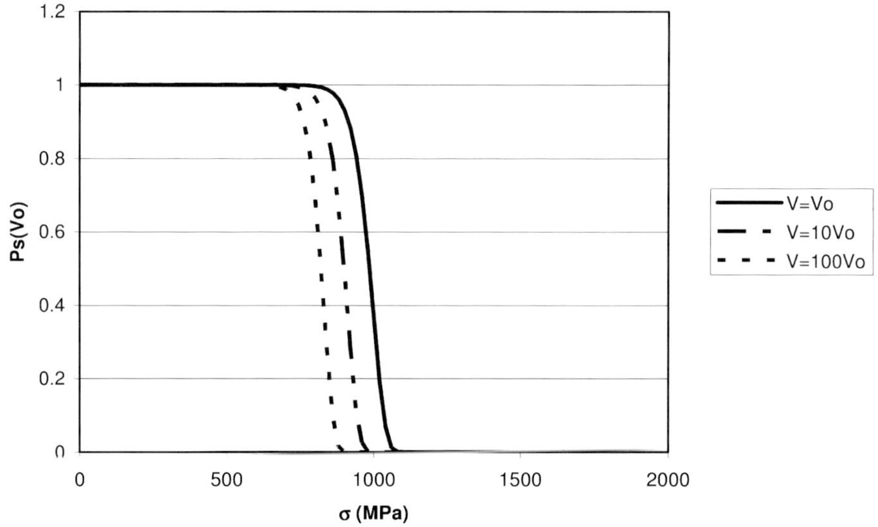

Fig. 7.16 Weibull probability to resist stress σ ($\sigma_o = 1$ GPa, $m = 25$). Three volumes have been considered ($V/V_o = 1$, 10 and 100).

fused silica optical communications fibres show m in the range 30–100! Some authors argued that the variation observed for $m > 30$ could result only from section variation of the studied fibre. The 'intrinsic' strength of silica fibre would then be about 12 GPa, close to the theoretical strength determined in Section 7.1.

Let us consider uniformly stressed specimens; we have

$$P_s(V, \sigma) = \exp\left[-\frac{V}{V_o}\left(\frac{\sigma}{\sigma_o}\right)^m\right] \tag{7.40}$$

When larger specimens are considered (size effect from laboratory to practical application) the probability that the sample resists stress σ is

$$P_s(V, \sigma) = P_s(V_o, \sigma)^n \text{ with } V = nV_o \tag{7.41}$$

Figure 7.16 plots the probability $P_s(V, \sigma)$ for increasing sizes (namely $V/V_o = 1, 10$ and 100) where mean strength is observed to decrease accordingly. Size effect is indeed inherent to brittle fracture. The load applied to lead the specimen to fracture decreases as the volume of the tested specimen increases because of the greater probability of containing dangerous flaws. A corollary to this is that higher average strength values are always observed in bending than in tensile tests, because the maximum stresses are applied to only a small part of the bent specimen (Sections 7.8.2.1 and 7.8.2.2). Here, we suppose that residual stresses are null. If not, the situation is then even more in favour of bending resistance since superficial residual stresses are compressive. Whenever critical flaws can be supposed to locate at the surface as for flat glass (because of contact damage; see Chapter 8) the difference becomes less. When a fibre is considered, longitudinal and transversal size effects are expected. This statistical behaviour has very important technological consequences:

i. Diameter size effect could provide a way of improving strength by obtaining finer filaments. Composites formed from fibre embedded in a polymer matrix are of interest in view of the outstanding fibre strength as compared to flat glass.
ii. If significant enough, size effect could be used to perform accelerated testing increasing size.
iii. Wide difference may arise from small-scale laboratory specimens to larger structures.

So far, quantitative extrapolation methods following the Weibull law are still questioned in the literature (Zinck et al., 1999).

8
Contact Resistance of Glass

8.1
Sharp and Blunt Contact

8.1.1
Introduction

As detailed in Chapter 7, glass strength depends mostly on flaws (defects) from which cracks develop. The existence of flaws is the reason why glass specimens show a mechanical strength much below the theoretical one. These flaws may be either in the glass bulk like small bubbles remaining after fining operations and crystalline phases (refer to Chapter 10) or at the surface like microcracks generated by the contact between the glass and tools and later in use (Fig. 8.1). These may be generated during the process itself or from handling and service conditions. We address in the present chapter the generation mechanisms of such contact-induced flaws.

During the production process the glass follows a complex thermomechanical route involving contact with tools at different temperatures (with the possibility of a thermal gradient between glass and tool), while in use the contact is generally isothermal (both the contactor and glass being at room temperature). As discussed in detail below, over a given threshold the contact is no longer reversible (excursion out of the elastic domain). Non-reversible deformation and cracks can be observed. All these phenomena lead to the introduction of flaws at the glass surface and glass strength is dramatically affected (Gy, 1997). We shall not address chemical reaction with the surface. It is well known that just touching lightly the surface of a pristine silica rod dramatically lowers its strength. For instance, halogen lamp operating notes often recommend no contact between the fingers and the lamp. Microanalysis of fracture origin shows alkali and alkaline earth contaminants. Figure 8.1 shows the dramatic scratches at the internal surface of a used plate. We review the mechanics of such phenomena. Glassmakers have been aware of these phenomena for a long time and coating technology was developed accordingly to protect glass surfaces from erosion and damage either for mechanical or aesthetics purposes (fibre sizing, container hot end treatments; Chapters 10 and 12). More fundamentally, tailoring contact resistance has attracted much scientific and technological interest and

Glass; Mechanics and Technology. Eric Le Bourhis

ISBN: 978-3-527-31549-9

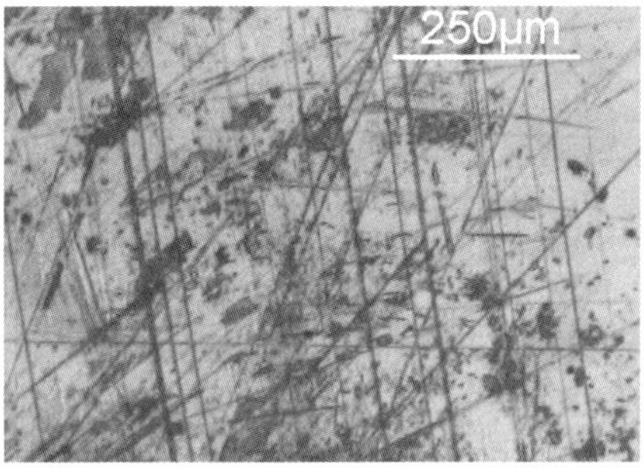

Fig. 8.1 Scratches on a used glass plate as observed by optical microscopy.

fracture mechanics described here was applied with success to many other types of materials showing brittle fracture (Le Bourhis et al. 2001b; Cook, 2006).

In the literature, two schematic contact conditions are generally considered. Blunt contact conditions consider that the material under a sphere deforms elastically up to brittle fracture. In contrast, under a sharp contact glass deforms elastically, plastically and finally fractures. The term 'plastic' is to be considered with caution for glass that is amorphous and 'plastic' means irreversible in this context. Glass may deform 'plastically' or inelastically below the glass transition temperature when small volumes are highly stressed. Under a sharp contact, pressure scales typically with hardness H that is commonly of the order of some gigapascals, while the deformed zone is elastically constrained by the surrounding volume of the material as discussed below.

Experimental tools have been developed using a spherical indenter to produce blunt contact while sharp contact is produced by conical or pyramidal tips as reviewed below. Also, scratch testing allows production of a lateral movement during the contact and hence a frictional force. These testing modes allow to approach real contact conditions.

8.1.2
Spherical Indentation

8.1.2.1 Elastic Loading

Figure 8.2 shows a diagram of the elastic loading of a glass surface by a spherical indenter. In such an experiment, a sphere of radius R is loaded under a force F, the contact radius being a. The specimen surface around the contact area also deforms elastically as shown schematically in Fig. 8.2.

Load (F)–penetration (h) relationship is known as the Hertz (1881) contact solution:

$$F = \frac{4}{3} R^{1/2} E^* h^{3/2} \tag{8.1}$$

where R is the radius of the tip and E^* the composite modulus, given by

$$\frac{1}{E^*} = \frac{1 - \nu_i^2}{E_i} + \frac{1 - \nu_s^2}{E_s} \tag{8.2}$$

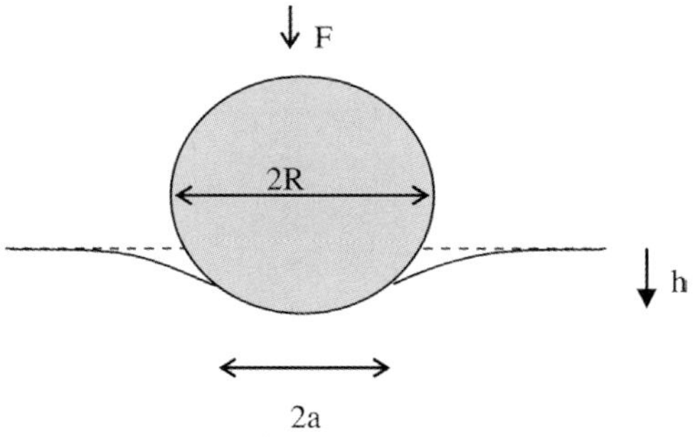

Fig. 8.2 Diagram of the elastic loading of a spherical indenter on a glass surface.

where E and ν are elastic modulus and Poisson ratio with subscripts i and s referring to the indenter and the specimen respectively.

For a diamond tip usually employed in the laboratory, $E_i = 1141$ GPa, $\nu_i = 0.07$. For materials much softer than diamond we have $1/E^* = (1 - \nu_s^2)/E_s$, and the indenter can be supposed perfectly rigid. In the general case, the glass and indenter properties are to be considered. When the two materials in contact are identical, a glass against another glass for instance, we have $1/E^* = 2(1 - \nu_s^2)/E_s$, and both the specimen and the indenter deform.

Figure 8.3 shows the calculated reversible loading curves of a glass under two spherical diamond indenters with different radii. The smaller the radius, the larger the penetration. The contact radius a is given by

$$\frac{a}{R} = \left(\frac{3}{4}\frac{F}{E^* R^2}\right)^{1/3} \tag{8.3}$$

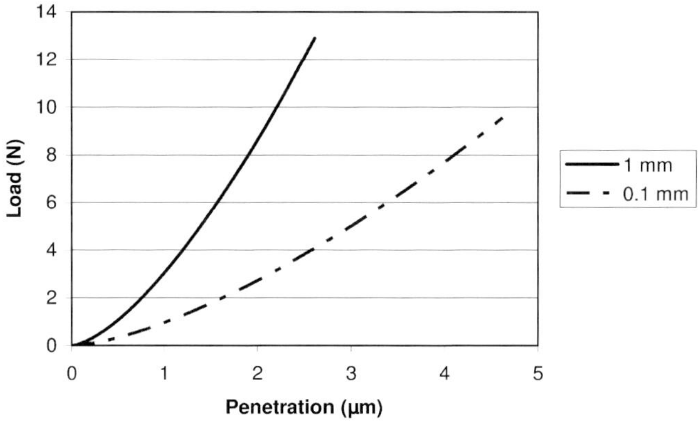

Fig. 8.3 Calculated load–penetration curve of a spherical diamond indenter into a standard silico-soda-lime glass surface ($E^* = 72$ GPa, $R = 1$ or 0.1 mm).

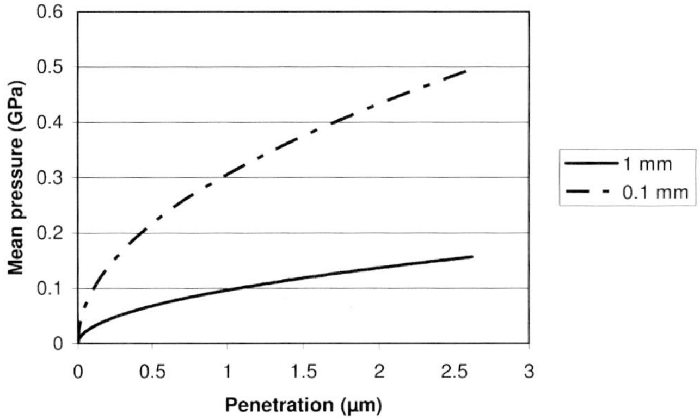

Fig. 8.4 Mean pressure as a function of penetration of a spherical diamond indenter into a standard silico-soda-lime glass surface ($E^* = 72$ GPa, $R = 1$ or 0.1 mm).

while the mean pressure p is

$$p = F/A_c = F/\pi a^2 \tag{8.4}$$

which is zero for $F = 0$ and progressively increases as load increases (Fig. 8.4).

Under usual loading conditions, the mean pressure is found to be well below the yield stress of the glass ($Y \sim H/3 \sim 2$ GPa for an SLS glass; Appendix 8) so that only elastic loading is considered up to fracture. Notably, we can observe that when the sphere radius decreases the mean pressure for a given penetration of the indenter increases (Fig. 8.4).

The maximum tensile stress in the specimen occurs at the contact circle:

$$\sigma_m = \frac{1}{2}(1 - 2\nu)\,p \tag{8.5}$$

while the maximum shear stress is located along the loading axis at depth $0.5a$ below the surface:

$$\tau_m = 0.48\,p \tag{8.6}$$

8.1.2.2 **Hertz Fracture and Indentation Toughness**

Under a spherical indenter, the mean pressure progressively increases when the applied load increases. Eventually, cone cracks may extend into the material as shown schematically in Fig. 8.5 and revealed experimentally in Figs. 8.6 and 8.7. Outside the contact area, pre-existing flaws are subjected to tensile stress, which at some point (depending on the flaw size) run around the contact forming a surface ring crack that grows downward following the principal stress conical trajectories, while on unloading the cracks close. The fracture usually remains visible because of the imperfect

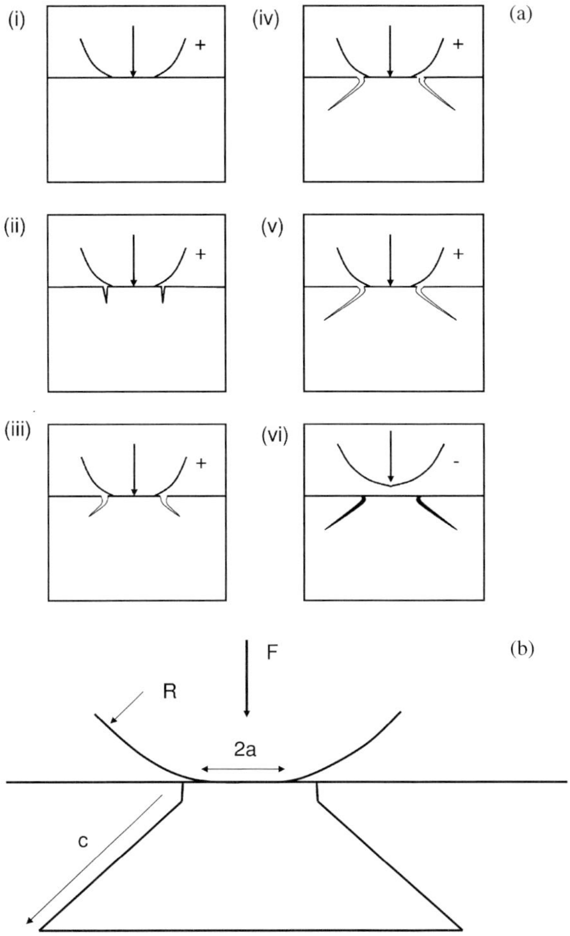

Fig. 8.5 Fracture under a blunt indenter: (a) cone fracture sequence on loading (+) and unloading (−); (b) cone fracture geometry (according to Lawn, 1993).

closure at the crack (Fig. 8.6). Notably, the residual stress field changes the crack geometry as revealed in Fig. 8.6b. Also, the load required for generation of cracks of similar size is superior in tempered specimens. In thermally tempered glazing, the skirt of the cone crack is almost parallel to the indented surface, while it is inclined to the surface in an annealed (stress-free) specimen. The crack evolution is depicted schematically in Fig. 8.5 (Lawn, 1993, 1998) with the plus sign indicating loading and the minus sign unloading. Next, let us consider a specimen with small residual stresses.

The mechanics of cone crack initiation started with Auerbach (1891). He showed empirically that the critical load to initiate a cone fracture in flat specimens was proportional to the radius of the indenting sphere $F_c \sim R$. For decades (Tillett, 1956; Lawn, 1993), Auerbach's law was debated since fracture theory suggests that cracks

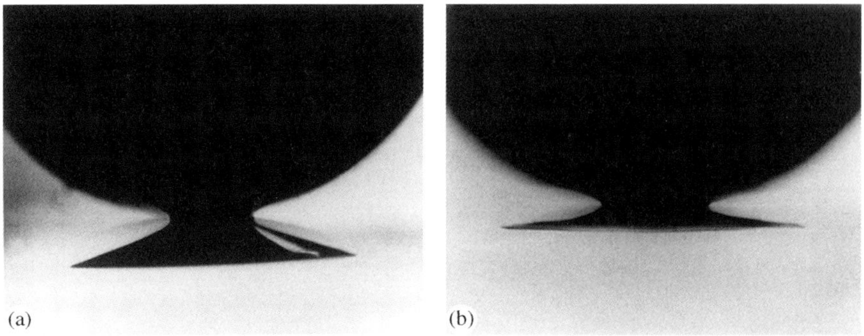

Fig. 8.6 Side views of the cone cracks under load in annealed and thermally tempered soda-lime glass formed under a 2.5 mm radius steel indenter. Note that the skirt of the cone crack in the annealed glass makes a significantly larger angle with the indented surface than that in the tempered glass (Reprinted from Chaudhri and Phillips, 1990, with permission from Taylor & Francis Ltd. http://www.informaworld.com). (a) load 793 N, cone base 2.4 mm; (b) load 2060 N, cone base 2.85 mm.

initiate at the points where maximum tensile stress in the Hertzian field equals the strength of the material. Using Eqs. (8.4) and (8.5) results in an $F_c \sim R^2$ relationship. The paradox was solved taking into account the statistical nature of fracture and showing the so-called Auerbach's range in a domain where F_c shows an insensitivity to flaw size. Then, the toughness of a glass can be determined using the critical load for crack generation (Auerbach's law) or the extension of the cracks (Franck and Lawn model) as reviewed by Geandier et al. (2003). The toughness and the critical load are related in Auerbach's range by (Lawn, 1993; Warren, 1995; Fischer-Cripps, 2004a)

$$K_c = \sqrt{\frac{E^* F_c}{CR}} \tag{8.7}$$

where F_c is the critical load for cone crack generation, C a constant that has to be calculated for a given material and indenter (Warren, 1995), R the indenter radius and E^* the composite modulus defined above. For glass contacting glass, $C = 3131$; it is 2790 for alumina contacting alumina (Warren, 1995).

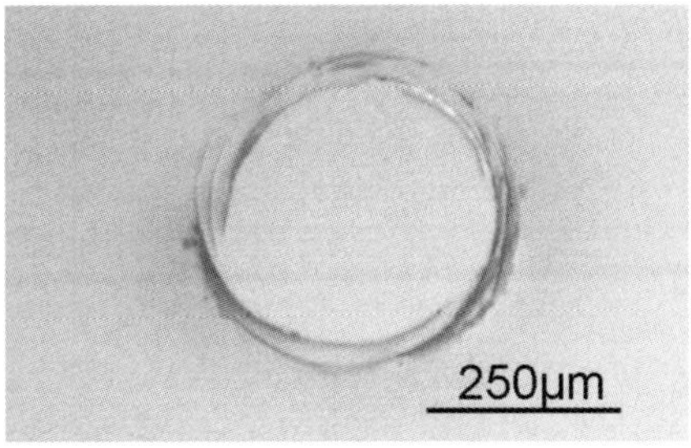

Fig. 8.7 Top view of ring cracks generated by a sphere (0.2 mm diameter) on the atmospheric side of a float glass.

In the Frank and Lawn (1967) model, only the cone crack is considered; the initial small circular shaped crack perpendicular to the surface is neglected, due to its small size compared to the cone length. It is assumed that the crack is propagating along the direction of the smallest principal stress (σ_3) perpendicularly to the maximum principal stress (Hertz stress field σ_1; we suppose no residual stresses) and that the crack presence and propagation do not modify the initial stress field. The critical stress intensity factor is then obtained by integration along the crack length c (Fig. 8.5; Section 7.5):

$$K_c = 2\sqrt{\frac{c}{\pi}}\int_0^c \frac{\sigma_1(x)}{\sqrt{c^2 - x^2}}\,dx \tag{8.8}$$

where $\sigma_1(x)$ is the Hertzian stress value along the crack (the opening stress normal to crack direction). Interestingly, Chaudhri and Kurkjian 1986 showed that cone cracks also form on dynamic loading with lower cone angles. The cone formed is 46°, that is, smaller than that of 62.8° formed under static loading (Chaudhri and Kurkjian, 1986). Moreover, the angle amplitude depends on the Poisson ratio of the studied glass and the residual stresses as discussed above.

A typical Hertz fracture obtained after a blunt impact is shown in Fig. 8.8 on a laminated (annealed) windscreen. Even small cracks develop under humid environments as discussed in more detail in Chapter 9 and should be repaired before they damage dramatically the glazing.

8.1.3 Sharp Indentation

8.1.3.1 Elastic Loading

Figure 8.9 shows a diagram of the elastic loading of a glass surface by a conical indenter. The approach can be extended to pyramidal tips (Vickers, cube corner, Berkovich 1951) and more generally self-similar tips taking $A_c = \pi a^2 = \alpha h^2$ and neglecting the influence of the indenter edges. A cone loaded under force F penetrates the specimen down to the penetration depth h, the contact radius being

Fig. 8.8 Hertz fracture on a laminated windscreen.

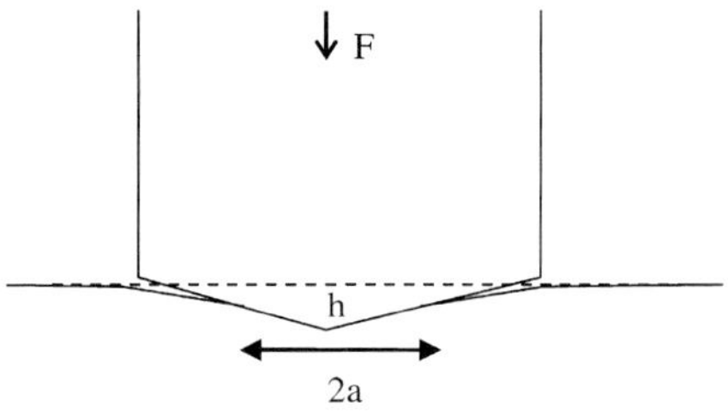

Fig. 8.9 Diagram of the elastic loading of a conical indenter on a glass surface.

a. Again as for a spherical tip, the specimen deforms elastically around the contact area as shown in Fig. 8.9 (see also Appendix 8).

Load (*F*)–penetration (*h*) relationship is known as the Sneddon (1965) contact solution and shows a square dependence on penetration (so-called self similarity or geometrical similarity of the contact):

$$F = \frac{2}{\pi \tan\beta} E_r h^2 \tag{8.9}$$

where β is the angle between the indenter flank and the specimen surface and E_r is the reduced modulus defined as

$$\frac{1}{E_r} = \frac{1 - \nu_s^2}{E_s} \tag{8.10}$$

where subscript s indicates the specimen.

The contact radius *a* is given by

$$a = \frac{2h}{\pi \tan\beta} \tag{8.11}$$

while the mean pressure *p* is

$$p = E_r \tan\beta / 2 \tag{8.12}$$

which interestingly is independent of the penetration (geometrical similarity). Mean pressure shows dependence on flank angle. For a sharper tip (larger β), a larger pressure is generated.

Using standard silico-soda-lime glass characteristics ($E_r = 76$ GPa) and a flank angle $\beta = 19°$ (a cone angle with the same area function as that of a Berkovich or Vickers tip) one finds a pressure $p \sim 13$ GPa well above the glass yield strength ($Y = H/3 \sim 2$ GPa; Appendix 8). In fact, glass is 'plastically' deformed as revealed by the indenter mark left on its surface (Fig. 8.10). The mark shows the symmetry of the indenter used (here four-fold symmetry since a Vickers pyramid tip was used). The image shows also fracture events that are discussed in more detail below. Most tips are not perfectly sharp but show tip rounding at their extremity. Under low loads the tip rounding yields

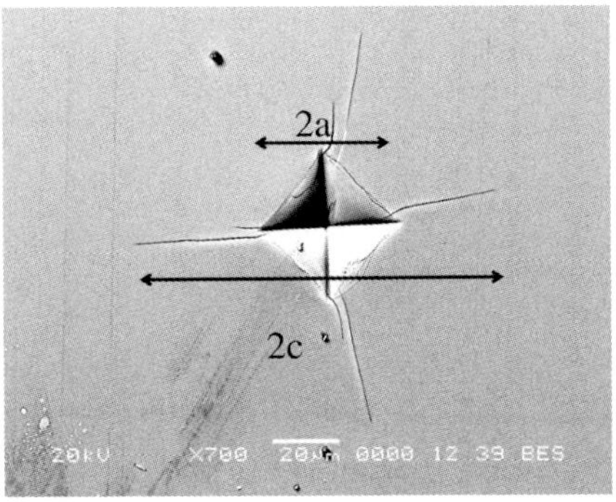

Fig. 8.10 Indented surface showing the permanent deformation and the radial cracks (scanning electron microscopy observation of standard glass indented under 5 N); *c* and *a* correspond to crack and mark lengths respectively.

a Hertz dependence between force and penetration with no plastic deformation being generated, while the response tends to Sneddon's law for elevated penetrations.

8.1.3.2 Elastic–Plastic Loading

As discussed above, a sharp contact yields elastic and irreversible deformations. Different approaches have been proposed to model the elastic–plastic response (Loubet et al., 1986; Sakai 1999). Basically purely elastic and purely plastic components are connected in series (Chapter 6; Oyen and Cook, 2003).

The total displacement of the indenter is the sum of the elastic h_e and plastic h_p displacements (a viscous h_v component may be added; Oyen and Cook, 2003; Mammeri et al., 2004; Appendix 8):

$$h = h_e + h_p \tag{8.13}$$

Because of the geometric similarity of the sharp indenter, all constitutive equations show the square of the respective penetrations:

$$F = AE_r h_e^2 \tag{8.14}$$

$$F = BHh_p^2 \tag{8.15}$$

where A and B (24.5 for a Vickers or Berkovich tip) are constant[1] for a bulk, E_r the reduced modulus and H the hardness. It should be noted that we assume here that the

[1] Parameter A may include mechanical properties (H/E ratio). Penetration can be written $h = h_s + h_c$, where h_s is the elastic deformation of the surface (Appendix 8) where H/E ratio appears (Loubet et al., 1986; Hainsworth et al., 1987; Malzbender et al., 2000).

sharp tip is a perfect pyramid or cone while tip rounding has to be considered as already discussed above. The refinement of the present approach can be obtained using numerical tools like finite element analysis (FEA; Appendix 11).

Combining Eqs. (8.13) to (8.15) yields

$$h = \sqrt{\frac{F}{AE_r}} + \sqrt{\frac{F}{BH}} = \sqrt{F}\left(\frac{1}{\sqrt{AE_r}} + \frac{1}{\sqrt{BH}}\right) \tag{8.16a}$$

Hence we obtain a square dependence of the force F on the total penetration:

$$F = \left(\frac{1}{\sqrt{AE_r}} + \frac{1}{\sqrt{BH}}\right)^{-2} h^2 = Kh^2 \tag{8.16b}$$

where K is an elastic–plastic prefactor. This result suggests that elastic–plastic response under a conical remains self-similar.

8.1.3.3 Hardness

Hardness measures the plastic resistance of a surface to contact loading, that is, its resistance to permanent deformation. Most experiments are carried out using a Vickers diamond pyramid. When a Vickers diamond (four-sided) pyramid is loaded over the plastic threshold, one can determine, after the test, the mean pressure or hardness H defined by the ratio between the applied load and the projected surface A_p of the permanent deformation left by the indenter. This area can be determined from the mark diagonals ($A_p = 2a^2$; Fig. 8.10) observed under elevated loads by means of optical microscopy. Hardness is

$$H = F/A_p \tag{8.17}$$

A large mean pressure (large hardness) means that the specimen shows an elevated resistance to plastic deformation under contact loading. Historically, the hardness ratio was defined using the actual surface of the indent and not its projected surface so that a geometric factor relates the hardness number (Vickers, Knoop, Brinell, etc.) and the mean pressure. For instance, for a Vickers tip the Vickers hardness H_v is $0.927H$ (92.7% of the mean pressure). According to the load range, macro-, micro- and nanoindentation domains have been defined. These domains should be borne in mind since indentation size effect (ISE, that is, a decrease of hardness with load in most cases) is reported for many materials including glasses, so that hardness values are to be compared within the same domain. It should be noted that these domains also refer to different ways to determine the surface of the indent. In the macro- and microindentation domains the indent surface is generally determined after unloading by optical microscopy. We shall refer to this area as A_p (projected area defined above). Going to the nanoindentation domain (Appendix 8), observation of the indent becomes tedious and hardness is determined under load and is defined as the ratio of the force to the projected contact area A_c that is extracted from the

Tab. 8.1 Microhardness of different glasses (Aben and Guillemet, 1993; Barton and Guillemet, 2005).

Glass	Microhardness, *H* (GPa)
Silica	7.0–7.5
Aluminosilicate	5.4–5.9
Borosilicate	5.8–6.6
Soda-lime-silica	5.4–6.6
Lead-silicate	4.5
Aluminophosphate	3.2

loading–unloading curves (Appendix 8):

$$H = F/A_c \tag{8.18}$$

Microhardness values are reported for different glasses in Table 8.1 and do not show strong difference according to the composition of the glass even though they are known to deform quite differently whether they are considered normal or anomalous glasses. The nature of the permanent deformation is still controversial however. It depends on the type of glass, on the temperature and also on the strain rate. At room temperature, so-called anomalous glasses like fused silica show a volume densification while normal glasses show shear flow (Mackenzie, 1963; Peter, 1964; Neely and Mackenzy, 1968; Arora et al., 1979). In amorphous semiconductors (a-Ge) phase transition to crystalline phases is even observed (Patriarche et al., 2004).

Normal glasses contain alkali and alkaline earth modifiers leading to non-bridging oxygen (NBO) atoms allowing for heterogeneous shear bands to form. Some authors suggested following Greaves' (1985) modified random network that shear would happen preferentially at the frontiers between modifier-rich and silica-rich volumes (Chapter 5). Heterogeneous flow yields non-Newtonian viscoplasticity. The same formalism is used by the plasticity community and this explains why the term 'plasticity' is used for a glass even though the dislocation concept has to be reconsidered in a non-periodical structure (G'Sell, 2001). Douglas (1958) argued that 'when the load is first applied' to the glass specimen 'the pressure under the point of the diamond will be very high and the viscosity will be reduced' (this is non-Newtonian flow), and 'flow will take place until the diamond has penetrated sufficiently so that the force is no longer great enough to cause sufficient reduction in viscosity for flow to continue'. Also, water incorporation at glass surfaces (Chapter 5) is suspected to play a role in glass response (Han and Tomozawa, 1990).

In contrast, anomalous glasses which contain very few modifying ions (silica being the archetype), present a much cross-linked network preventing heterogeneous flow. Thereafter, indentation induces compaction (or densification) of the glass. It occurs by bending and twisting but not breaking the silicon–oxygen bonds in the glass network. In fact, when the glass is heated up sufficiently the mark recovers (Neely and Mackenzy, 1968; Yoshida et al., 2001).

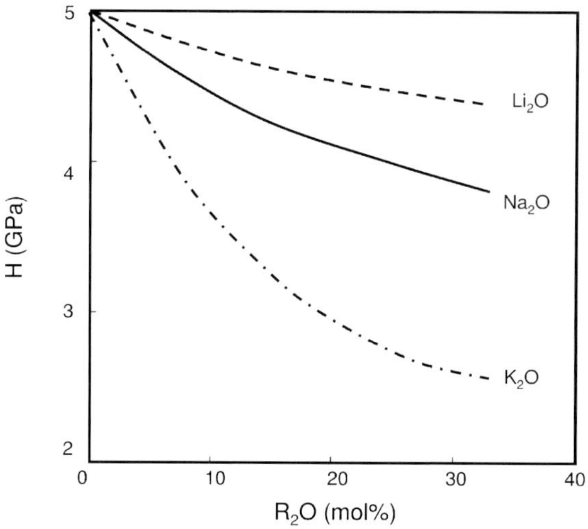

Fig. 8.11 Hardness *H* of silicate glass as a function of modifier concentration (according to Kennedy et al., 1980).

Kennedy et al. (1980) reported changes in the hardness of silica as modifiers are progressively introduced. As observed for Young's modulus *E* changes with alkali (Chapter 7), softening of the silica glass is more pronounced for a modifier having a lower Dietzel field intensity, that is, for a modifier that is expected to form weaker interaction with oxygen (Chapter 5; Fig. 8.11). While heterogeneous shear flow is volume-conservative, compaction, as the word suggests, leads to decrease in volume. This is the reason why the residual stresses caused by indentation into an anomalous glass are smaller than those in a normal glass. This has been clearly demonstrated by Arora et al. 1979 using photoelasticity (Chapter 10) to reveal differences in stresses around indent sites for normal and anomalous glasses. More recently, Perriot (2005) carried out cross-sectional Raman investigation on indented silica and revealed compaction gradient in silica (Fig. 8.12). Also, Yoshida et al. (2005) showed that compaction contribution to indentation deformation in normal glass could be greater than generally accepted. In fact, the separation of anomalous and normal glasses according to the presence of modifiers is still questioned. Rouxel et al. (2006) showed recently that SLS glasses show substantial compaction (obviously less than silica) that recovers upon annealing. Furthermore there seems to be a close relationship between densification and Poisson's ratio.

The mechanical behaviour of a glass becomes more and more time dependent as the temperature is increased and as the viscous behaviour becomes predominant over densification process (Westbrook, 1960; Wiederhorn and Hockey, 1980; Le Bourhis and Metayer, 2000; Le Bourhis and Rouxel, 2003). Under large strain rates, the supercooled liquid shows non-Newtonian behaviour with an apparent viscosity decreasing with the strain rate amplitude (Simmons et al., 1988; Appendix 10). Although this behaviour makes the process operations more difficult, it can yield

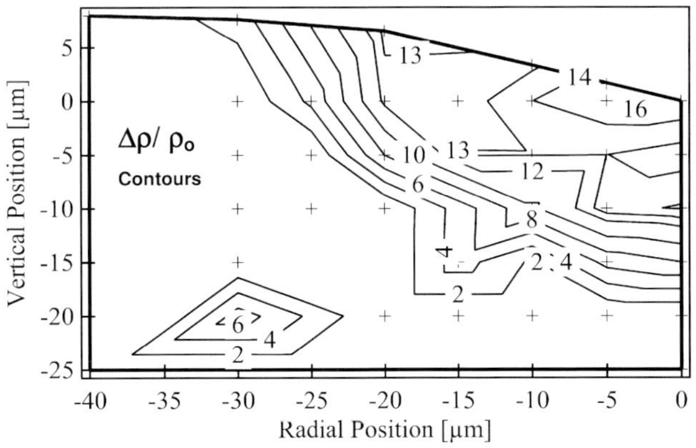

Fig. 8.12 Cross-sectional compaction map in silica under a Vickers indentation (Perriot, 2005).

advantages since the operations can be carried out under lower stresses (for the same flow, rheofluidification) once high strain rates can be achieved.

Hardness depends strongly on the temperature as shown in Fig. 8.13 for a soda-lime-silica glass. As the temperature approaches T_g, viscous flow (volume conservative) comes into play. Strangely, very few data existed before the 1990s while this behaviour is very important from the process point of view. These studies are in fact not easy to carry out and discrepancies appear between authors (Westbrook, 1960; Watanabe et al., 2001; Le Bourhis and Rouxel, 2003; Shang and Rouxel, 2005). These have been attributed to the differences in loading time, hold period and surface roughness employed by different authors. Also, modelling of the behaviour has to consider both viscosity and elasticity (Chapter 6, Appendix 8). Viscous deformation is

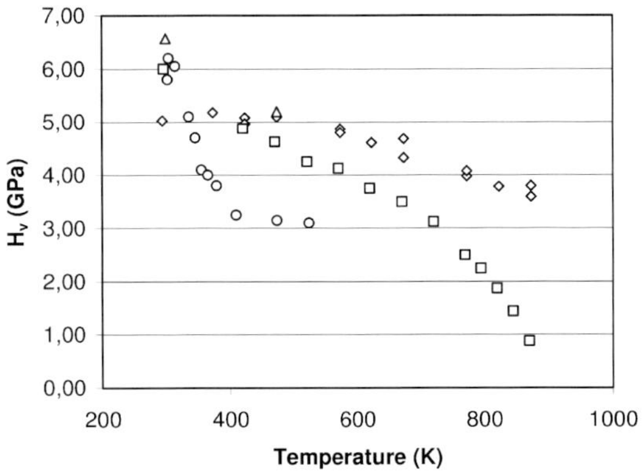

Fig. 8.13 Mean hardness as a function of temperature for soda-lime-silica glass: □, Watanabe et al. (2001); Δ, Smith and Zheng (2000); ◇, Le Bourhis and Metayer (2000);○, Westbrook (1960). (Reprinted from Le Bourhis and Rouxel, 2003, with permission from Elsevier.)

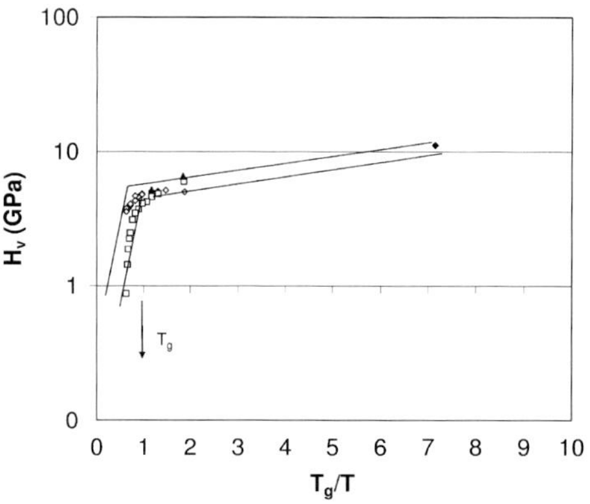

Fig. 8.14 Hardness (log scale) as a function of reciprocal temperature for soda-lime-silica glass: □, Watanabe et al. (2001); ▲, Smith and Zheng (2000); ◇, Le Bourhis and Metayer (2000); ◆, Kurkjian et al. (1995). (Reprinted from Le Bourhis and Rouxel, 2003, with permission from Elsevier.)

time-dependent and depends on history. Also, well below the glass transition, a decrease of the hardness is observed on increasing the temperature. At such low temperature ($<300\,^{\circ}$C), non-Newtonian flow is expected (Chapter 6) and makes the analysis more difficult. It is of much interest to show the hardness of a studied soda-lime-silica glass on a log scale as a function of the reciprocal temperature where the data obtained at liquid nitrogen temperature can be added (Kurkjian et al., 1995). Figure 8.14 summarizes the results over a wide range of temperatures (70–900 K) and reveals two deformation regimes: (i) at low temperatures where lower variations in hardness are observed and (ii) in the vicinity of T_g where higher hardness variations are observed. Interestingly, the data at liquid nitrogen temperature align quite well with the data at intermediate temperatures. These results are very similar to those obtained for the same materials by mechanical spectroscopy for the elastic modulus where a transition was also observed at or near T_g on the Young's modulus versus T/T_g curves (Gadaud and Pautrot, 2003; see also Chapter 6). This transition mainly reflects changes occurring in the material when the temperature increases from the glassy state associated with a slow weakening (softening) regime to the supercooled liquid state, corresponding to a fast weakening regime. Also, it was previously observed that the permanent deformation changed in nature from compaction dominated at liquid nitrogen temperature to shear flow dominated when the temperature increases (Kurkjian et al., 1995; Yoshida et al., 2001) and this also correlates with the progressive change in slope observed in Fig. 8.14.

8.1.3.4 **Radial–Median Cracking**

Sharp indenters like Vickers diamond pyramids produce two basic types of crack pattern: (i) the radial–median system that appears first and (ii) the lateral one that appears under further increasing loads (Figs. 8.15 and 8.16).

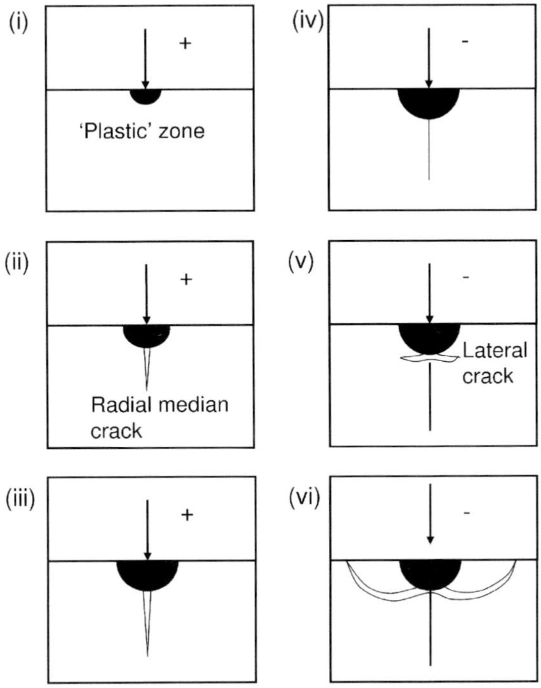

Fig. 8.15 Radial median fracture under a sharp indenter. On loading (i, ii, iii) median cracks form and extend on unloading (iv, v, vi), to form radial cracks at the surface. When large enough loads are used lateral cracks form on unloading and extend parallel to the surface (according to Lawn, 1993).

The radial–median cracks appear first. At a critical load, one or more nascent flaws within the deformation zone become unstable and pop in to form a subsurface median crack. On increased loading, the cracks propagate downward, while on unloading the median cracks close up below the contact zone while they simultaneously open up in the residual tensile stress field at the surface that is generated by the plastically deformed zone. This residual stress field yields subcritical crack growth as detailed in Chapter 9. After indenting under elevated loads, a lateral crack may appear while unloading the indenter. They propagate parallel to glass surface. When such cracks emerge at the specimen surface, chips are produced.

The proposed sequence has been verified and/or altered on different types of glasses thanks to *in situ* observations (Chaudhri and Phillips, 1990; Cook and Pharr, 1990; Chandrasekar and Chaudhri, 1993; Fig. 8.16). It has been also shown that glass type has some influence. In particular the behaviour is modified when the glass densifies (Cook and Pharr, 1990; Sglavo and Green, 2001).

In anomalous glasses (showing densification under sharp contact) there is a blunting effect on the indenter apex (Sglavo and Green, 2001; Bertoldi and Sglavo, 2004) and the stress state in the elastic matrix surrounding the indentation site is

Fig. 8.16 Side views of the crack pattern in an annealed glass under a Vickers indenter during quasi-static loading–unloading. A full-penny median crack is shown in frame 1. As the load increases the full-penny crack increases in size and then breaks through to the surface to form a half-penny crack (frame 2). As load decreases a lateral crack forms (frame 4). Frame 6 shows the top view after unloading (Reprinted from Chaudhri and Phillips, 1990, with permission from Taylor & Francis Ltd. http://www.informaworld.com). Maximum load is for frame 2 and is 387 N.

similar to the Hertzian stress field generated beneath blunt indenters. As discussed above, the residual stress intensity is also reduced because of less constraining effect that the displaced material in the deformation zone exerts on the remaining elastic matrix (Bertoldi and Sglavo, 2004). These phenomena obviously change crack

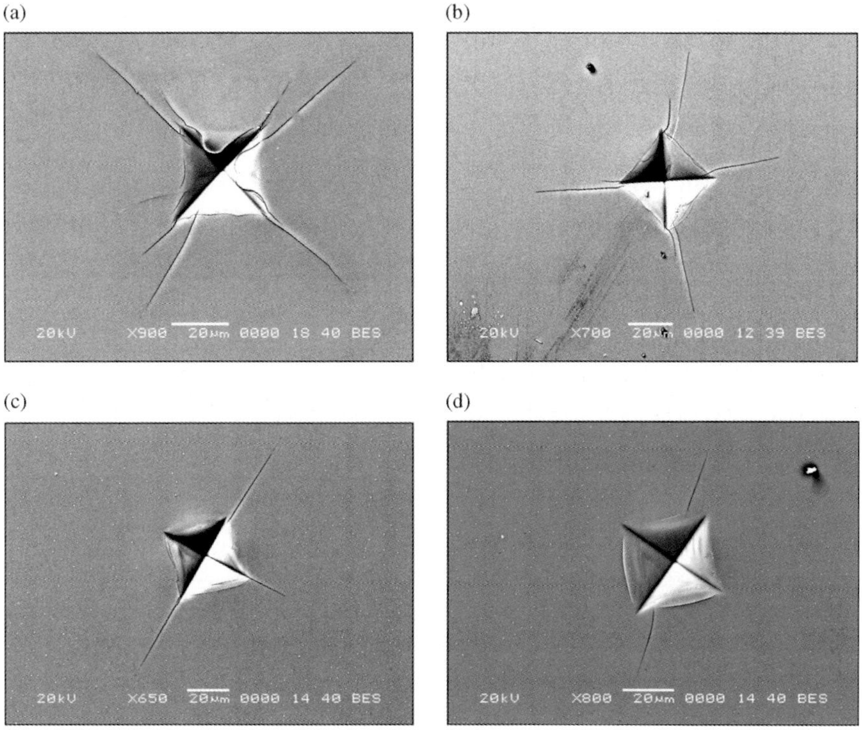

Fig. 8.17 Glass surface after indentation test: (a) at room temperature; (b) at 300 °C; (c) at 500 °C; (d) at 600 °C. (Reprinted from Le Bourhis and Metayer, 2000, with permission from Elsevier.)

propagation. The resulting main crack system produced by indentation resembles the Hertzian cone (Hagan, 1979; Tandon and Cook, 1992, 1993; Sglavo and Green, 2001), though some shorter median, lateral and radial cracks are also present.

Observation of the glass surface indented at different temperatures is also instructive (Le Bourhis and Metayer, 2000). Figure 8.17 shows a series of micrographs of a soda-lime-silica glass surface after indentation at increasing temperatures. A progressive decrease in the number of cracks is observed when the temperature is increased (only two of them are observed on the micrograph obtained after testing at 600 °C). In contrast, other glasses like GeAsSe chalcogenide glasses show a constant number of radial cracks for all indentation temperatures (Le Bourhis et al., 2001). These results can be further analysed in terms of indentation toughness as presented in detail in Section 8.1.3.5. As under a spherical contact, indentation toughness can be extracted from sharp indentation response.

8.1.3.5 **Indentation Toughness**

Using the dimensions of radial cracks, the apparent indentation fracture toughness K_c of a glass can be determined as proposed originally by Anstis et al. (1981). The relation between K_c and the half of the mean size of the two cracks c is given by (Fig. 8.10)

$$K_c = \chi \frac{F}{c^{3/2}} \tag{8.19}$$

where F is the load applied on the indenter and χ is a function of the elastic–plastic ratio E/H that informs on the intensity of indentation residual stress field (proportional to χF):

$$\chi = 0.016\sqrt{\frac{E}{H}} \tag{8.20}$$

The indentation procedure may be very useful to determine fracture toughness, in particular when only a small volume of glass is available. It has to be used carefully. In fact, as mentioned above, the radial cracks propagate in the residual stress field induced by the 'plastic' zone. Subcritical crack growth or ageing (i.e. time between test and measurement of crack length) may happen and should be taken into account when the indentation procedure is used since cracks extend with time (stress corrosion; Chapter 9; Fig. 8.18).

Geandier et al. (2003) insist on indentation toughness discrepancies between authors. These are attributed to different procedures (ageing, contact geometry). Also, Rizkalla et al. (1996) reviewed eight different formulae used in the literature to extract toughness from the observed crack length, Eqs. (8.19) and (8.20) being the most usual. Therefore it is important to consider indentation toughness values with care and the data in view of experimental procedures and analysis. However, the method is revealed to be useful also at elevated temperatures since another important

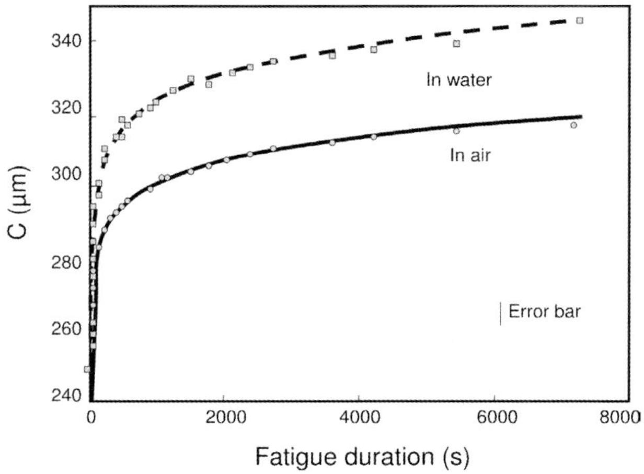

Fig. 8.18 Radial crack length (c) as a function of the fatigue duration (t) in a window-type glass (according to Dériano et al., 2004). Indentation load of 49.05 N (loading time $t' = 15$ s). Note the effect of water (see also Chapter 9).

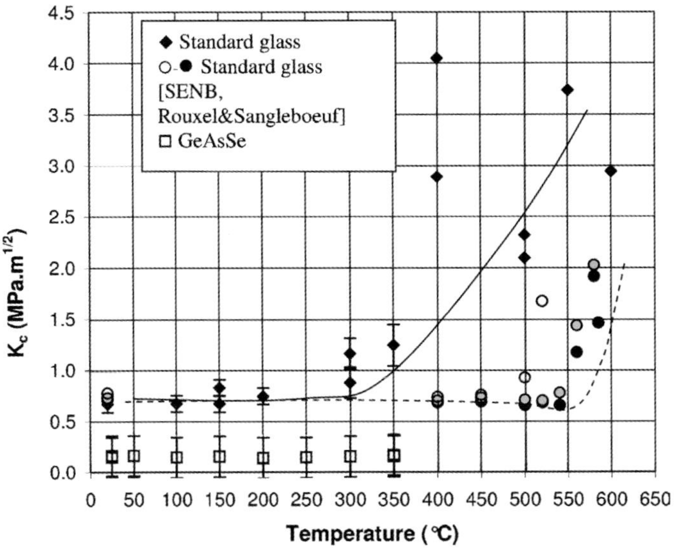

Fig. 8.19 Mean fracture toughness as a function of temperature for soda-lime-silica glass and GeAsSe glass (error bar $= \pm$SD). The indentation values for soda-lime-silica glass are compared to values obtained on chevron-notched bars (Rouxel and Sangleboeuf, 2000) under three different loading rates (open circle, $2\,\mathrm{MPa\,m^{1/2}\,s^{-1}}$; shaded circle, $4\,\mathrm{MPa\,m^{1/2}\,s^{-1}}$; filled circle, $8\,\mathrm{MPa\,m^{1/2}\,s^{-1}}$). Lines represent guides for the eyes. (Reprinted from Le Bourhis and Rouxel, 2003, with permission from Elsevier.)

point for glass processing is to characterize glass contact resistance at working temperatures where fracture toughness measurements with conventional techniques are rare and difficult. Double-torsion (DT) or double-cantilever beam (DCB) techniques although suited for investigations under various environments (Chapters 7 and 9) cannot be adapted easily to measurements at elevated temperatures. In soda-lime-silica glass, fracture toughness was reported to increase very slowly at low temperatures (Fig. 8.19), while approaching the glass transition K_c was observed to increase strongly with temperature in good agreement with values obtained from the loading of chevron-notched bars (Rouxel and Sangleboeuf, 2000). In contrast to this behaviour, chalcogenide glasses show very small changes in toughness at temperatures ranging from room temperature to transition temperature (Le Bourhis and Rouxel, 2003). The indentation method allows one to determine whether a glass shows brittle–'ductile' transition. This field is attractive from both the technological and fundamental point of view. More understanding of glass behaviour under contact loading and particularly its time dependence is required (Appendix 8).

8.1.3.6 Lateral Cracking and Chipping

After median-crack generation, when loading to larger indentation loads, lateral cracks may appear eventually. These propagate parallel to the indented surface and are circular in form. They are observed easily with optical microscopy since they produce optical interferences due to the imperfect closure at the crack interface.

Under elevated loads the cracks may emerge at the surface and then form a chip which is later removed from the glass surface. These cracks are responsible for glass erosion and abrasion (see Section 8.4).

8.1.3.7 **Brittleness Index**

When contacted, a glass may deform (over some threshold load) irreversibly, that is, plastically and/or with fracture. It is of utmost importance to understand, control and tailor glass response. In this perspective, brittleness index introduced by Lawn and Marschall (1979) as the ratio of hardness and toughness is of great importance:

$$B = H/K_c \tag{8.21}$$

This ratio shows competition between plastic and brittle deformation modes. Notably, this index shows no dependence on the elastic deformation mode (dependence on Young's modulus; see Section 8.4). B values calculated for different glasses and one metal are shown in Table 8.2.

The brittleness index unit is $\mu m^{-1/2}$ so that B^{-2} is to be considered as a length scale (in fact as detailed below $120B^{-2}$ is the relevant length scale).

Lawn and Evans (1977) showed that brittleness is closely related to the critical load to initiate a median–radial crack (see also Ishikawa and Shinkai, 1982). The threshold load is

$$F_c = \lambda_o K_c B^{-3} \tag{8.23}$$

while the threshold crack size is written

$$c_c = \mu_o B^{-2} \tag{8.23}$$

λ_o and μ_o being geometrical constants ($\lambda_o = 1.6 \times 10^4$, $\mu_o = 120$). Using Table 8.2 data, critical force is determined as about 0.02 N and 800 kN for glasses and Fe respectively. Hence decreasing B yields a dramatic increase in critical load. Brittleness happens to be a very important property for glass manufacturers since this parameter controls surface damage. Thereafter, the glazing strength and performance in use should be positively affected with decreasing B. Much work has been carried out to address this issue during the past decade.

Tab. 8.2 Hardness, toughness and brittleness of different glasses. Data for a metal (Fe) are shown for comparison.

Glass	H (GPa)	K_c (MPam$^{1/2}$)	B ($\mu m^{-1/2}$)
Fused silica	7.0	0.7	10
Soda-lime-silica	5.8	0.7	8.2
Borosilicate	6.2	0.7	8.8
Fe	5	50	0.1

As discussed above hardness and toughness are related to the mark size a and the crack size c (see Fig. 8.10) by

$$H = \frac{F}{2a^2} \tag{8.24}$$

and

$$K_c = \chi \frac{F}{c^{3/2}} \tag{8.25}$$

where χ is a function of the elastic–plastic ratio E/H, define above. These equations combine into

$$\frac{c}{a} = \left[2^{3/4} \frac{H}{K_c} \frac{\chi}{H^{1/4}}\right]^{2/3} F^{1/6} = \left[2^{3/4} \frac{\chi}{H^{1/4}}\right]^{2/3} F^{1/6} B^{2/3} \tag{8.26}$$

$$\frac{c}{a} = \varpi F^{1/6} B^{2/3} \tag{8.27}$$

Sehgal et al. (1995) suggested that the prefactor ϖ varies in a small range (2.2 to 2×10^{-4} $Pa^{-1/4}$) for common glasses. They carried out extensive studies to determine the c/a ratio for a large number of glasses and used the c/a ratio as a brittleness index. In fact, both indices are related by

$$B = \kappa F^{-1/4} \left[\frac{c}{a}\right]^{3/2} \tag{8.28}$$

When the study can be carried out in a small range of loads, the dependence on load can be neglected. As detailed in next paragraph, Sehgal et al. (1995) revealed that glass brittleness can be changed dramatically by varying the glass composition (in particular the silica content).

8.2 Sharp Contact Resistance

An important fact is that the contact resistance of a glass will be dependent on the tempering treatment, the contact temperature and the glass composition. Tempering introduces compressive stresses at the article surface. These contribute negative stress intensity factor when they are compressive ($K_r < 0$; Chapter 7). Equation (8.19) can be rewritten under such conditions (Tandon and Cook, 1992, 1993) as

$$K_c = \chi \frac{F}{c^{3/2}} + K_r \tag{8.29}$$

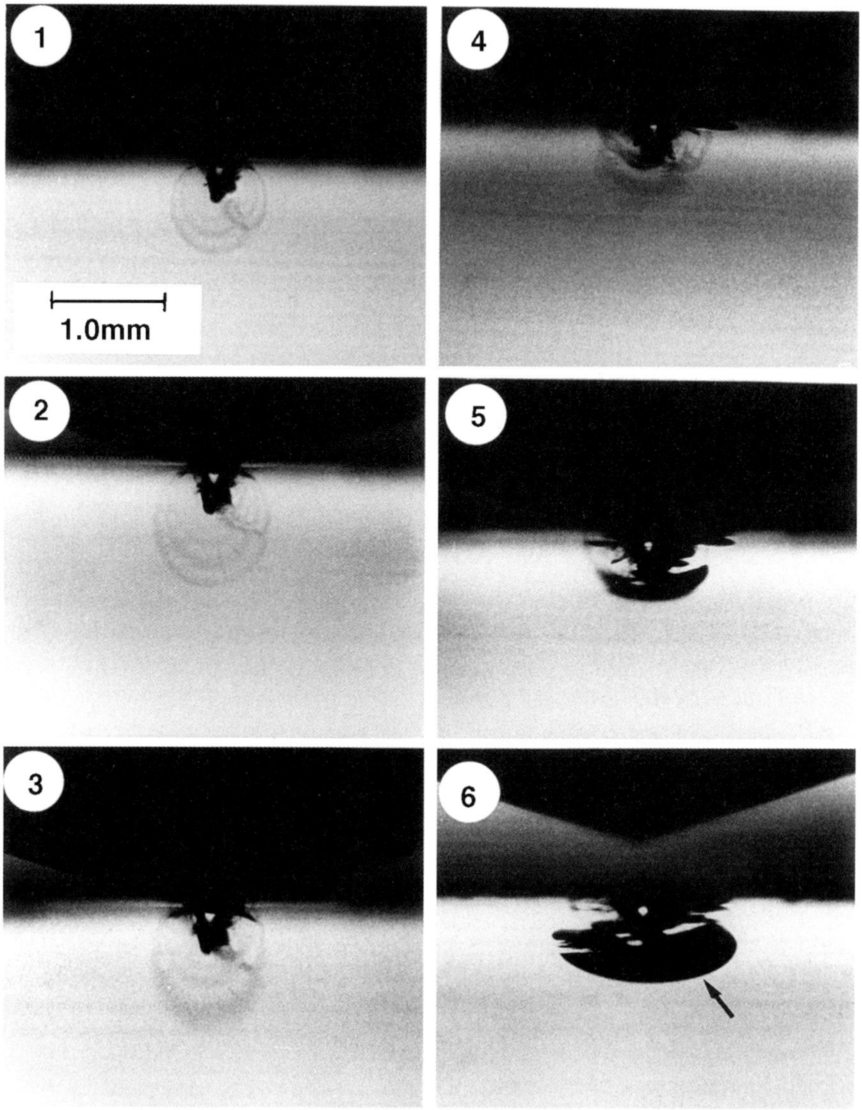

Fig. 8.20 Side views of the crack pattern in a thermally tempered glass under a Vickers indenter during quasi-static loading–unloading. A full-penny median crack appears in frame 1 and breaks through to the surface on unloading (frame 3). As load decreases lateral cracks form (frame 5) (Reprinted from Chaudhri and Phillips, 1990, with permission from Taylor & Francis Ltd. http://www.informaworld.com). Maximum load is for frame 2 and is 470 N.

As a matter of fact, the load necessary to generate a crack of the same radius is more important in a tempered glass than in an annealed glass (compare Figs. 8.20 and 8.16). Chandrasekar and Chaudhri (1993) suggested a method to determine the residual stresses in a tempered glazing comparing the brittle responses of an annealed and a tempered specimen under contact loading. The relationship proposed by these authors is

$$\sigma_s = \wp \frac{F^* - F}{c^2} \tag{8.30}$$

where σ_s is the surface residual stress amplitude, c the crack size, F^* and F the critical load to produce the crack in a tempered and a annealed specimen and $\wp$ a geometrical factor ($\wp = 1/[1.16\pi^{3/2}\tan(68^\circ)]$ for a Vickers indenter). This phenomenon is emphasized in chemically tempered glass as illustrated in Fig. 8.21 with Eq. (8.30) being modified to take into account the rapidly varying residual stress (Morris et al., 2004). Under a 150 N load chemically tempered glasses show no radial cracking while untreated ones show well developed median–radial crack systems. This is the demonstration that compressive surface stresses not only have the capability to strengthen the glass with an existing flaw population, but also help reduce susceptibility to further strength degradation due to handling and service. This is an essential issue as regards brittle materials. Not only are the intrinsic properties to be selected and improved, but also the extrinsic characteristic, here the flaw size distribution, can be tailored.

As discussed above, the contact response of glass depends on the temperature. Viscous flow and toughness may increase as temperature increases as observed for soda-lime-silica glass approaching the glass transition. Little literature has been

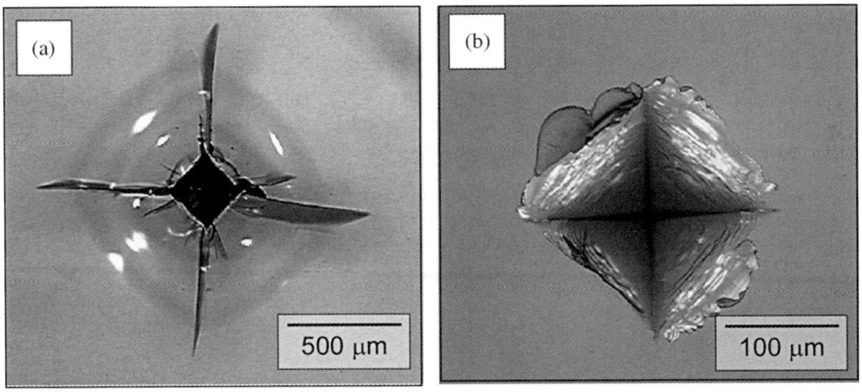

Fig. 8.21 Optical micrographs of 150 N indentations on (a) unexchanged aluminosilicate (ALS) glass (Code 0317 Corning Glass Works, Corning, NY) with radial cracks emanating from the impression corner and (b) ALS ion-exchanged at 400 °C for 30 min. Radial cracking is suppressed by the compressive surface stresses generated by ion-exchange process. (Reprinted from Morris et al., 2004, with permission from Springer Science and Business Media.)

devoted to this subject even though contact damage may happen throughout the industrial process. Notably, coatings are employed in most glass processes (flat glass, containers, fibres) to protect the glass surface as detailed in Chapter 10.

Also, brittleness index changes with the composition of the glass since hardness and toughness are both affected (Sehgal et al., 1995; Le Bourhis and Rouxel, 2003). Sehgal et al. (1995) proposed developing new glasses showing better contact resistance to fracture. They fabricated a large amount of glasses and measured their performance in terms brittleness index (Section 8.1.3.7). Figure 8.22 shows the indentation response of some of the studied glasses at room temperature. Some glasses show no radial cracks in Fig. 8.22c and d while others show a developed radial-crack system under the same loading conditions (Fig. 8.22a and b).

The overall data of their glass series are ordered according to density and normal–anomalous character in Fig. 8.23 (Sehgal et al., 1995). Normal glasses show a linear

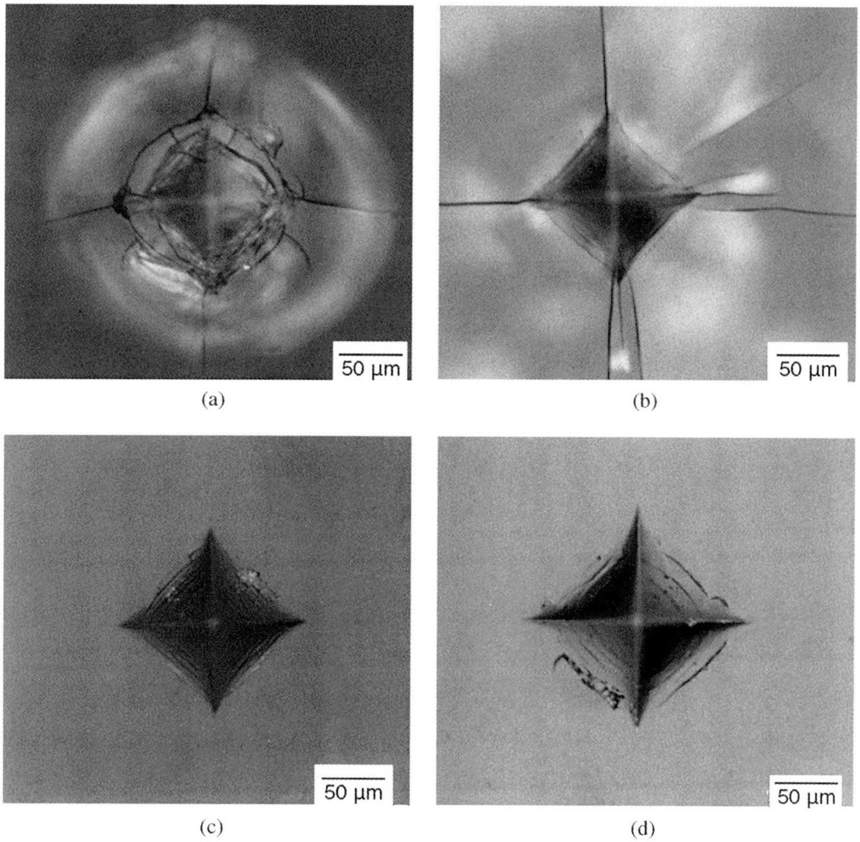

Fig. 8.22 Contact response of glass drastically changed by changes in composition: (a) low-density SiO_2; (b) large density; (c, d) intermediate density. (Reprinted from Sehgal et al., 1995, with permission from Elsevier.)

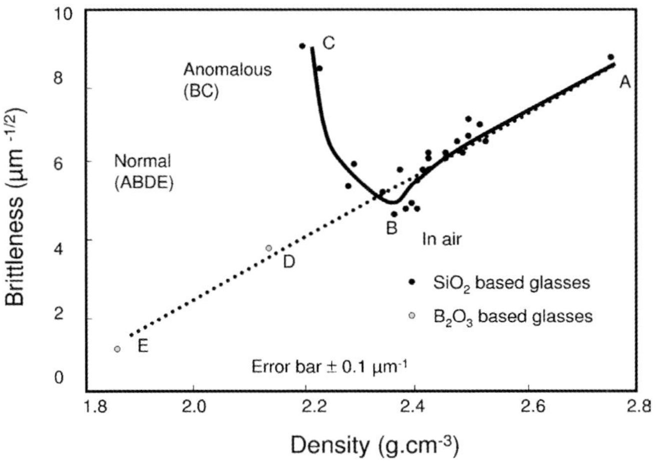

Fig. 8.23 Brittleness versus density for SiO_2- and B_2O_3-based glasses. ABC: SiO_2-based glasses; DE: B_2O_3-based glasses. Lines are drawn as guides for the eye. (Reprinted from Sehgal et al., 1995, with permission from Elsevier.)

variation of the brittleness index with the density (Fig. 8.23). The best performance is obtained for B_2O_3-based glasses. SiO_2-based glasses show a minimum of the brittleness index for a density around 2.4 g cm^{-3}. This is due to the anomalous character of SiO_2-rich glasses that show compaction as extensively discussed in Section 8.1.3.3. Sehgal et al. (1995) suggested that brittleness is dependent on densification and plastic flow modes of deformation before crack initiation. In fact, not only density is important, but also packing density should be considered (Chapter 5). As a matter of fact, B_2O_3-based glasses would be chosen for optimized resistance to contact loading. While contact resistance is to be optimized to limit the damaging of a glass surface, cutting of glass requires (Section 8.5) radial crack generation and propagation. Therefore, a compromise has to be found.

8.3
Scratch Resistance

The indentation techniques have been extended to sliding contact adding a lateral movement to the tip. These techniques are referred to as scratch testing and frictional force can be monitored. As for the indentation technique, two contact geometries are considered. Spherical contact allows elastic–brittle transition (Fig. 8.24) to be studied while sharp sliding contact generates elastic–plastic–brittle behaviour. Also, a single contact or multiple contacts can be used. Under a spherical indenter one observes a series of unclosed ring cracks on the indenter track. Experimentally, glass response is sensitive to the environment and in particular to moisture (Le Houérou et al., 2003). It is in fact very well known by manufacturers that oil should be employed with the roller when marking the glass before cutting (see Section 8.5).

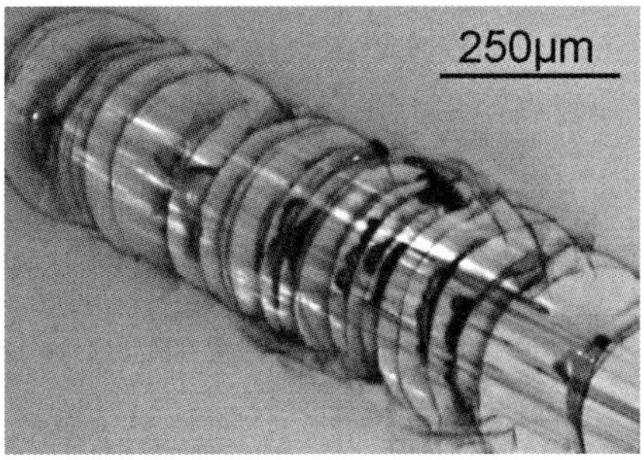

Fig. 8.24 Top view of cracks generated by sliding a sphere (0.2 mm diameter) on the atmospheric side of a float glass. Note the partial formation of the ring surface crack. Indenter movement from top left to bottom right corner.

An important factor for industrial application is the response to repeated scratch. This behaviour will be important for glass edge grinding. Change from ductile grooving to chipping is observed when the number of repeated passes increases. The critical number of passes depends both on the type of glass and applied load (Thonggoom and Funkenbusch, 2005). Also water diffusion into the glass is expected to change the surface behaviour (Chapter 5, Appendix 8).

8.4 Abrasion Resistance

The contact performance of a glass has implication also for the abrasion performance (Malkin, 1989; Tawakoli, 1993; Buijs, 1994; Le Bourhis 1998a). Buijs (1994) proposed a model for abrasion taking into account the dynamic loading of impacting particles.

Considering sharp impactors generating lateral cracks the volume of removed material for each impact is (Marshall et al., 1982; Buijs, 1994; Fig. 8.25)

$$V = \pi c_l^2 h_l \tag{8.31}$$

where h_l is the depth of the lateral crack and c_l its radius. Marshall et al. (1982) showed that

$$c_l = \alpha_0 \frac{E^{3/8}}{K_c^{1/2} H^{1/2}} F^{5/8} \tag{8.32a}$$

and

$$h_l = \alpha_1 \frac{E^{1/2}}{H} F^{1/2} \tag{8.32b}$$

where α_i are geometrical factors depending on the shape of the particle (Buijs, 1994), so that

$$V = \pi\alpha_0^2\alpha_1 \frac{E^{5/4}}{K_c H^2} F^{7/4} \tag{8.33}$$

The load achieves its maximum F_m at maximum penetration h_m. This can be calculated assuming the kinetic energy of the abrading particle is transmitted to the sample (taking into account the restitution coefficient e):

$$\frac{1}{2}mv^2(1-e^2) = \int_0^{h_m} F(h)dh \tag{8.34}$$

Neglecting the elastic contribution like Buijs 1994 (see Section 8.1.3.2), the force is

$$F(h) = \alpha_2 H h^2 \tag{8.35}$$

where α_2 is a geometrical factor ($\alpha_2 = 24.5$ for a Vickers contact) and we find

$$\frac{1}{2}mv^2(1-e^2) = \frac{1}{3}\alpha_2 H h_m^3 \tag{8.36}$$

so that

$$F_m = \alpha_2 H h_m^2 = \alpha_2 H \left[\frac{3}{2}\frac{mv^2(1-e^2)}{\alpha_2 H}\right]^{2/3} \sim H^{1/3} \tag{8.37}$$

Finally the removal volume per particle is determined to be

$$V \propto \frac{E^{5/4}}{K_c H^{17/12}} \tag{8.38}$$

Equation (8.38) shows dependence on all deformation modes (elastic, plastic and brittle modes) while brittleness index introduced in Section 8.1.3.7 shows only

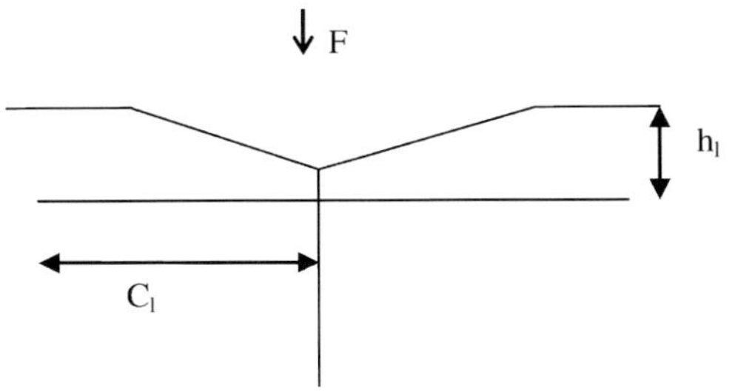

Fig. 8.25 Medial and lateral cracking under a conical indenter. When the lateral crack emerges at the surface a chip forms.

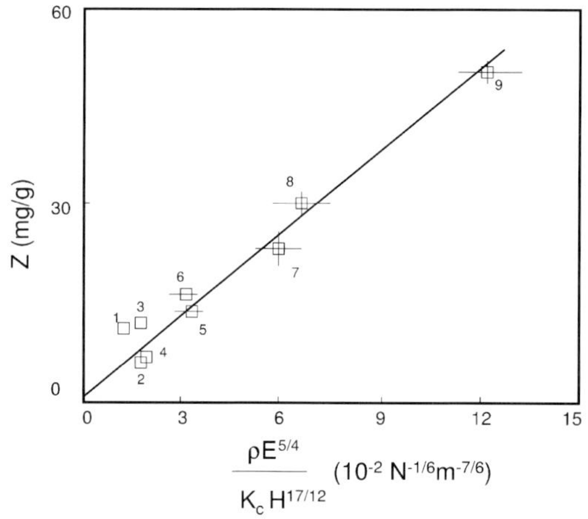

Fig. 8.26 Abrasion performance changes with elastic–plastic–brittle ratio. 1, Herasil (fused silica); 2, BK7 (crown); 3, SLS (float glass, Glaverbel); 4, ZKN7 (crown); 5, F2 (flint); 6, LF5 (flint); 7, SF55 (flint); 8, SF6; 9, SF58 (flint). (Reprinted from Buijs, 1994, with permission from Blackwell Publishing.)

competition between 'plastic' and brittle modes. The removal rate *Z*, that is, the thickness of the layer removed per unit of time, is then expected to be proportional to the volume *V* and the rate of incident particles. In erosion studies the removal rate is mostly expressed in terms of the mass of the workpiece removed per mass of the abrasive and is plotted in Fig. 8.26 for different glasses. The plot shows very good agreement of the model with experimentally extracted rates indicating that the proposed approach by Buijs (1994) allows good prediction of the erosion process. The plot indicates clearly that glass abrasion can be tailored according to its elastic, plastic and brittle properties. Glass composition has not been optimized for abrasion; rather, glass manufacturers prefer to get further insight into sanding, grinding and cutting operations in order to improve glass quality and performance (Malkin, 1989; Tawakoli, 1993; Boccaccini, 1997). As mentioned above it is of utmost importance that a glass surface is protected against abrasion (glass fibre sizing, container HEC and CEC; see Chapter 10).

8.5 Cutting and Drilling of Glass

Another important aspect of contact behaviour of glass is the cutting and drilling of glass sheets.

Cutting is divided into three operations: (i) scribing, (ii) flexuring and (iii) breaking. Scribing of the glass under a sharp tip (in fact a glass-cutter wheel) allows the

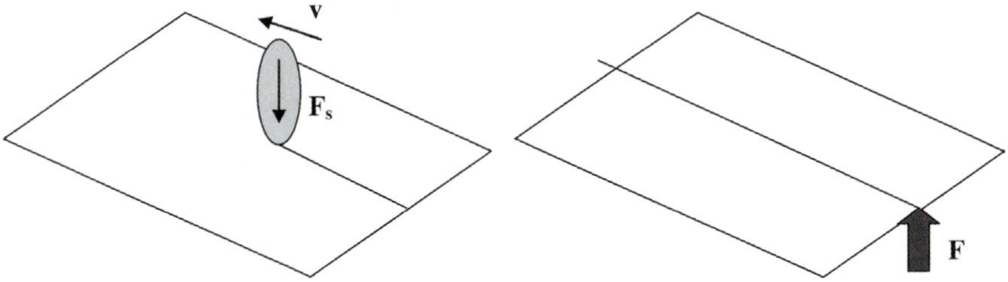

Fig. 8.27 Glass cutting principle (scribing, flexuring).

generation of median cracks similar to those described under a sharp indenter (one observes that for scribing a lateral movement is to be added; see Section 8.3). The applied vertical force F_S while scribing has to be controlled carefully to introduce median cracks while avoiding the generation of lateral ones that yield chips and degrade the glazing edge quality. Subsequently, the introduced median cracks are driven through the glass thickness applying tensile stresses on the indented side by bending the glazing (Fig. 8.27).

Scribing of good quality is achieved using sharp rollers to mark the glass and cutting oil to protect the mark from the atmosphere and so from ageing (see Chapter 9). Adler SA (France) recommends different roller angles according to glass thickness, namely 125°, 135° (standard) and 145° for 1.5–2, 3–4 and 4–6 mm thick glazing respectively. It also recommends using a mixture of 90% kerdane and 10% SAE 20 oil. Cutting oil is applied on the glass surface with a pad in front of the cutting head. The plate is then bent setting the crack tip under tensile stresses to force its propagation. These operations are carried out automatically on float lines (Fig. 8.28) and manually in distribution centres.

The success of this procedure depends on the control of crack generation and flexure stress, and also on the amplitude of residual stresses (see Chapter 6) that may

Fig. 8.28 Glass cutting machine (Bystronic Maschinen AG/ marketing brochure 2003) showing the scribing tool on a translating arm. (Courtesy of Bystronic Maschinen AG.)

appear during the cooling of the glass (Guillemet, 1990). Because of the development of thin glass (Chapter 10) and fibres, laser cutting has been developed. Thermal shock induced by the laser allows for crack formation and further development. For very thin glass (also fibres) laser-induced melting permits the direct separation (cutting) of the two pieces. One advantage of laser cutting is that in principle operations like grinding can be avoided.

Drilling of glass may be done using a rotating tube fed with a mix of loose abrasive grains and water. These are complex conditions involving several contact points and repeated loading. Also, the environment plays an important role in the results. The tube may be of steel or soft metals. However when fine dimensional control and rapid drilling rate are necessary, tungsten carbide and diamond core drills are used. Their high initial cost is balanced by drill life and production rate and quality.

9
Ageing of Glass

9.1
Fatigue in Glass

Fatigue in glass means that its strength is not constant over time but deteriorates. This phenomenon may happen when a glass is stressed in an atmosphere containing water (which is mostly the case). This is a major drawback since design strengths are to be considered three or four times lower than the initial one (so-called safety factor).

9.1.1
Static Fatigue

Fatigue is observed when testing glass under stress below catastrophic failure. An important observation is that failure eventually happens after a period of time that increases when decreasing the stress. Figure 9.1 shows the failure times of borosilicate glass rods in bending. Because of the large excursion in time and stress the data are plotted on logarithmic scales. Glass strength is observed to decrease dramatically with increasing time excursion. The strength is divided by that observed in nitrogen at 77 K (−196 °C) where fatigue is avoided as detailed below. Note that for the shortest test time, fatigue is already observed. Fatigue is also reported for soda-lime-silica and silica fibres. Silica fibres have received much attention as they are employed for waveguides. They are coated with polymers in order to protect their surface from moisture and hence to improve their resistance to fatigue.

9.1.2
Testing Methods

Methods for testing at the shortest timescale are quite difficult, while testing over long times leads to long delays before one gets the overall behaviour. The simplest way to measure the susceptibility to fatigue is to carry out static tests in which a constant load is applied to a series of fibres. The tests can be carried out in simple tension or bending. The time at rupture is then measured and determined to vary with loading

Glass; Mechanics and Technology. Eric Le Bourhis

ISBN: 978-3-527-31549-9

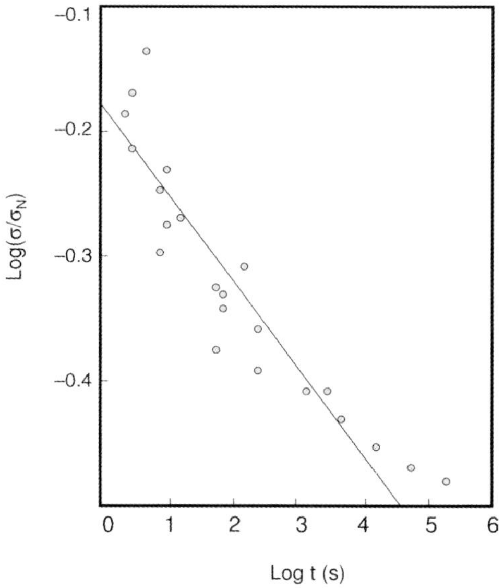

Fig. 9.1 Static fatigue of borosilicate glass rods (σ_N is the mean fracture stress in the same test – four-point bending – at 77 K, according to Friedman et al., 1982).

amplitude (Fig. 9.1). Another way is to test the specimens under different loading rates. This method requires fewer samples and is thus often preferred.

9.2
Stress Corrosion

Fatigue results from the fact that cracks may propagate and develop although the stress intensity factor (SIF) is less than the toughness (so-called stress corrosion). Accordingly SIF increases progressively and may eventually reach the critical value for breakage.

9.2.1
Introduction (Domain III)

Figure 9.2 shows the typical features of the velocity v of a running crack as a function of SIF, K. Alternatively, v is sometimes plotted as a function of the strain energy release rate G (Chapter 7). An important feature is that below toughness value K_c crack velocity is not zero. Around and above fracture toughness K_c the crack speed does not depend on the environment: it increases steeply with SIF and saturates at a characteristic speed that is around half the speed of sound (1500 m s^{-1} for a soda-lime glass). At some critical value (about 3 MPa m$^{1/2}$ for a silico-soda-lime glass) crack branching happens into two or more cracks (so-called multiplication of cracks). In a

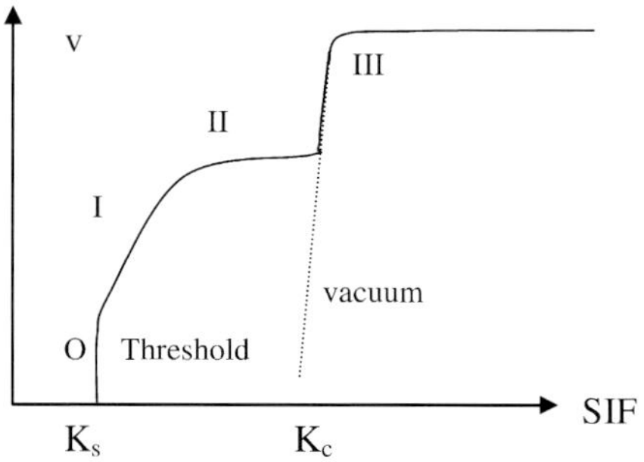

Fig. 9.2 Crack velocity ν as a function of stress intensity factor (SIF) K.

narrow range around K_c the crack velocity ranges between 10^{-3} and $1\ \mathrm{m\,s^{-1}}$ (domain III). The slope of the curve is very steep. In the absence of any water (in vacuum), this curve would extrapolate linearly to lower crack speed (dotted line in Fig. 9.2). The brittle behaviour of glass is related to the relatively low toughness value (K_c, ranging between 0.6 and $1\ \mathrm{MPa\,m^{1/2}}$) as discussed in Chapter 7. The actual behaviour of glass in ambient conditions is much more complex since subcritical crack growth may happen. In domains I and II the tip of a surface crack slowly runs even though stress concentration is below the toughness value. This phenomenon is often called static fatigue since it does not yield immediate fracture but decreases glass strength to subsequent fracture. This is a well-known phenomenon to insurance companies that recommend protecting a new impact on a windshield from weathering until it is repaired. Reparation consists in pumping moisture from the impact and protecting it in order to reduce the subcritical growth of impact cracks which indeed depends strongly on the environment as discussed below (liquid or gas, moisture or not).

9.2.2
Domains O and I: Reaction Controlled

At low driving forces, crack velocity is determined by the kinetics of the chemical reaction responsible for breaking bonds at the crack tip (Section 9.3). Threshold at O corresponds to fracture equilibrium in the related humid environment, $K_s = \sqrt{2E\zeta_e}$, where ζ_e is the glass surface energy in the same environment. With $\zeta_e = 0.4\ \mathrm{J\,m^{-2}}$, one gets $K_s = 0.25\ \mathrm{MPa\,m^{1/2}}$ for a silico-soda-lime glass. Below this threshold there is no detectable movement of the cracks. Over this threshold, cracks will propagate with velocity increasing with SIF. Fracture propagation is determined by the rates of chemisorption of reactive species to the strained bonds at the crack tip and subsequent rupture of these bonds. This is a process that is clearly sensitive to environmental parameters (Fig. 9.3) and to glass composition (Fig. 9.4).

In this regime, crack velocity increases rapidly with driving force. Starting from a very low velocity (down to $10^{-10}\ \mathrm{m\,s^{-1}}$), ν is generally represented by an empirical

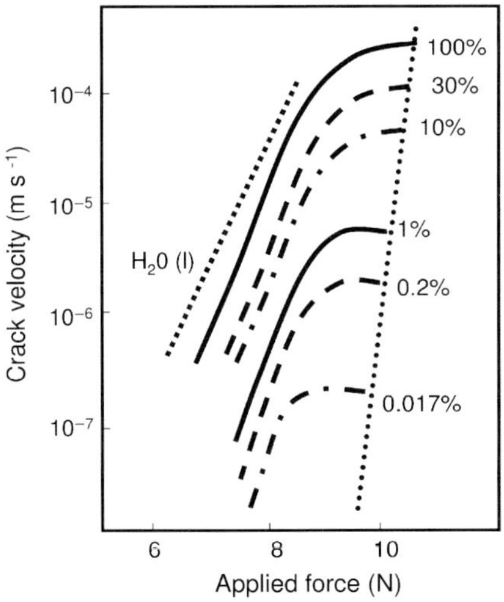

Fig. 9.3 Crack velocity as a function of applied load for different moisture contents (according to Wiederhorn, 1967).

power law dependence on the SIF (also named Evan's law):

$$v = AK^n \tag{9.1}$$

where n, the slope of the curve in a logarithmic diagram, is called the fatigue parameter. The smaller n is, the larger the susceptibility to fatigue. The range of variation of n is very broad, from 12 to 50 (Fig. 9.4). Region I and region III are

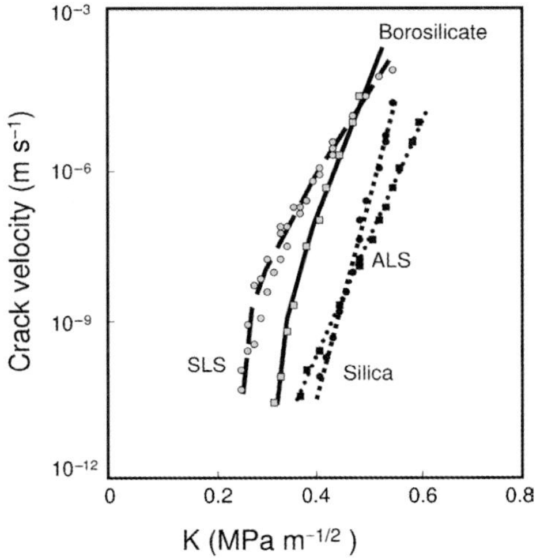

Fig. 9.4 Crack velocity as a function of stress intensity factor (SIF) K for different glasses Data obtained in water at 25 °C. (Reprinted from Wiederhorn and Bolz, 1970, with permission from Blackwell Publishing.)

connected together by region II in which the velocity does not depend very much on the SIF, but depends on the amount of water in the environment (Section 9.2.3).

The effect of an increasing amount of water in the environment (in the atmosphere or in another inert liquid medium) is a shift of region I towards lower SIF, without changing the slope *n* (Fig. 9.3). It can be shown that the absolute amount of water does not matter, but rather the ratio of the partial pressure of water to that at saturation is important (humidity ratio; Wiederhorn, 1967; Freiman, 1985).

9.2.3
Domain II: Transport Controlled

In the second regime (domain II), crack velocity is limited by the transport of reactants to the crack tip. In fact, when the driving force for crack growth increases, the transport of reactive species to the crack tip becomes the rate-limiting step. In this regime, crack growth rates are nearly independent of the SIF, but depend strongly on the environment. The effect of an increasing amount of water in the environment (in the atmosphere or in another inert liquid medium) is a shift of region II towards higher crack speeds (Fig. 9.3). At even higher driving forces, catastrophic failure takes place as detailed in Section 9.2.1 and Chapter 7, and the crack velocity no longer depends on environmental factors.

9.3
Charles and Hillig Theory

The classic theory due to Charles and Hillig 1962 to account for stress corrosion involves the chemical reaction of a water molecule with silica, taking place at the tip of the crack:

$$\mathrm{Si-O-Si} + \mathrm{H_2O} \rightarrow \mathrm{Si-OH} + \mathrm{HO-Si} \qquad (9.2)$$

This reaction shows the transformation of a siloxane bond into two silanol groups that leads to the development of the flaw (Fig. 9.5). Hydrolysis of the glass surface happens following the same reaction and a glass gel is formed (Chapter 5). Notably at the crack tip hydrolysis is enhanced by stress concentration.

It should be noted that this ‘classical’ interpretation involving a chemical reaction at the very tip of a crack is still questioned by Tomozawa (1998). As the diffusion of

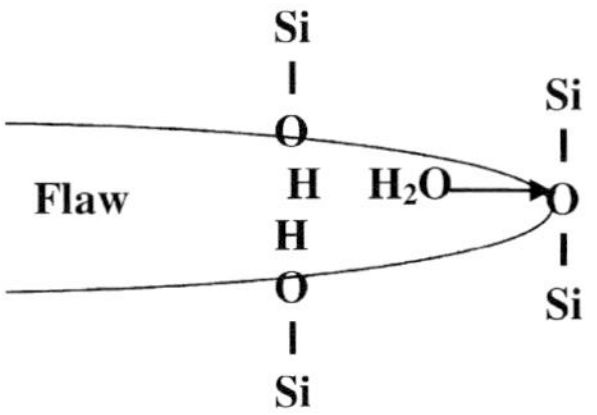

Fig. 9.5 Schematic representation of flaw extension after Si–O–Si reaction with water at a tip extremity.

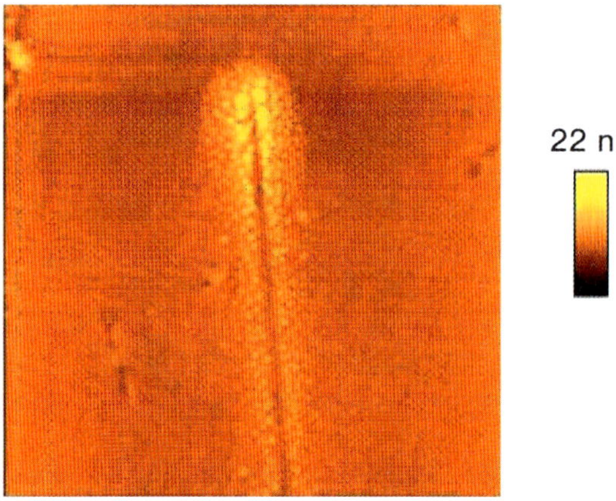

Fig. 9.6 Crystallites at the surface of glass close to the tip of an aged indentation crack. AFM view, size of field: 8.5 μm × 8.5 μm. (Courtesy of B. Nghiêm).

molecular water within glass is activated by stress, he suggests that this diffusion process and the corresponding modification of glass properties in the crack tip area might explain the subcritical growth. Koike and Tomozawa (2006) showed also that surface structural relaxation is enhanced as compared to that of the bulk in the presence of water. Recently, Nghiêm (1998) showed that Na^+ diffusion process is actually enhanced at the crack tip under the presence of the high stress concentration. Fig. 9.6 shows a subthreshold crack extremity that reveals Na^+-rich crystallites resulting from Na stress-enhanced diffusion. In fact, crack propagation involves complex physical and chemical phenomena that are to be taken into account to get further insight into critical crack growth.

9.4 Lifetime under Static Fatigue

These related phenomena have important consequences since a specimen submitted to $K < K_c$ (domain I) eventually breaks (delayed fracture) because of the development of a crack up to its critical value (crack also sharpens; Doremus, 1994).

Let us consider a crack with initial length a_i and note the initial SIF, K_i. We have $K_i < K_c$. At time to rupture t, the crack length is a_c and the SIF is K_c. Meanwhile the SIF is

$$K = \sigma Y \sqrt{a} \tag{9.3}$$

where σ is the applied stress supposed to be constant. The SIF variation with time is written

$$\frac{dK}{dt} = \frac{1}{2}\sigma Y \frac{1}{\sqrt{a}}\frac{da}{dt} = \frac{\sigma^2 Y^2}{2K}\nu \tag{9.4}$$

Considering again Evan's law we have for crack velocity v

$$v = AK^n \tag{9.5}$$

and so

$$dt = \frac{2K^{1-n}}{A\sigma^2 Y^2} dK \tag{9.6}$$

which after integration from initial state to fracture and supposing $K_c^{2-n} \ll K_i^{2-n}$ since $n \sim 10$–50 yields

$$t = \frac{2K_i^{2-n}}{A\sigma^2 Y^2 (n-2)} \tag{9.7}$$

where $K_i = \sigma Y \sqrt{a_i}$. We have also $K_c = \sigma_i Y \sqrt{a_i}$, where σ_i is the intrinsic strength (without ageing effect), then $K_i = K_c \sigma / \sigma_i$, which we introduce into Eq. (9.7) to get

$$t = \frac{2}{n-2} \frac{1}{AY^2 K_c^{n-2}} \frac{\sigma_i^{n-2}}{\sigma^n} \tag{9.8}$$

Hence time to rupture t varies as σ_i^{n-2}/σ^n with a strong dependence on applied stress σ and inert strength (Figs. 9.1 and 9.7).

Supposing the operative flaw origins in fatigue are the same as those in an inert environment, one should be able to express lifetime distribution in terms of a Weibull function (Zarzycki, 1982; Lawn, 1993).

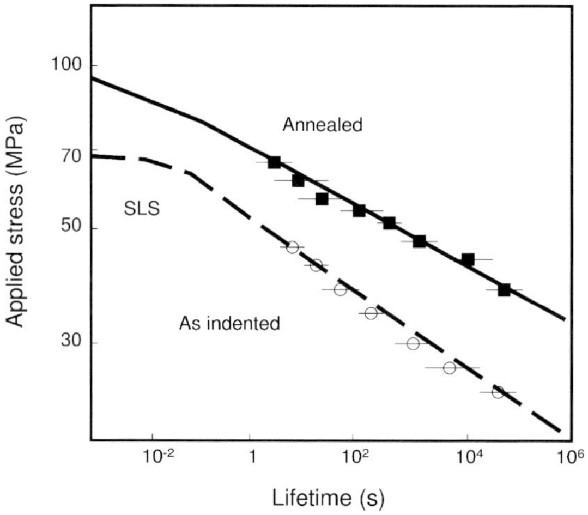

Fig. 9.7 Increased susceptibility to static fatigue due to indentation residual stresses. Vickers-indented soda-lime glass (load: 5 N), tested in water (according to Lawn, 1993).

As detailed in Chapter 7, fracture can be described statistically. Using Weibull statistics we have for the intrinsic strength σ_i (without ageing effect under inert environment):

$$P_s(V, \sigma_i) = \exp\left(-\int_V \frac{F(\sigma_i)}{V_o}\,dV\right) = \exp\left[-\frac{1}{V_o}\int_V \left(\frac{\sigma_i}{\sigma_o}\right)^m dV\right] \tag{9.9}$$

Equation (9.9) combines with Eq. (9.8) to give

$$P_s(V, \sigma) = \exp\left[-\frac{1}{V_o}\int_V \left(\frac{1}{\sigma_o}\right)^m \left(\frac{(n-2)AY^2K_c^{n-2}\sigma^n t}{2}\right)^{m/(n-2)} dV\right] \tag{9.10}$$

which can be rewritten in a simple and usual form

$$\begin{aligned} P_s(V, \sigma) &= \exp\left[-\frac{1}{V_o}\int_V \left(\frac{(n-2)AY^2K_c^{n-2}\sigma^n t}{2\sigma_o^{n-2}}\right)^{m/(n-2)} dV\right] \\ &= \exp\left[-\frac{1}{V_o}\int_V \left(\frac{t}{t_o}\right)^{m/(n-2)} dV\right] \\ &= \exp\left[-\frac{1}{V_o}\int_V \left(\frac{t}{t_o}\right)^{m^*} dV\right] \end{aligned} \tag{9.11}$$

with

$$m^* = \frac{m}{n-2} \tag{9.12}$$

and

$$t_o = \left(\frac{2\sigma_o^{n-2}}{(n-2)AY^2K_c^{n-2}\sigma^n}\right) \tag{9.13}$$

It is concluded that statistical analysis can be employed for delayed fracture prediction using the Weibull modulus m^* and timescale t_o modified by Evan's parameter n.

For flat glass and containers we have $n-2 > m$ (this is also the case for most ceramics) and hence $m^* < 1$ corresponding to a large spread of lifetime values. In contrast, glass fibres show much higher Weibull modulus so that $n-2 < m$ and $m^* > 2$ allowing for a better prediction of fatigue strength.

Equation (9.11) can be written in the form

$$t = \sigma^{-n} f[P_s(V, \sigma)] \tag{9.14}$$

where f is a function of $P_s(V,\sigma)$. Interestingly, this result shows again a strong dependence on applied stress σ as $1/\sigma^n$.

9.5 Applications

Delayed fracture has to be determined and known in order to predict glazing performance. This is a complex task since it involves environment-dependent parameters, while glass fracture prediction supposes flaw population statistics to be known. As mentioned above, for glass articles showing low Weibull modulus, a large spread of lifetime would be obtained. Proof testing is a way to eliminate weak members. The elements are subjected to a short-term stress cycle similar to that employed in service while in excess of that required. The articles with the largest flaws fail and are eliminated from the distribution which is biased towards the high-strength region. As a matter of fact, the articles can be used under service conditions with a higher degree of confidence.

Alternatively, there exist ways to prevent stress corrosion. One of these is to temper the glass. Then, the cracks that are generally present at the surface (Chapter 8) are submitted to compressive stresses provided that the glazing has been designed in such a way that the service tensile stress does not exceed the residual compression (Chapters 6 and 7). Another way is to take advantage of the development of hydrophobic thin films which might also yield new solutions to fatigue issues (Chapter 12). Such films can be used to protect the glass from moisture. As detailed in Chapter 12, one problem of such coatings is their physical and mechanical stability with time (poor resistance against UV and contact loadings; see Chapter 8).

9.6 NiS Phase Transformation

Ageing of glass may also be observed because of the physical transformation of internal defects like metastable nickel sulphide (NiS) inclusions. These undergo a slow transformation at room temperature from a denser to a lower density phase. These inclusions can be of different compositions, namely NiS, $Ni_{1-x}S$ or Ni_7S_6, with sizes in the range 1–10 μm (Barry et al., 1998). S comes from the raw materials and combustion, while Ni might appear because of the abrasion of the tools used for batching (Chapter 10). In fact, it remains impossible to eliminate these NiS inclusions. Their content is estimated at about 1 particle for 10 tons of glass.

The transformation takes place around 335–380 °C and is written

$$\alpha\text{-NiS(high temperature)} \rightarrow \beta\text{-NiS(low temperature)} \qquad (9.15)$$

with a relative increase in volume of the inclusion given by

$$\varepsilon = \Delta V / V \sim 2 - 4\% \tag{9.16}$$

Transformation strain can be accommodated over T_g by viscous flow. On the contrary, below T_g transformation strain remains in the glass article. It may drive the slow growth of the cracks up to the point where they become unstable in the residual tensile stress field in the core of the tempered glass yielding spontaneous fracture (Swain, 1981; Aben and Guillemet, 1993). In fact, the inspection of fragmented windows reveals that NiS particles are present at the origin of the fracture (at the centre of the mirror zone; Chapter 7) and located at the glass centre. This does not mean they are not uniformly distributed in the glass but, rather, fracture happens only when the particles are in the tensile zone. Phase-transformed NiS inclusion can be observed in a polarizing microscope using photoelasticity to reveal the induced stress field (Aben and Guillemet, 1993; for details on photoelasticity see Chapter 11).

The radial displacement at a distance r is a function of the increase in volume ε:

$$u_r = \frac{\varepsilon}{3}\left(\frac{R}{r}\right)^3 \tag{9.17}$$

where R is the particle size. Radial displacement increases considerably with particle size (as R^3). The maximum stress resulting from phase transformation (when NiS volume change is not accommodated at elevated temperatures) scales with

$$\sigma \sim \varepsilon G \tag{9.18}$$

where G is the shear elastic modulus. At 800 °C (in the glass melt) only the α phase exists. When the cooling rate is low, there is time enough for all NiS particles to transform into the β form. The change in volume can be accommodated when the glass is still viscoelastic. The result is that all NiS inclusions are in the β form and the glazing is stabilized. This is the case for a float glass at the lehr exit (Chapter 10). On the contrary, when the glazing is tempered the float glass is to be re-annealed at about 650 °C while NiS inclusions transform from the β into the α form again. On tempering the glass, the change back to the β phase cannot take place at elevated temperatures because of the very fast cooling rates being employed but continues at room temperature (Chapter 6). It is accelerated by solar radiation absorption that increases the temperature of the glazing. This transformation follows kinetics over a long time frame; the volume mismatch induces stresses into the glass so that fracture may actually happen a long time after production (some months or years). Obviously there are several parameters that yield a more rapid fracture (Barry et al., 1998):

i. the size of the inclusion,
ii. the presence of a flaw at its proximity,
iii. the intensity of the tensile stress, and
iv. the tempering degree of the glazing.

Glassmakers report fracture induced by 40–100 μm particles for glazing tempered at compressive surface stresses of 100–120 MPa. Thereafter, it has been of utmost importance to find a treatment to transform the NiS particles into their β stable phase, in particular in tempered glazing. Therefore, a thermal treatment (heat soak test, HST) of these glazings has been added to accelerate the transformation of NiS inclusions. This modification of the manufacturing process of tempered glass ('heat soaking') can prevent the phase transformation from taking place during service and thus efficiently guarantees the product against such spontaneous failures (Barry et al., 1998; Gy, 2003). The temperatures during such a treatment are assumed to be low enough as compared to the transition temperature so that we may neglect changes in glass thermal strengthening. In fact, heat soaking treatment is carried out at about 300 °C. At this temperature, NiS transformation is completed after 2 h (Barry et al., 1998) and thus any products containing particles of size larger than the critical dimension break, thus eliminating the risk of breakage during use (Aben and Guillemet, 1993). Moreover, since a properly tempered glass has a compressive residual stress everywhere in the surfaces, tempered glazings designed in this way are protected against ageing since static fatigue is avoided as detailed above. Nonetheless, this heat soak test involves industrial added costs since (i) it requires heat-treating furnaces, (ii) it increases the operation times and (iii) it may affect the tempering whenever deviation of maximum treatment temperature happens in the HST processing.

9.7 Crack Healing

All the ageing phenomena mentioned above are room temperature events. Obviously, thermal treatments may considerably change the behaviour of glass since viscous flow may progressively appear. In fact, crack healing is reported with related 'closing' of the cracks. Thereafter, the mechanical performance may be improved (Fig. 9.7). Nonetheless after the heat treatment is completed, ageing is observed again in the annealed glass as its stress at rupture decreases with the lifetime as detailed in Section 9.3 (see also Chapter 7).

10
Mechanics of Glass Processes

10.1
Introduction

Industrial processes are numerous and depend on the application (glazing, containers, fibres, etc.). It is beyond the scope of this book to review all of them. Instead, we shall illustrate by some examples how glass mechanics is to be considered throughout the industrial process. We have selected widespread industrial tools regarding float glass, containers and fibres. This chapter can also be nicely illustrated by a video produced by France Télévision (FR3) entitled 'C'est pas sorcier se met au verre'. All processes require a furnace with a high capacity, a long production line for upstream operations (batching, melting, fining, forming and annealing), while downstream operations (reheating, forming, moulding, coating, tempering, laminating) can be performed on-site or off-site (Fig. 10.1).

All these operations require a deep understanding and application of mechanics. At elevated temperatures, viscosity controls the homogeneity of the glass melt, the flatness of the glass while the thermomechanical properties determine the strength of the refractories used to fabricate the tools. At lower temperatures the tribological resistance of glass products will control surface quality and strength of the manufactured objects (flat glass, containers). Industrial production numbers show that tools have achieved great efficiencies since the industrial furnaces for float line and bottle line produce an average of 500 and 300 tons of glass per day respectively. A unique float line produces per day glazing surfaces required for one building as high as Montparnasse tower (Flessel, 2003). The production frequency of bottles is about 10 per second in a conventional production line. Further evolutions are carried out to meet increased capacity, flexibility, availability and glass quality. In fact, maintenance, cleanliness and fabrication practices have been considerably modified. Four primary operations for glass manufacturing are

i. batching,
ii. melting,
iii. fining and
iv. forming.

Glass; Mechanics and Technology. Eric Le Bourhis

ISBN: 978-3-527-31549-9

Fig. 10.1 Fibreglass production unit at Ladenbourg. (Copyright Saint-Gobain.)

Batching, melting and fining operations are common to all glass manufacturing processes with some variations according to the furnace type. The forming and post-process depend on the end product.

10.1.1 Batching

Batching encompasses raw material selection based on chemistry, purity, uniformity and particle size. The batch comprises sand for silica, while modifiers are introduced as carbonates instead of oxides to reduce energy costs (Table 10.1). The batch selection is adapted to the end product, for instance fibre production requires more selection and finer raw material particle size than container production. Also raw material humidity is controlled using IR analyser on modern production lines; also, impurity concentrations (Fe, Ni) are checked. Recycling of the glass has become an important parameter at this step and allows energy gains since a lower energy is required to melt the raw materials mixed with the recycled glass. Flat glass manufacturers recycle up to 30 % of their own cullet while domestic recycling of the containers has become widespread (60 to 90 % of the domestic packaging is recycled in Europe for instance). Recycled containers are separated according to their

Tab. 10.1 Raw material composition for standard soda-lime-silica glass production.

Raw material	Sand	Sodium carbonate	Calcium carbonate	Metal oxides
Concentration (%)	54	24	16	2

colour to optimize the achieved product. Also, recycling may introduce ceramic contaminants that undergo reactions with the glass melt and are present as inclusions in the finished product. Metal and organic contaminants create instability during glass processing (through reduction/oxidation reactions) and degrade the quality of the glass. Delivery, mixing and sizing processes are highly abrasive, and equipments contain metals and ceramic-coated wear surfaces. Therefore, contamination risks exist from the tools and, actually, nickel sulphide particles are believed to form from nickel contained in such tools and from sulphur impurities introduced by combustion and are responsible for the delayed fracture of tempered glazing (Chapter 9).

10.1.2 Melting

Melting consists of complex chemical and physical phenomena. A large energy above the thermodynamic value 2.7 MJ kg^{-1} is required to fabricate soda-lime-silica glass. Out of 1.2 kg of raw materials, 1 kg forms the glass while 0.2 kg forms the gas, mainly CO_2. These large volumes of gas help in the fining of the glass as discussed in the next section.

Sand and sodium carbonate grains react as follows:

$$SiO_2 + Na_2CO_3 \rightarrow Na_2SiO_3 + CO_2$$

While calcium carbonate transform as

$$CaCO_3 \rightarrow CaO + CO_2$$

Glass furnaces are used for melting the raw material particles and for transforming these into glass. The low melting constituents (alkali oxides) melt and dissolve the higher melting constituents such as quartz and alumina. Different furnaces are used for producing containers, fibreglass, flat glass and speciality glass. They can be divided into those heated electrically and those heated primarily by combustion. Often electrical heating is used in combination with fuel firing (so-called electric boost) to improve heating uniformity and melt efficiency and to reduce gas consumption and emissions that is a matter of prime importance in the context of global climate changes. Recycling glass allows emissions to be reduced, which were at a maximum in 1980 before decreasing up to the present (Flessel, 2003). Other polluting emissions (NO_x, SO_x) are generated and are reduced improving combustion (using oxygen instead of air) and fuel quality. Furnaces are all made of refractories that allow for long lifetimes (of ~10 years for float and container furnaces) before requiring rebuild costs. The cost of rebuilding a float or container furnace is in the region of €1 million.

Electrical heating is used extensively and exclusively in smaller speciality and fibreglass melting units because of its lower initial cost and low emissions as compared to combustion furnaces even though energy costs remain high. This drawback is compensated by flexibility in particular when small production volumes are required for speciality products. Glass conductivity plays an important role in this process: while being electrically insulating at room temperature, the glass melt can be heated by the Joules effect (see also Appendix 13).

Most furnaces are combustion heated which can be further divided according to the method used to recover exhaust waste heat and the way fuel is burnt (with air or oxygen). Oxygen fuel technology offers several advantages even though requiring pure oxygen. Regenerators can be avoided, eliminating furnace superstructures. Heat recovery is of utmost importance since only 10 % is used for the melt while 70 % is lost through exhaust. Exhaust waste heat recovery is performed using regenerators that alternately store and recover heat, the shift being about every 20 minutes.

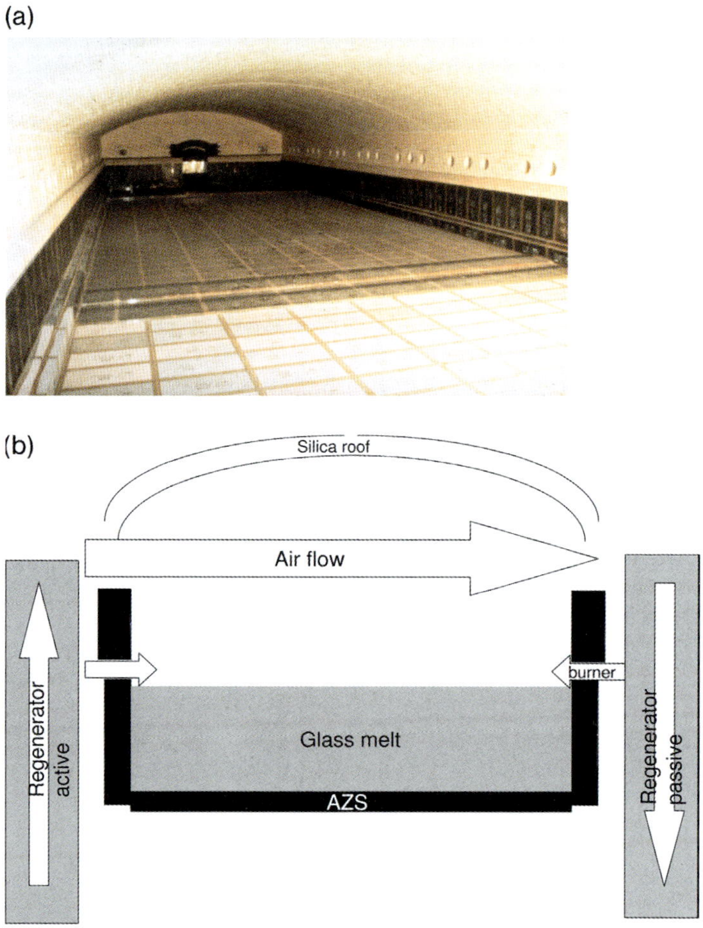

Fig. 10.2 (a) Glass furnace interior. (Copyright Saint Gobain.) (b) Glass furnace schematic. Burning and air flow alternate, regenerators being either active or passive accordingly.

In float furnaces, several regenerators (their number depending on the furnace capacity) are placed on each side of the furnace while in container furnaces, two regenerators are placed both on the back side (the gas flow forming a U loop). Regenerators consist of enormous thermal capacity refractories that alternately either store energy or recover warm air to the active burners. In a float furnace, burners are located on opposite sidewalls (Fig. 10.2). Firing alternates from side to side, the regenerators being alternately passive (stocking) or active (restitution). Such a procedure allows for production of glass with only twice the thermodynamic number (5.4 MJ kg^{-1}; Flessel, 2003). Alumina-zirconia-silica (AZS) refractories are used to contain the glass melt and to resist corrosion. The furnace roof is made of silica which presents low thermal expansion, which is important since the roof experiences elevated temperature (~1600 °C), the melt being annealed from its top. Also when dislodged pieces fall into the melt, they are more easily dissolved. Downstream, dislodged pieces cannot be dissolved any longer and then dedicated materials are selected: pure alumina is used for the exit of float glass furnaces and feeders used for delivery of automatic blowing machines. Large thermal gradients exist in the melt because raw materials are introduced progressively. Convection flow (velocity of the order of millimetres per second) is induced and favours thermal exchange and glass homogeneity. Of much interest for the glassmaker, two principal convection flows are generated in the float furnace, one in the melting zone and another in the fining zone. This allows preservation of the fining zone from the batch. In container furnaces, separation is obtained by a barrier placed at the bottom of the furnace. In contrast, radiative transfer is the dominant exchange mode between the roof and the glass melt surface.

10.1.3
Fining

Glass fining produces a molten glass that is uniform in terms of composition and temperature, and bubble free. In fact, the temperature of the glass delivered from the melting furnace, and its composition, has to be regulated with extreme care. This requirement influences the next operation, that is, glass forming, and as seen in Chapter 6, the latter operation is dependent on viscosity, that is, on the way the glass has been prepared. Consider for instance that a glass gob is to fall under its own weight in order to become correctly placed in a mould. If the viscosity is too high it will not do so in time for the next operation (say the introduction of compressed air to blow the glass out to the shape of the mould) but if it is too low the glass will fall too far and the finished article may be too thick at its base. Fortunately, large volumes of gas are rejected from the raw materials before transformation in a glass melt and this helps homogenize the melt considerably. Before glass delivery, bubbles have to be eliminated. Bubbles contain trapped air (oxygen, nitrogen) and decomposition products of the raw materials, mainly carbon dioxide. Large bubbles rise to the top of the melt and participate in melt homogenization. In contrast, small bubbles would not eliminate because of increasing required time as their size diminishes (Appendix 4). There are then two alternatives: either gas dissolves into the melt or bubbles are increased in size so that they rise more rapidly to the surface. These operations are accelerated by fining agents that are introduced in small quantities in the batch and hence in the glass melt. Fining agents like arsenic and antimony increase dissolution rates. The mechanism by which these agents enhance dissolution of gas from bubbles is still debated; a change in oxidation state of the melt or increasing the rate of oxygen diffusion is suggested. Sodium or calcium sulphate fix on small bubbles that are favourable nucleation sites for fining agents and release gas, increasing the related bubble size and this hence facilitates their elimination (Appendix 4):

$$2Na_2SO_4 \rightarrow 2Na_2O + O_2 + 2SO_2$$

The fining zone is generally gas fired; an electric boost can also provide increased efficiency in getting uniform temperature.

10.1.4
Forming

As stated above, the forming operations have been developed for different applications. Flat glass, container and fibre tools present different viscosity, stress and strain rate regimes. The tools have been developed accordingly. Some of these are described in the continuation of this chapter.

10.2
Float Process

Let us consider the fabrication of flat glass which is produced nowadays by the float process. This process was introduced by Pilkington Brothers Ltd in the 1950s (see Chapter 3). Firstly, the raw materials (sand, soda, lime) are continuously introduced into the furnace, melted at $\sim$1500 °C, homogenized by convection and fined to eliminate bubbles (Fig. 10.3). The furnace contains typically 2000 tons of glass and produces every day 500 tons of glass. The viscous liquid travels onto the float at a temperature of $\sim$1100 °C under a nitrogen atmosphere in order to prevent corrosion of the tin bath. Under these conditions the equilibrium thickness of the glass sheet is about 6 mm so that the sheet has to be expanded or contracted by top rolls (or rollers working from the top of glass) to produce thinner or thicker glass sheets respectively. The equilibrium equation is written (Fig. 10.4; Langmuir, 1933)

$$\frac{g t_H^2 \rho_g}{2}\left(1 - \frac{\rho_g}{\rho_b}\right) = \zeta_{ga} + \zeta_{gb} - \zeta_{ba}$$

where g is the gravitational acceleration, t_H the sheet thickness, ρ_g and ρ_b the density of glass and bath, ζ the surface energy and subscripts g, b and a indicate glass, bath and atmosphere respectively. Taking $\rho_g = 2500\,\mathrm{kg\,m^{-3}}$, $\rho_b = 6500\,\mathrm{kg\,m^{-3}}$, $\zeta_{ba} = \zeta_{gb} = 0.5\,\mathrm{J\,m^{-2}}$ and $\zeta_{ga} = 0.35\,\mathrm{J\,m^{-2}}$, one finds t_H=6.8 mm.

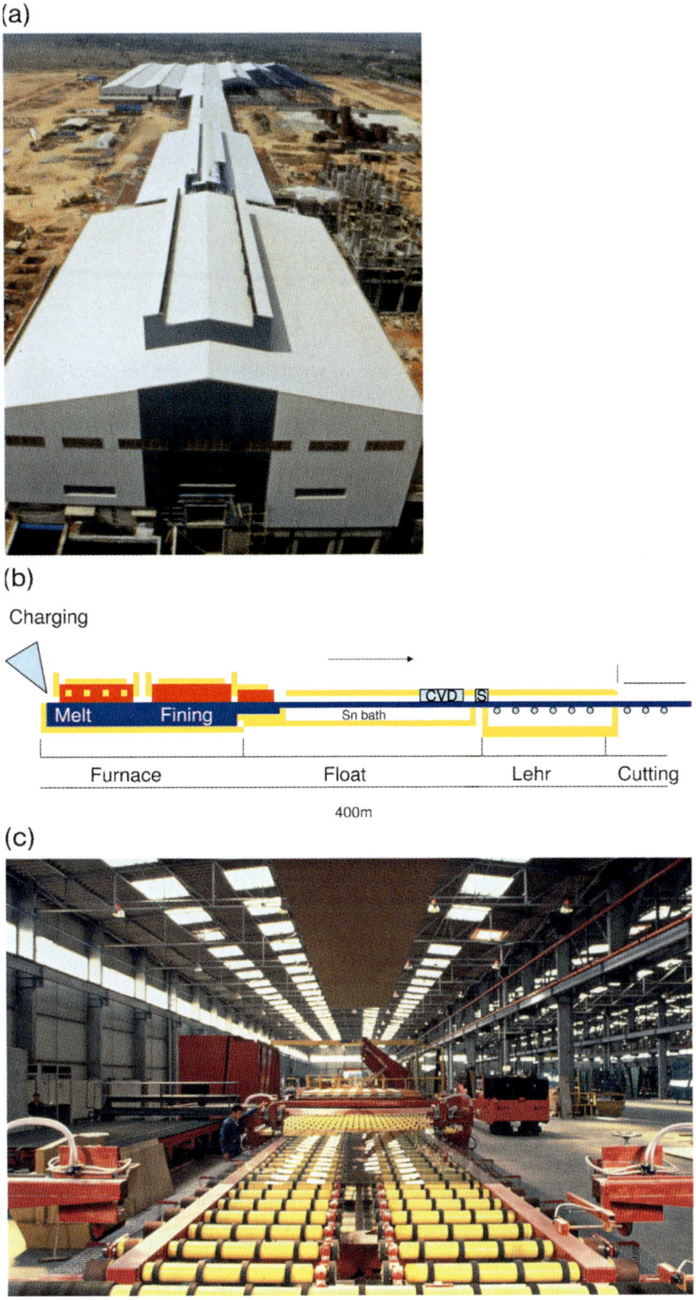

Fig. 10.3 (a) Float line at Chennai. (Copyright Saint-Gobain.) (b) Schematic representation of a float line (CVD and spray unit locations are indicated and optional). (c) Float glass line end showing transiting glass on rollers. (Copyright Saint-Gobain.)

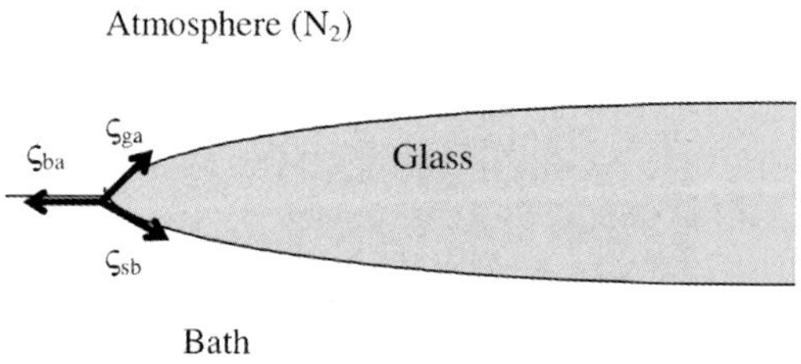

Fig. 10.4 Float ribbon mechanical equilibrium on a tin bath.

Several pairs of top rolls are used; they are made of steel and are water cooled. Their rotation axis is horizontal and shifted out of the float axis to draw the glass ribbon. The range of commercial thickness is between 2 and 19 mm. Glass with lower thickness ($<$2 mm) is difficult to produce and fusion draw is preferred (Section 10.4). The glass sheet is extracted from the float at ~600 °C and transferred to the lehr. At such temperature the glass ribbon is viscous enough to be drawn upward out of the tin bath. From this point, tin side of the glass (the bottom side that was in contact with the tin bath) is submitted to several contact loadings by rollers at temperatures ranging from 600 °C to room temperature while it is transferred. The use of an SO_2 atmosphere allows reduction of the friction between the glass sheet and the rollers. In fact, the glass surface reacts at elevated temperatures with SO_2 to form a lubricant layer that helps protect the glass surface. In the lehr, the thermal history is carefully controlled to prevent glass distortion (in particular for thin glass) and to design the residual stresses within the glass since it passes through the glass transition (~550 °C). The glass is then transferred to the control, cutting and scoring sections (Chapter 8) to produce 6 m × 3 m panels (Ridealgh, 2006). The glass edges that were in contact with the top rolls have to be removed. Here the on-line control of membrane stresses all over the process is essential for cutting operations (Chapter 11). The sheets are transferred with care avoiding sharp contacts thanks to the use of suction cups (Chapter 8). The panels are then stored with soft separators in between and transported in a tilted stack.

This process has been a revolution for the glass industry because it offers very good optical quality without requiring any further operation, while polishing was necessary before the invention of the float process (Chapter 3). Therefore, most flat glass is produced through this process. This implies that to produce curved, tempered, laminated glazing it is necessary to heat the glass again close to its transition temperature (Fig. 10.5; Chapter 6). These further operations can be carried out on-site or off-site. However, in view of the efficiency of float lines the out-of-line operations are accepted.

When glass panels are to be transferred to the transformation units specially designed trucks (called Floatliners, with improved shock absorbers) are used. Once there, the panels are cut into smaller pieces (Fig. 10.6). Distributions over the panels are optimized for cutting the different pieces (windshields, back windows, etc.) generating less glass waste (these wastes are recycled however). According to the

Fig. 10.5 Schematic representation of horizontal tempering furnace. (Courtesy Tamglass Ltd, http://www.tamglass.com.)

application, the final shape is cut, edge abraded, formed at elevated temperatures and either tempered when strengthened glass is to be produced or cooled slowly when laminated glass is to be produced. All these operations are carried out carefully to avoid glass distortions and surface damage. In fact, the evolution from vertical to horizontal tempering has improved the process capacities but increased the risk of roll-wave distortions (Bartoe and McDanel, 2003). All the operations realized at elevated temperatures and horizontally require optimization of the viscosity regime to the spacing of the rollers (to limit glazing sagging) and also require improvement of the glass–roller interface. Ceramic rollers are to be polished, clean of abrasive pieces, to remain unbowed at elevated temperatures, and show low eccentricity.

An important characteristic of float glass is the intrinsic difference between the surface in contact with the bath (tin side) and the opposite side (atmospheric side).

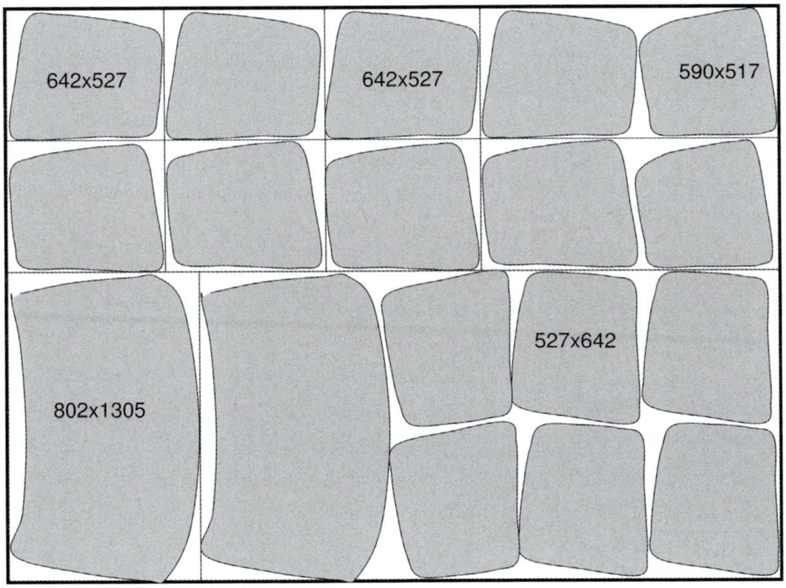

Fig. 10.6 Glass panel subdivisions into glazings (automotive laterals and backlights).

The tin side of the glass is in contact with the bath (and enriched with tin) and it is contacted by the lehr rollers. None of this happens to the top side. Concentration and penetration of tin have been measured extensively on both faces (tin and atmosphere sides; Bruckner and Navarro, 1971; Pantano et al., 1993; Stella and Verita, 1993; Moseler et al., 2002; Frischat, 2002). However, the reported tin profiles are complex with the tin in different oxidation states. For the tin side, a satellite peak has been reported by some investigators and varies according to the thermal and oxidation conditions within the bath chamber and during the subsequent treatments (thermal conditions in the lehr, SO_2 treatment, tempering; Suscavage and Pantano, 1983; Wang, 1997; Williams et al., 1997). However, it is accepted that a tin gradient of about 10 μm in depth is present on the tin side of float glass (for common thicknesses of 2–6 mm). Consequently, the physical properties of the two faces of float glass are different. For example, it is easy to propagate an evanescent optical wave into the outermost layers of the tin side while it is difficult to do so at the atmospheric side. This mirage effect is used to measure the surface residual stresses at the tin side of float glass, determining polarization changes induced by the residual surface stresses (Chapter 11). However, work still needs to be carried out to know better the physical properties of this superficial glass since it is characterized by a non-uniform, rapidly vanishing tin concentration which makes the investigations more difficult (Le Bourhis, 1998). Interestingly, some studies report different strengths for the two sides (Entwistle 1993) and this could be in part attributed to different physical properties and surface damage.

A major change to the float process was the introduction of on-line coatings. As the glass ribbon moves continuously, high deposition rates are required (60–100 $nm\ s^{-1}$; McCurdy 2000) which to date are only obtained using chemical vapour deposition (CVD). The first manufactured coatings (low-E glazing) were tin oxide (SnO_2) which was deposited from an organotin compound ($(CH_3)_2SnCl_2$) and oxygen. The on-line CVD process is fast and cheap because the glass is coated as it is properly annealed minimizing post-processing. Also as compared to sputtered PVD coatings (Chapter 12), CVD coatings adhere more strongly to the glass since they are deposited at higher temperature (600 °C). Unfortunately, CVD does not allow the broad application range obtained by magnetron sputtering that is more flexible in terms of materials and compositions (Chapter 12).

10.3 Fusion Draw

Fusion draw was invented by Corning in the 1960s for automotive applications (windshields). With this process (Fig. 10.7), the glass sheet is formed from a continuous glass flow delivered by a refractory piece called an 'isopipe'. Glass flows out laterally along the two sides of the pipe. These two flows later join downwards to form the glass sheet. The control of the thickness is carried out with rollers that do not contact the central part of the sheet (as for the float process in the tin bath section). The damaged edges are later removed while scoring. An important characteristic of

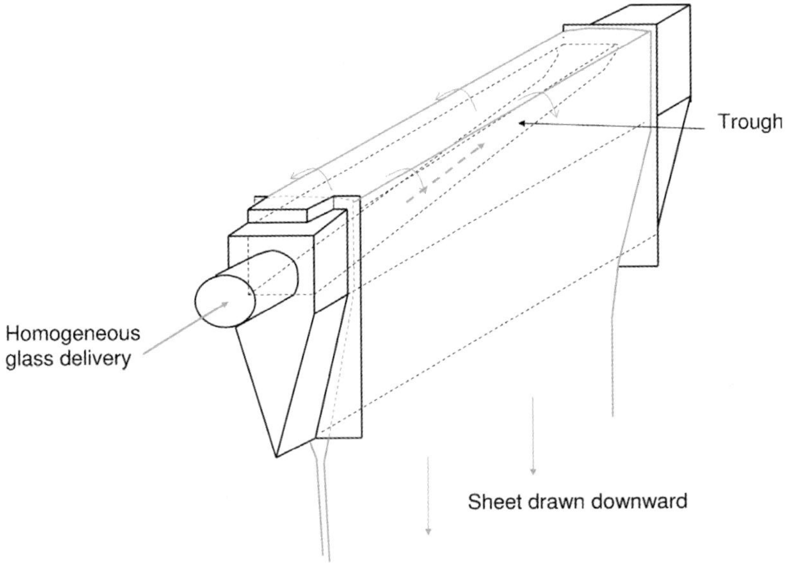

Fig. 10.7 Fusion draw process.

this process is that the external faces of the sheet are not mechanically damaged since they are not in contact with any tool. In fact, the two glass surfaces that were in contact with the refractory are fused together when they join. This process allows production of sheets of 1 m^2 with an excellent control of thickness down to 0.5 mm. This is an important point to reduce flat display panel weight (display applications; see Chapter 12). Moreover, no polishing of the glass is necessary since the process delivers excellent glass surface quality.

10.4 Container Process

Containers are formed on a large scale (Fig. 10.8) using different processes listed in Table 10.2, among which we shall focus on pressing, press-and-blow and blow-and-blow processes.

Tab. 10.2 Processes and associated products.

Process	Articles
Pressing	Culinary and table vessels, pots, lenses, bricks
Centrifugal casting (or spinning)	Plates, cathodic tubes
Blow-blow	Bottles
Press-blow	Goblets, light bottles
Press-blow-spin	Goblets, laboratory tools

Fig. 10.8 Fabrication line for bottles. (Copyright Saint-Gobain.)

All processes start up with batching and melting, and after conditioning to the desired temperature, the hot gob (molten glass) is transferred from the furnace to the forming operations through gob feeders (Fig. 10.9). When glass moves from the melting tank to the forming machine, the glass has the appearance of a thick and red-orange syrup. The gob feeder controls the weight, temperature and shape of the gob, all of which are critical to container quality. Gob feeder rates can approach 300 gobs per minute. Several forming machines are distributed by one gob feeder. The gob

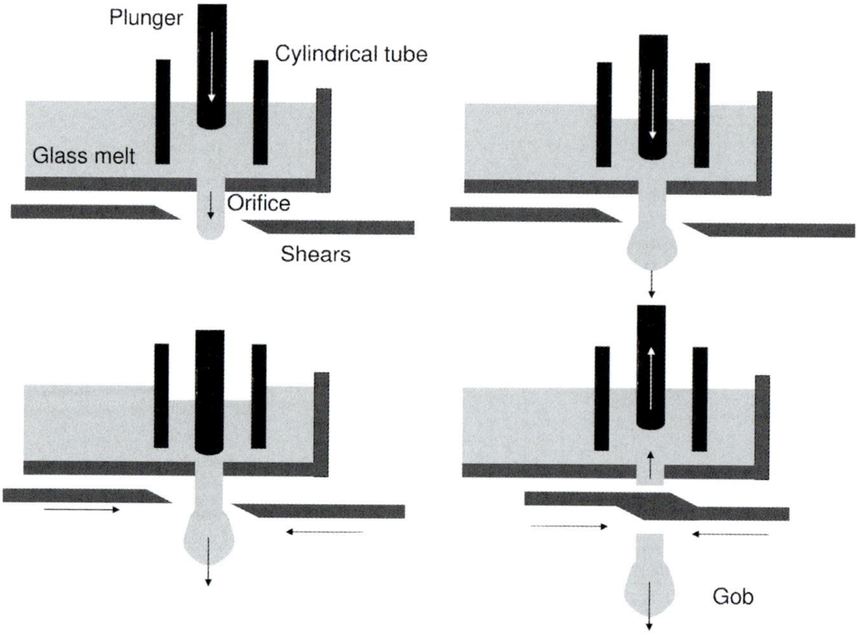

Fig. 10.9 Four-step gob feeder process.

feeder comprises a plunger that drives the glass melt to the orifice and shears that cut the gob at the desired volume. The cutting of the gob is facilitated by the cooling of the glass melt resulting at the shear contact (and hence an increase in viscosity) while backward movement of the plunger induces the upward displacement of the glass melt. The gob thus formed, next enters the forming operations.

While forming, heat is extracted from the glass in a controlled manner. The moulds have to be made of good thermal conductors with a smooth surface aspect and lubricated. These properties determine the forming step and external surface quality. The process divides into three steps:

i. parison forming,
ii. container finishing (only for press-and-blow and blow-and-blow), and
iii. container annealing and coating.

10.4.1 Pressing

Pressing is used for widened containers (plates, cups and glasses) and also for producing thicker cross-sections for automotive lenses. The gob of molten glass is loaded in a mould, the plunger is lowered and forces the glass to spread and fill the mould (Fig. 10.10). Pressing can be performed with free sides (without a ring) when dimension tolerance is acceptable. The size of the plate or glass is then determined by the parison temperature and the pressing force. Patterns on the mould surface allow for imprinting the glass object and make it look more attractive and more expensive than it is (Chapter 4; see also Chapter 12 on value issue). For more dimensional control, the mould is closed by a ring allowing for production of automotive headlight lenses with high-precision shape. An automatic machine comprises 6 to 16 different moulds allowing for high production yield.

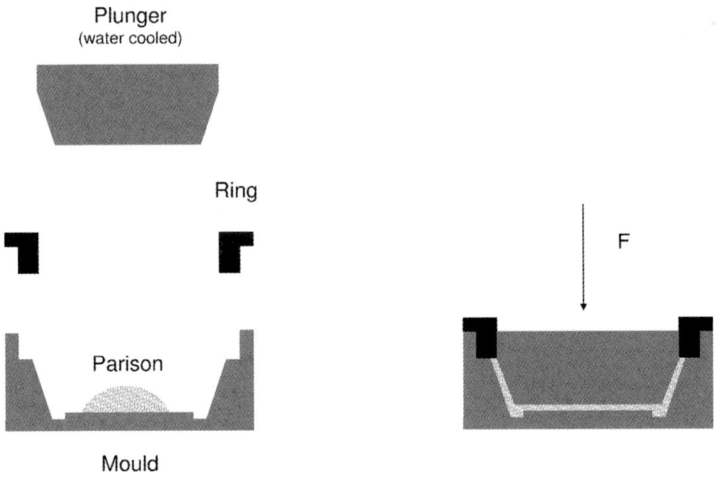

Fig. 10.10 Pressing process.

Fig. 10.11 Forming of flasks. (Copyright Saint-Gobain.)

10.4.2
Press-and-Blow, Blow-and-Blow Processes

These processes offer the possibility of producing narrow necks for bottles (Fig. 10.11). Three moulds are used: (i) the parison mould, (ii) the finishing mould and (iii) the neckring (Fig. 10.12). The latter is used throughout the process and allows for forming the neck and transferring the parison from the parison mould to the finishing one. Moulds are made of two parts which open to release the parison and the glass object. As a matter of fact a slight line is marked (so-called mould mark) on the containers.

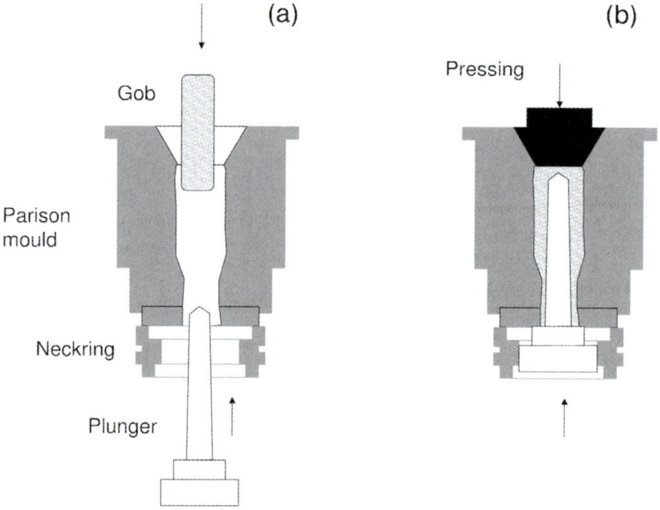

Fig. 10.12 Press-and-blow process: (a) gob feeding, (b) parison pressing, (c) parison mould opening and plunger release, (d) transfer to finishing mould, (e) sagging, (f) blowing, (g) annealing and HEC.

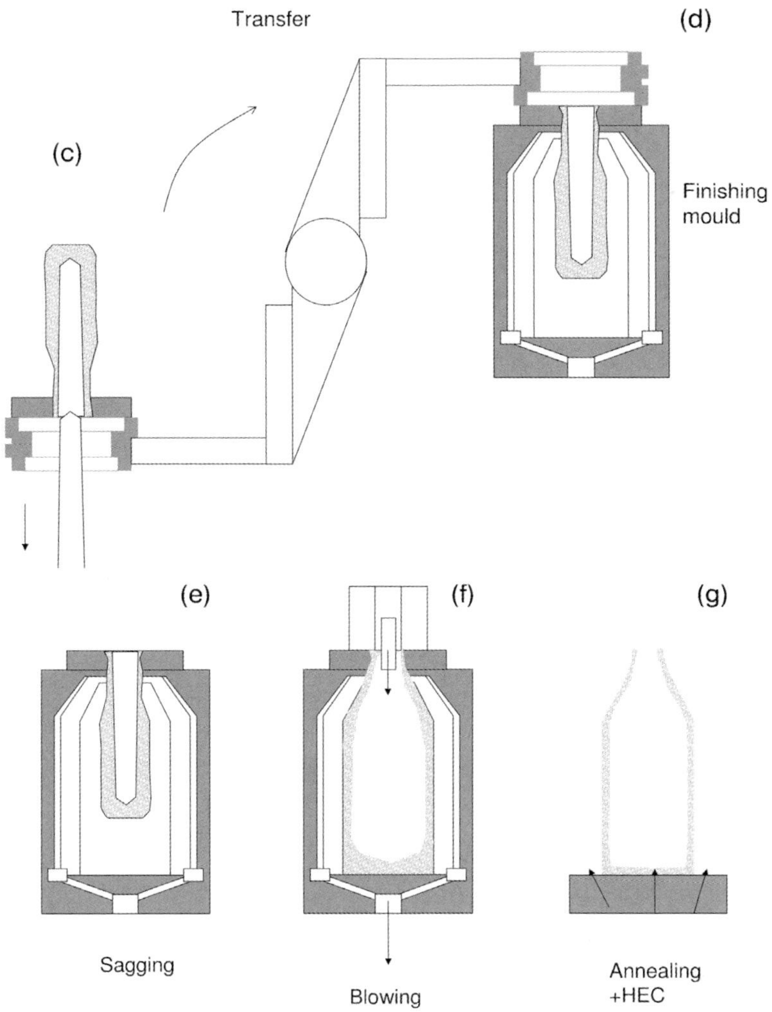

Fig. 10.12 *(Continued)*

There are two important processes to fabricate bottles named after the way used to produce the parison and to finish the container, namely pressing and then blowing or both steps blowing. Both processes start with gob feeding (Fig. 10.12a).

In the press-and-blow process a plunger (mandrel) is used to press the parison (Fig. 10.12b) against the parison mould. The plunger progressively enters the gob while it is pressed at the other extremity. Next the parison is transferred into the blowing section (Fig. 10.12c and d). The final shape is obtained after blowing the parison into the forming mould (Fig. 10.12e and f). The container is then transferred on a chain belt (which may mark the container bottom), annealed and treated (Fig. 10.12g; HEC and CEC discussed later). Annealing of the glass allows removal of the residual stresses from containers that are formed rapidly from elevated temperatures as discussed above (Fig. 10.12g). The blow-and-blow process mostly differs by the way the parison is formed. Instead of being pressed, the gob is blown in the parison mould. Next, as in the press-and-blow process, the parison is then transferred and blown into a finishing mould to achieve the final shape. Temperature control is essential to control the glass flow and distribution. It should be noted that the glass deformation under its own load (sagging) is to be considered to achieve optimized distribution of the glass. This has become a technological challenge when fabricating light bottles. Also, the control of the adhesion between the metal tools and the hot glass requires careful control of the working temperature. The combination of all these parameters is complex and finite element modelling is of great help to get further understanding of the process (Laevsky, 2003; Appendix 11).

The press-and-blow process allows for a better control of the shape of the parison and hence of the glass distribution as compared to the blow-and-blow process. For obvious economical reasons the glass distribution is optimized to allow for production gains. Light bottles are produced with the press-and-blow process while heavier bottles with more complex shape are produced with the blow-and-blow process. Notably, in the press-and blow process internal defects are introduced by the plunger when it comes in contact with the glass parison (Lornage, 2004). These deleteriously affect the mechanical performance of the bottles. Also, glass manufacturers lubricate the external surface to limit contact damage in the production lines and in use.

Coating technology is routinely used to protect the external surface of the containers. Firstly, coatings are deposited at elevated temperatures (so-called hot-end coatings (HECs) $\sim 500\,^{\circ}C$). HECs are usually 10 nm of a hard ceramic material. The ceramic material is applied at the hot end using CVD and is either tin oxide (SnO_2) or titanium oxide (TiO_2), precursors being tin tetrachloride ($SnCl_4$ treatment) and titanium tetrachloride respectively. These materials are fabricated immediately after forming when the container surface is still at an elevated temperature. Hence, CVD coatings offer good adhesion and strength. These reduce further contact damage (Chapters 8 and 12). Such HECs have several critical requirements that make difficult and challenging the development of new coatings: rapid deposition (1–5 ms), efficient, low cost, uniform, safe and non-corrosive (McKown, 2000).

At lower temperatures cold-end coatings (CECs; $150\,^{\circ}C$) are further deposited. They are 50 nm of organic material, namely partially oxidized polyethylene, which is spray-deposited from a dilute aqueous emulsion. The coating is applied when the glass temperature is about $150\,^{\circ}C$, after the container is annealed. This temperature is critical since when it is too high an anhydrous film forms and delaminates.

10.5 Fibre Process

Glass fibre production tools are very different when considering optical fibres, rovings or glass wool. Processes differ in the way drawing force is applied and three drawing modes are distinguished:

Tab. 10.3 Processes and associated products.

Process	Articles
Tensile drawing	Optical fibre (using a preform), fibres for reinforcement
Centrifugal drawing (or spraying)	Fibrous glass for thermal insulation, acoustical insulation, air filtration

i. tensile drawing,
ii. gas friction drawing, and
iii. centrifugal drawing.

Nowadays, processes employ either tensile or centrifugal drawing (Table 10.3), gas friction being combined with centrifugal drawing.

10.5.1 Tensile Drawing

The process starts with molten glass delivered through a platinum heated orifice plate. The number of orifices varies from 200 to 8000 with diameters between 1 and 2.5 mm. Glass flows through each orifice and is drawn downwards. The drawing speed determines the glass filament diameter. Common speeds are between 10 and 50 $m\,s^{-1}$ allowing one to achieve fibre diameters ranging between 5 and 25 μm. Rovings are produced by combining several hundred fibres (range 100–2000 filaments). Several fibres are drawn in parallel and gathered as shown in Fig. 10.13. Anti-abrading treatment is then essential for keeping excellent mechanical resistance. Surface treatment also called sizing is required for adhesion to the matrix in the case of composites. Figure 10.13 illustrates the manufacturing of multi-fibres and the application of sizing. In a typical fibre-forming machine (Fig. 10.14), the coating applicator is located just below the fibre-forming bushing. Glass fibres leave the

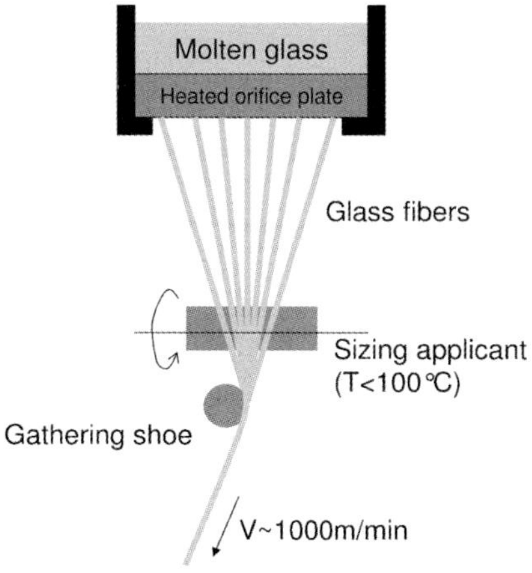

Fig. 10.13 Fibre drawing process.

Fig. 10.14 Fibre drawing. A multipurpose aqueous dispersion (sizing) is applied to the glass filaments before they are pulled together to form a reinforcement fibre. Among other functions, the sizing lubricates the fibre contacts to reduce surface flaw generation. (Copyright Saint Gobain.)

forming die at a temperature over 1000 °C and are rapidly quenched to deposition temperature (at 100 °C) by the aqueous sizing solution. Since the fibres are drawn under high speeds only short times (about 5 ms for a drawing speed of 1000 m min^{-1}) are available for applying the sizing. The filaments are then formed into a multi-filament strand and wound into a package.

Coatings play an essential role in the manufacturing of products made from glass fibre. Because of its abrasiveness, glass fibre would be useless if not coated during the sizing operation. In fact, without the protective coating the fibre loses its strength and destructs (see also Chapter 8). Sizing not only increases durability, but also adds significant value to the product. Although glass composition is tailored according to applications, sizing is the most important factor in differentiating one fibre product from another. The wide variety of fibre glass applications leads to more than 100 sizing compositions on the market, each containing from 2 to 10 components. All sizings are aqueous chemical solutions containing 0.05–10 % solids. Water has a key role in the technology. In fact, while it carries and dilutes solid components, it also cools the hot fibre allowing for the deposition of solids. These are composed of the film former (polymer), the coupling agent allowing adhesion of the film with the fibre, lubricants, antistatic compounds and antioxidants.

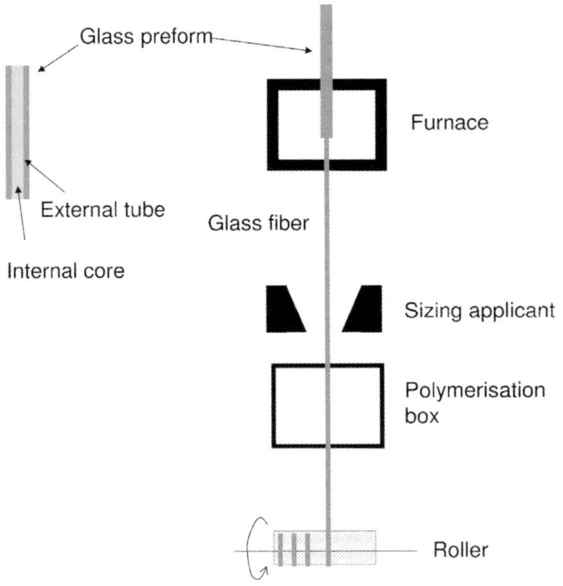

Fig. 10.15 Optical fibre drawing process.

Optical fibres require refractive index gradient necessary to propagate the optical beam. Different routes exist to generate a fibre with either discontinuously or continuously varying index. When a discontinuously changing index is to be fabricated, a preform can be made assembling an internal core of larger refractive index and an external tube made of a glass of lower refractive index. This preform is then drawn to produce very fine fibres (Fig. 10.15). Again the fibres are protected against abrasion by a polymer membrane. Alternatively, a double crucible can be used to produce the two glasses that are further drawn to form the fibre. When a continuously changing gradient is to be produced ion exchange is used. In the exchanged layer, the glass composition and refractive index are modified by the treatment. The SELFOC® process developed by Nippon Sheet Glass comprises a borosilicate glass fibre treatment in a K^+ bath.

10.5.2 Centrifugal Drawing

Fibres used in glass wool and mats have to be chopped. Also, the surface treatment will control the performance in use. The process again starts with the glass delivery from a furnace. In 1955, Johns-Manville Company developed a process for low viscosity melt. Glass flow with viscosity in the range 3–10 Pa s is dropped onto rotating cylinders spinning at an elevated velocity around their horizontal axis (Fig. 10.16). Glass drops are projected from one cylinder to another before producing fibres of different lengths and reaching a moving belt downwards where remaining

Fig. 10.16 Wool fibre wheel process. (Copyright Saint-Gobain.)

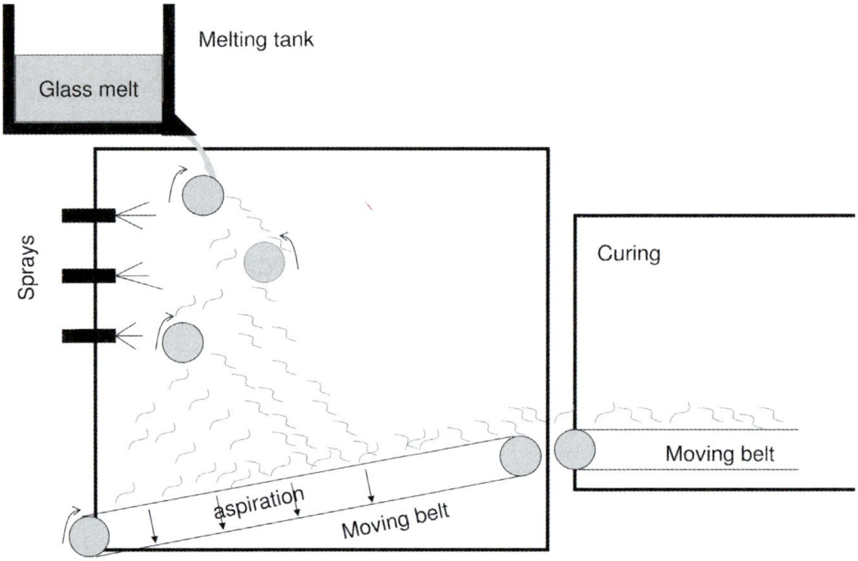

Fig. 10.17 Wool fibre wheel drawing process.

Fig. 10.18 Moving belts towards curing oven. (Copyright Saint-Gobain.)

glass drops are aspired. Meanwhile the fibres are sprayed with organic binders with curing happening in the polymerization furnace downstream (Figs. 10.17 and 10.18).

At about the same time (1957) Saint Gobain Company developed the TEL process which supplies most of its glass wool production. The main element is the centrifuge that spins around a vertical axis at 3000 rotations per minute (Fig. 10.19). It is supplied by a unique glass stream. Its flank presents large holes through which glass

Fig. 10.19 TEL process. (Copyright Saint-Gobain.)

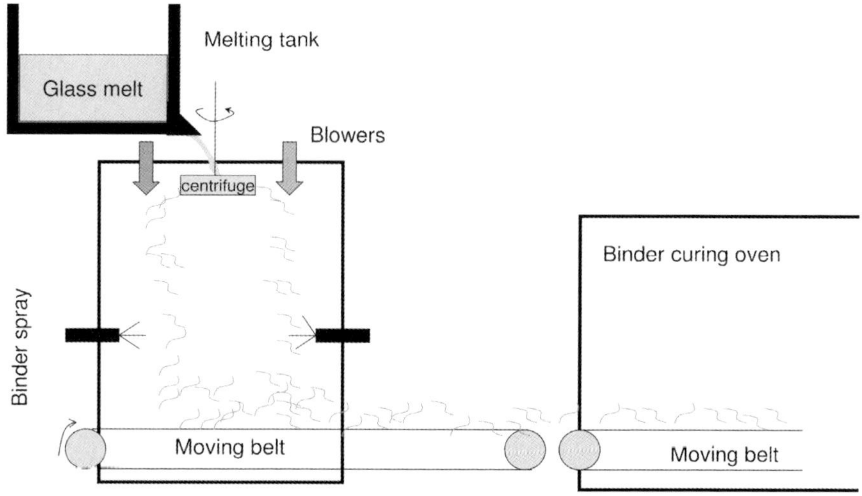

Fig. 10.20 Wool fibre TEL drawing process.

drops are projected against a circumferential and inclined wall presenting numerous smaller holes (~10 000 about 1 mm in diameter). Centrifugal drawing is assisted by annealed air blowers (Fig. 10.20). On reaching the moving belt the fibres are sprayed by polymers (protecting the fibres from erosion) further cured downstream. Moving belts allow for a continuous process delivery.

11
Production Control of Residual Stresses

11.1
Introduction

As in all industrial production lines, quality is controlled throughout the glass process on-line or off-line. As in Chapter 10, we shall not review all means for controlling the production but focus on some of them.

Glass is transparent and allows for various inspections (Fig. 11.1). Occasionally a bubble is not removed during fining, a sand grain refuses to melt, a tremor in the tin puts ripple into the glass ribbon. Automated on-line inspection allows two things: it reveals upstream process deficiencies that might be corrected and it enables computers downstream to steer the cutter round flaws or reject the glass article. In the latter case the rejected glass (also called cullet) is introduced back into the furnace.

In addition to transparency glass shows an interesting photoelastic property that allows for the determination of the stress levels. As mentioned in Chapter 7, a destructive method allows control to a high level of the residual stresses and consists in measuring the density of fragments in broken articles (R43 standard). In fact, residual stresses are very important when tempering (strengthening) is concerned. The most important way to introduce residual stresses is thermal tempering with level of stresses reaching 40 to 180 MPa. For aeroplane and train windshields (requiring high safety level) chemical tempering may be used to achieve higher level of stresses in the range 500–600 MPa (Chapter 6).

Cutting of glass however requires low residual stresses levels. In fact, after scribing, the glass is set under flexure (Chapter 8). The crack propagation path is then determined by the applied stress and residual stresses.[1)] When residual stresses are too large, glass cutting is of poor quality. The destructive method mentioned does not apply for

[1)] Noticeably, stresses can be either transient (appearing when the glass is submitted to elastic loading and recovering after unloading) or permanent (residual stresses, Appendix 7).

Glass; Mechanics and Technology. Eric Le Bourhis

ISBN: 978-3-527-31549-9

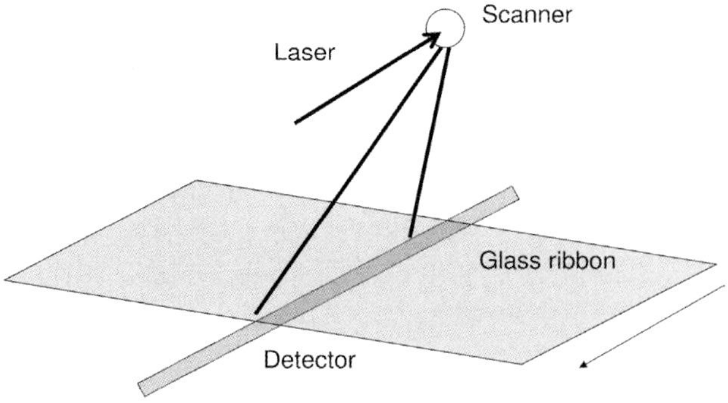

Fig. 11.1 Glass ribbon optical inspection.

quantifying such low residual stress levels since only a few fragments are generated and hence uncertainty is too large. Instead photoelastic methods are employed.

11.2
Residual Stresses in Flat Glass

In the following we shall assume the stress field is biaxial (normal residual stress $\sigma_3 = 0$; Fig. 11.2; Appendix 3) since the glass thickness is much smaller than the glazing dimensions. The stress tensor can be reduced to a symmetrical one with two diagonal terms called principal stresses ($\sigma_{ij} = 0$; Fig. 11.3), the principal directions

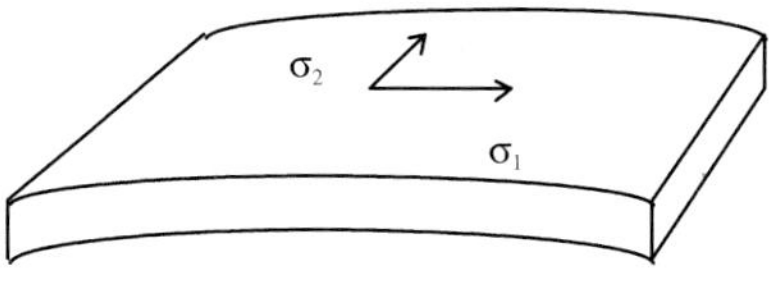

Fig. 11.2 Biaxial stress field in a glass sheet.

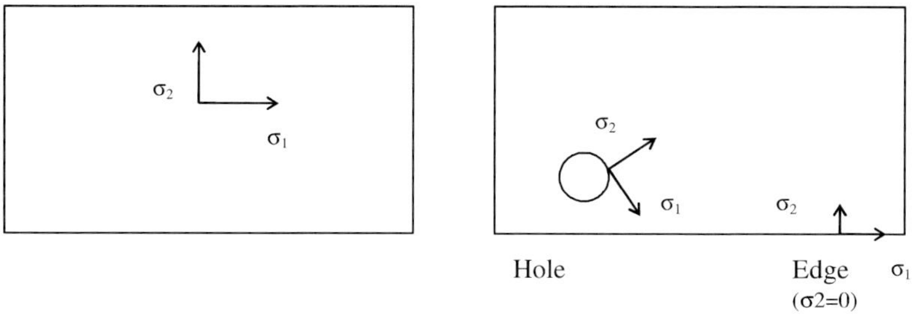

Fig. 11.3 Principal residual stresses in a glass sheet.

Table 11.1 Types of stresses and profiles in a glass sheet.

Thickness stresses	**Membrane stresses**	**Flexural stresses**
$\sigma_1 = \sigma_2$ (isotrope, equibiaxial)	$\sigma_1 \neq \sigma_2$	$\sigma_1 \neq \sigma_2$
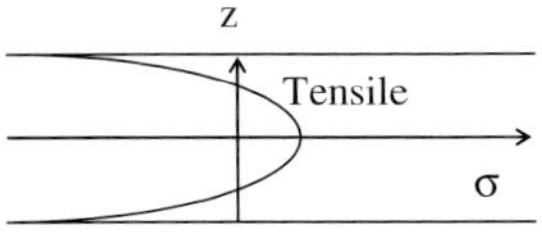	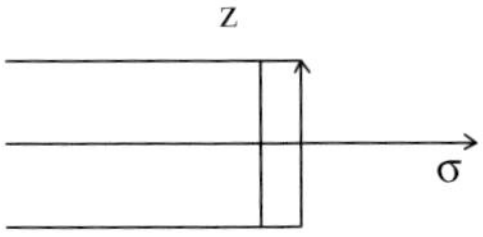	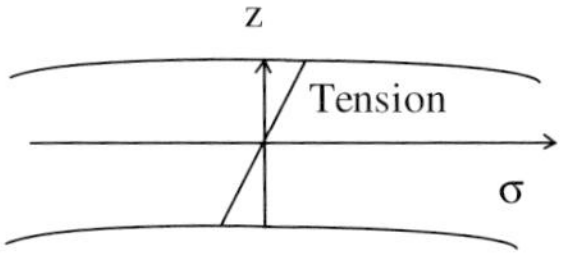
Compressive Thermal tempering	Compression ex: edge	Compression
$\int \sigma(z) dz = 0$	$\int \sigma(z) dz \neq 0$	$\int \sigma(z) dz = 0$

corresponding generally to directions of symmetry of the glazing or article as shown in Fig. 11.3 (Appendix 9).

The relations yielding principal stresses σ_1 and σ_2 are given by

$$\sigma_1 = \frac{\sigma_x + \sigma_y}{2} + \sqrt{\left(\frac{\sigma_x - \sigma_y}{2}\right)^2 + \tau_{xy}^2} \tag{11.1}$$

$$\sigma_2 = \frac{\sigma_x + \sigma_y}{2} - \sqrt{\left(\frac{\sigma_x - \sigma_y}{2}\right)^2 + \tau_{xy}^2} \tag{11.2}$$

where subscripts x and y refer to the respective axes. Stresses in a glazing are generally classified into three categories related to their origin (namely thickness, membrane and flexural stress; Table 11.1) with a general situation corresponding to a combination of those stress fields (Aben and Guillemet, 1993).

11.3 Basics of Photoelasticity in Flat Glass

Let us consider a polarized light; it is characterized by its wave vector $\boldsymbol{k}$ and polarization (light vector) $\boldsymbol{E}$ or $\boldsymbol{B}$ (they are perpendicular; Fig. 11.4). We shall consider plane, circular and elliptical polarization as defined in Table 11.2.

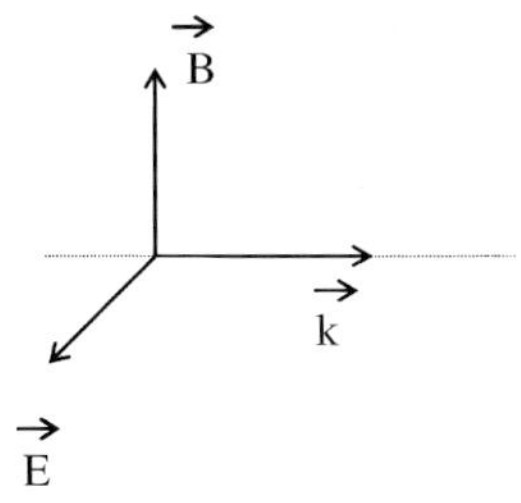

Fig. 11.4 Electromagnetic wave structure.

Table 11.2 Electromagnetic wave polarizations.

Plane polarized	Circularly polarized	Elliptically polarized
Phase shift between E_x and E_y, $\phi = 0$	Phase shift between E_x and E_y, $\phi = \pi/2$	Phase shift between E_x and E_y, $\phi \neq 0, \pi/2$

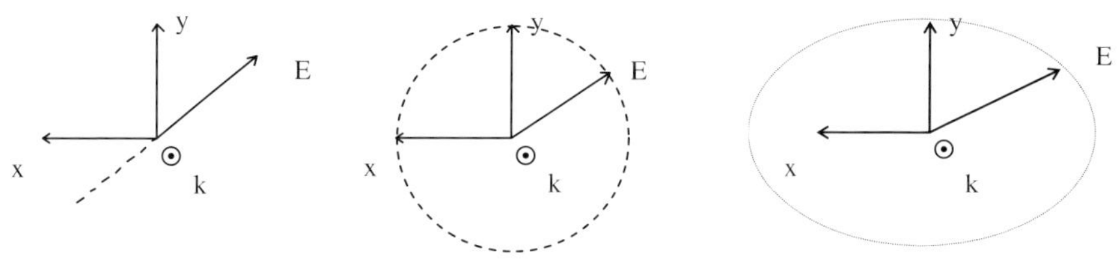

After passing through a glass specimen the light polarization is modified because glass becomes anisotropic when submitted to stresses (residual or transient; Appendix 7). The refractive index is not unique (Chapter 4, Appendix 1) but depends on the stress tensor. The indices along principal directions are

$$n_1 = n_o + C_1\sigma_1 + C_2\sigma_2 + C_2\sigma_3 \quad (11.3a)$$

$$n_2 = n_o + C_1\sigma_2 + C_2\sigma_3 + C_2\sigma_1 \quad (11.3b)$$

$$n_3 = n_o + C_1\sigma_3 + C_2\sigma_1 + C_2\sigma_2 \quad (11.3c)$$

where n_o refers to the refractive index without application of any stresses (Appendix 1). Using the plane stress hypothesis Eqs. (11.3a) to (11.3c) become

$$n_1 = n_o + C_1\sigma_1 + C_2\sigma_2 \quad (11.4a)$$

$$n_2 = n_o + C_1\sigma_2 + C_2\sigma_1 \quad (11.4b)$$

$$n_3 = n_o + C_2\sigma_1 + C_2\sigma_2 \quad (11.4c)$$

Suppose that light enters the glass plate normally (along 3-axis, Fig. 11.5) then

$$n_1 - n_2 = (C_1 - C_2)(\sigma_1 - \sigma_2) = C(\sigma_1 - \sigma_2) \quad (11.5)$$

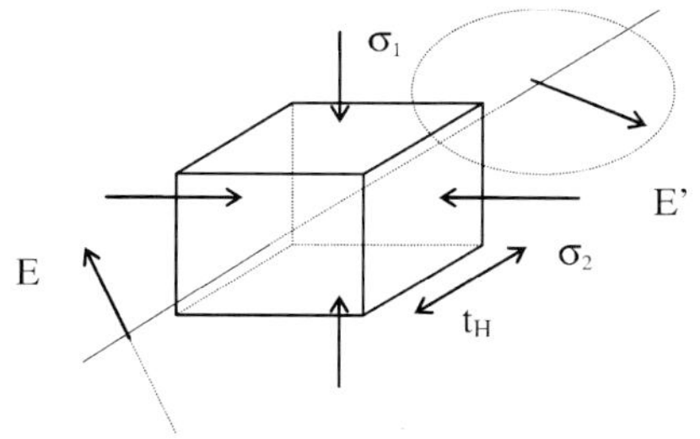

Fig. 11.5 Light path in glass under biaxial stress field.

Table 11.3 Brewster constant for different glasses (Redner, 1998).

Glass	Brewster constant, *C* (B)
Lead silica SRM 708	2.857
Borosilicate	3.65
Corning glass 704, 706	3.77
Soda-lime glass	2.6

with the phase shift being

$$\beta = \frac{2\pi}{\lambda}\delta \tag{11.6}$$

where the relative retardation is

$$\delta = t_H(n_1 - n_2) = Ct_H(\sigma_1 - \sigma_2). \tag{11.7}$$

It should be noted that the phase shift is proportional to the path length through the specimen (here the thickness t_H) and to the difference in principal stresses $(\sigma_1 - \sigma_2)$. C is the Brewster constant that depends on the type of glass (Table 11.3; $C = 2.6$

(a)

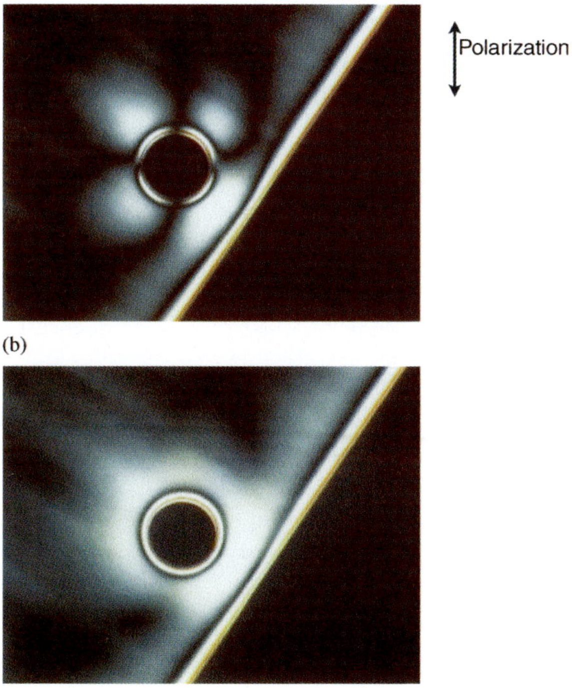

(b)

Fig. 11.6 Photoelastic patterns of an automobile side window.

$B = 2.6 \times 10^{-12}\,m^2\,N^{-1}$, $22\,MPa\,cm^{-1}\,\lambda^{-1}$ for silico-soda-lime glass). Photoelastic contrasts are revealed using polarized light. This is illustrated in the photoelastic photographs of Fig. 11.6 obtained with a plane polarized light (a) and a circularly polarized light (b).

A plane polariscope is formed by two normal polarizers (Fig. 11.7a) while circular polarized wave is produced using the scheme in Fig. 11.7b with additional ¼ wave plates with 45° relative rotation from the polarizers (Dally and Riley, 1965; Aben and Guillemet, 1993; see also Appendix 7), the plane polarized light transforming into circular polarized light after crossing the 1/4 wave plate (Appendix 1).

The principal directions σ_1 or σ_2 are detected as dark bands (isoclinics) when a plane polariscope is used and polarization being parallel to either σ_1 or σ_2. The isoclinics depend on the relative orientation of the sample, polarizer and analyzer, while zones where retardation δ is constant showing the same colour (black included) are isochromatics independent of the relative orientation of the sample, polarizer and analyzer. Hence, isochromatics inform on stress intensity (in fact, principal stress difference amplitude). In a circular polariscope, only isochromatics can be detected.

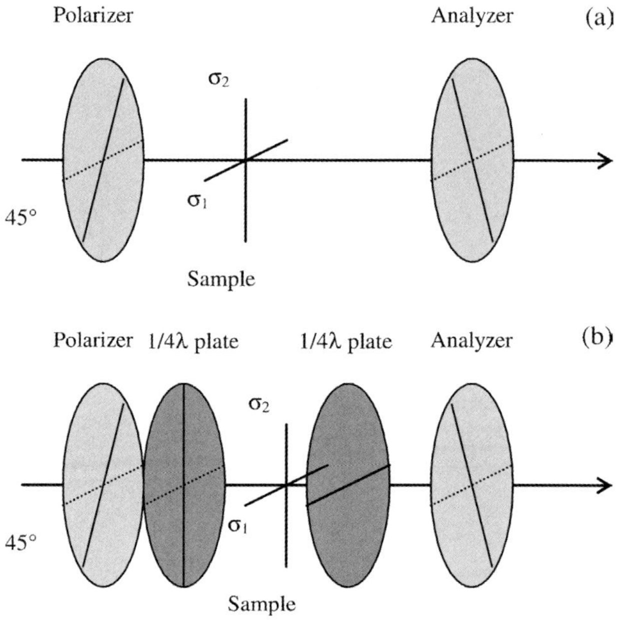

Fig. 11.7 (a) Plane and (b) circular polariscopes. Polarizer directions are normal, in (b) 1/4 wave plates are rotated by 45° with respect to polarizers.

11.4 Stress Meters

The quantitative determination of stress requires a calibrated relative retardation δ' to compensate the measured one δ ($\delta + \delta' = 0$) and we shall use from here onward the term 'compensator'. As stated above, generally we measure a stress difference that reduces to one component only when the other one is null. This assumption is valid, at the edge of a glazing (edge stress meter, shown below), and when the light beam propagates at the surface of the glass sheet (surface stress meter). Both instruments are described in the two next sections.

11.4.1 Edge Stress Meters

Edge stresses appear because of the more rapid cooling of the glass in the presence of the edge surface. Compressive stress is maximum at the edge and balanced by a tensile stress overshoot at some distance from the surface depending on the thermal strengthening (or annealing) conditions (Fig. 11.8). Edge stress measurement is carried out using a calibrated compensation obtained from a Babinet or a Senarmont compensator. We shall describe the latter, which has been commercialized by Sharples Stress Engineers Ltd (Strain Optics Company uses a Babinet compensator; Fig. 11.9). Polarized light enters and crosses the glass sheet as shown in Fig. 11.9a. Because of glass photoelasticity a phase shifted light emerges and is analysed by a ¼ wave plate and analyser. The ¼ wave plate transforms the elliptically polarized light into plane polarized light rotated by an angle $\beta/2$ (Appendix 1) that is written

$$\beta/2 = \pi/\lambda\delta \tag{11.8}$$

$\beta/2$ is determined by rotating the analyser to darken the image (compensation) at the considered point. The measurement procedure then consists in checking that

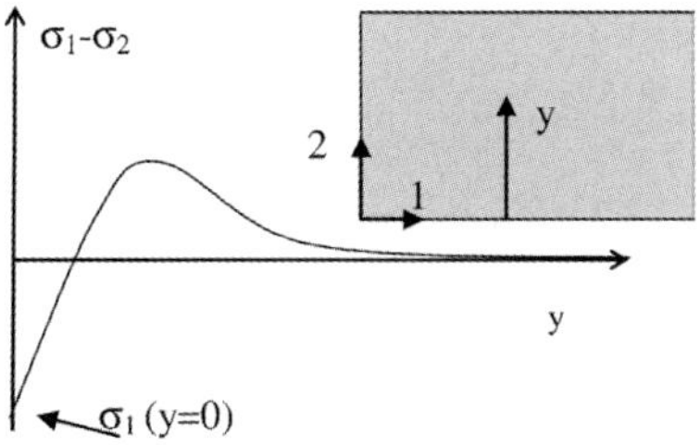

Fig. 11.8 Edge stress profile, showing compression at the edge; overshoot in the interior and then decaying stress at long distance from the edge. Note that only at the edge does principal stress difference reduce to stress component along glass edge (σ_1).

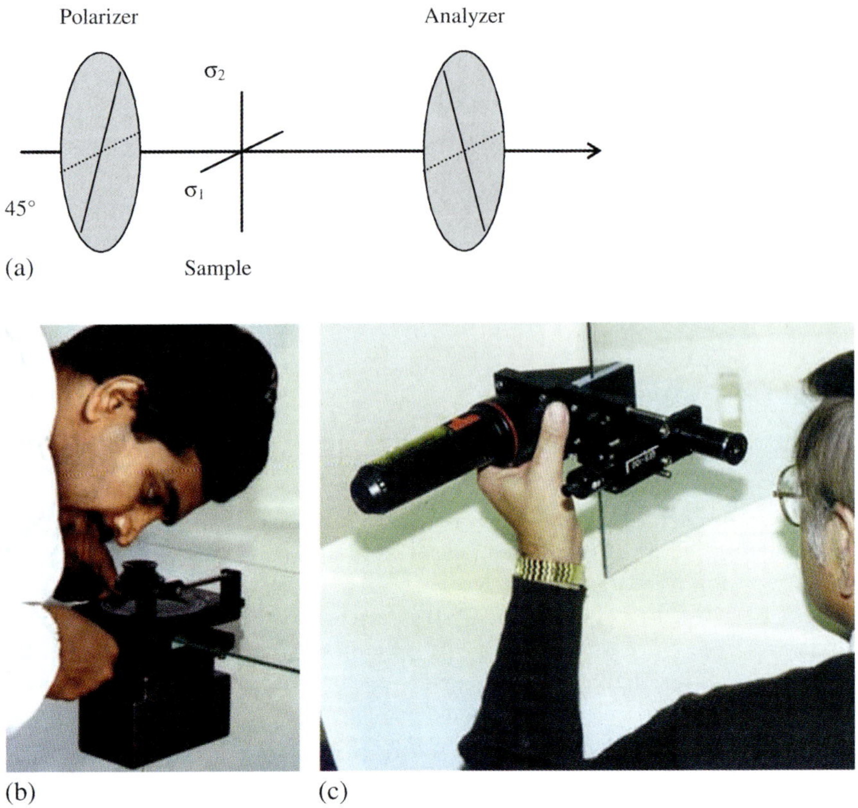

Fig. 11.9 (a) Edge stress meter principle with a Senarmont compensator. (b) Sharples edge stress meter. (Photograph courtesy of Sharples Stress Engineers Ltd, http://www.sharplesstress.com/.) (c) Strain optics Inc. edge stress meter. (Photograph courtesy of Strainoptics Inc., www.strainoptics.com.)

without any glazing no compensation is required (β is zero), then in determining β for different positions from the edge using monochromatic light and extracting the stress profile (Fig. 11.8). Sharples apparatus (Fig. 11.9) comprises a green filter and the related calibration table with the stress values for a related thickness and phase shift (Eq. (11.8)). It is important to note that stress and position are determined together since edge stress increases at the proximity of the edge. In thermally tempered glazing edge stress may be relatively high while edge stress meters allow determination of β between 0 and 2π only. Therefore, if the expected value β is higher than 2π, only the complementary fraction is measured. Moreover, only at the edge, one stress component (normal to the edge) reduces to zero and the stress measurement corresponds to stress component along glass edge.

This measurement can be extended to the ribbon control in the float process (so-called form and edge stresses). Automatic stress meters have been

Fig. 11.10 Strainoptics Inc. on-line stress scanner. (Photograph courtesy of Strainoptics Inc., www.strainoptics.com.)

commercialized and allow for scanning the ribbon from side to side and extracting the stresses map of the ribbon (Fig. 11.10). The information throughout the glass ribbon is then used to control the production and correct the process upstream.

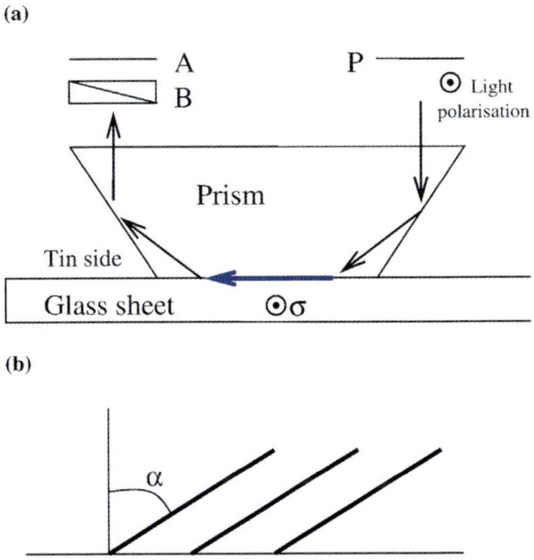

Fig. 11.11 (a) Surface stress meter principle. A coupling optical liquid is used to favour light transmission from the prism to the glass sheet at the tin side. P, polarizer; B, Babinet compensator; A, analyser. Stress component σ normal to optical path is determined. Blue arrows show the optical path at glass surface. (b) Dark interference fringes whose inclination increases with surface stress level.

Fig. 11.12 Surface stress meter, Strainoptics Inc. apparatus (GASP® surface polarimeter). (Photograph courtesy of Strainoptics Inc., www.strainoptics.com.)

11.4.2 Surface Stress Meters

Another important apparatus is the surface stress meter that allows measurement of the surface stresses (since the stress component normal to the surface is zero). Using the mirage effect due to the tin concentration gradient at the tin side of a float glass, the surface stress meter allows introduction of polarized light at the critical incidence thanks to a designed prism (Aben and Guillemet, 1993). While this method works well at the tin side, it is difficult on the atmospheric side that does not present strong tin concentration gradient. Subsequently, light propagates at the glazing surface being phase shifted by surface stress-induced photoelasticity. Light polarization is analysed after emergence from the glazing using the same prism (Fig. 11.11). The commercially available apparatus uses a Babinet compensator to determine the phase shift by measuring the slope of dark interference fringes (using previously calibrated tables; Fig. 11.12). Surface stress component σ normal to the optical path is determined. Stress amplitude is written

$$\sigma = \frac{(n_e - n_o)\tan\theta\sqrt{1-(n/N)}}{C}\tan\alpha \qquad (11.13)$$

where n and N are the glass and prism indices, n_o and n_e the Babinet compensator indices (ordinary and extraordinary, $n_e - n_o = 0.009$ for quartz), θ the Babinet compensator angle, C the glass Brewster constant and α the fringe inclination (Aben and Guillemet, 1993). The surface stress meters are adapted to the measurement of thermally induced stresses in float glass. Apparatus has also been developed on the basis of a superficial wave propagation in order to measure surface stresses in chemically tempered glazing (Aben and Guillemet, 1993; Abrams et al., 2003).

12
High-tech Products and R&D

12.1
Market Trend-Driven R&D

R&D has led to much development of glass applications improving continuously performances and devices. Marketing is becoming more important in R&D decisions. A new product is designed to answer customers' expectations. Although out of the scope of this book, product cost is of utmost importance in technological design. Ashby and Johnson (2003) reviewed the basics of the industrial approach that is controlled by a 'simple' relationship:

$$C < P < V \tag{12.1}$$

where C is the cost of the product comprising manufacture and marketing, P is the price paid by the customer while V is the value or a measure of what the customer thinks it is worth. If $C > P$ the manufacturer will loose money while if $P > V$ the customer will not buy it. Therefore, according to the manufactured object the rule of Eq. (12.1) has to be carefully checked.

However, the value V is difficult to quantify; it encompasses subjective factors. It is often said for instance that 'glass is class', and one expects champagne be delivered in a (intentionally made) heavy glass bottle and nothing else. Moreover customers are becoming more and more favourable to environmentally friendly industries. Recyclability has to be considered from the beginning of the R&D procedure. Also safety is a key parameter to customers and it has been considerably improved using laminated glazing for windshield or tempered glazing fragmenting into small harmless pieces.

Glass has become a structural part in automobiles and building (see also Chapter 2). Glass assembly with point-fixings provides flexibility for design (Fig. 12.1). A car nowadays shows glazing with very complex shapes so that they merge into the automotive design and aesthetic. These shapes are obviously produced thanks to the increasing insight into glass science and technology.

More recently, the development of nanotechnology has yielded new products and applications for the glass industry. Display panels and coated glasses have been developed as reviewed below and offer new opportunities for the glass industry.

Glass; Mechanics and Technology. Eric Le Bourhis

ISBN: 978-3-527-31549-9

Fig. 12.1 Glass assembly with corner fixing used for modern buildings (here Palais des Congrès, close to the author's university; note glass edges).

12.2
Flat Display Panels

During the last 15 years new panel technologies have spread out for information and image diffusion (mobile phones, computers (Fig. 12.2), televisions, etc.). These flat panels have modified strongly the glass industry. Synergy is necessary because only a combination of thermomechanical, chemical and electronics know-how allows for the production of liquid crystal displays and plasma display panels. These technologies are often called macroelectronics since they concern large devices

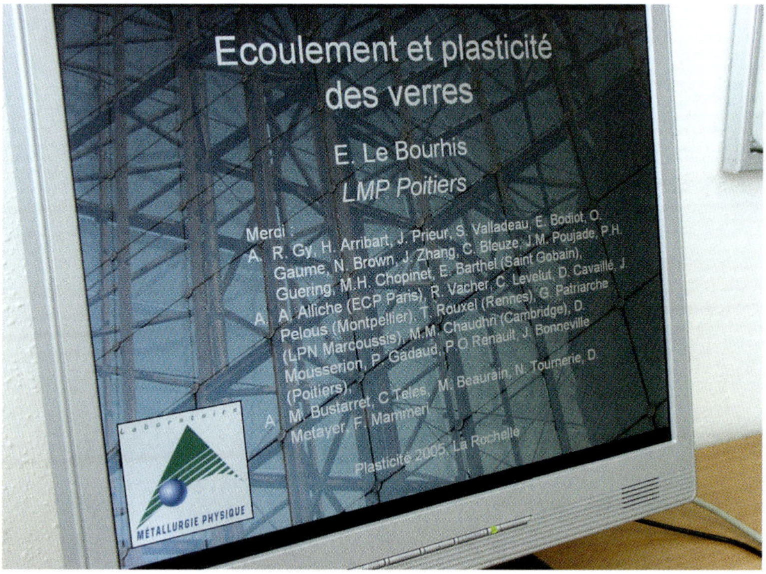

Fig. 12.2 Liquid crystal flat panel used for computers.

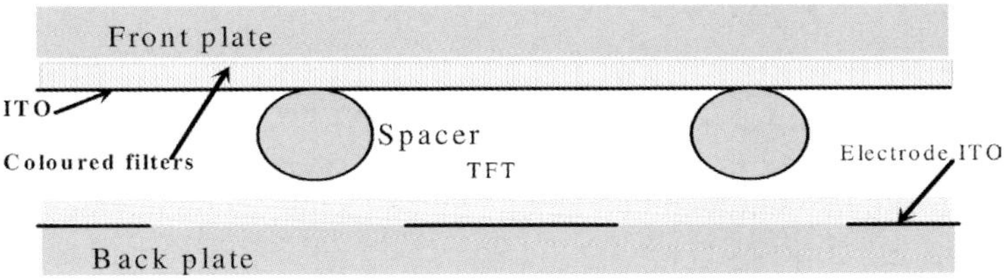

Fig. 12.3 Diagram of active matrix liquid crystal displays (AMLCDs).

over the largest possible area as opposed to integrated circuits in the microelectronics industry that makes the smallest possible devices in the smallest possible area. This is a very important and growing market: sales of active matrix liquid crystal displays have superseded the conventional glass cathode-ray tube in both revenue (2002) and units (2004) to become the dominant display technology (Reuss et al., 2006).

12.2.1 Active Matrix Liquid Crystal Displays

Active matrix liquid crystal displays (AMLCDs) represent the most important market for flat panels. Panels are realized today with 2 m diagonals: a Samsung AMLCD with 2.08 m diagonal with 2 megapixels was presented at the Society for Information Display exhibition in May 2005 at Boston. AMLCDs utilize semiconductor technologies. A network of thin-film transistors (TFTs) is realized on the active panel (Fig. 12.3). A second glass substrate (passive substrate) is used to fabricate coloured filters (Caillot et al., 1997). TFT technology comprises photolithography, dry etching, plasma-enhanced chemical vapour deposition (PECVD) and lift-off. Furthermore, heat treatments of the deposited structures have to be carried out at elevated temperatures and the thermomechanical stability of the glass substrates has to be considered carefully since alkali ion leaching and spatial deformations affect the panel performance (Section 12.2.3). In search of load gain, fusion draw (Chapter 10) has been used to produce very thin glass substrates while float technology has been pushed towards thinner glasses and new compositions.

12.2.2 Plasma Display Panels

A plasma display panel (PDP) structure is shown in Fig. 12.4. This technology has required synergy between glass and ceramics developments (see Section 12.3), electronics and optics design. Electrodes, dielectrics and phosphors have to be coated and selectively etched. Therefore, photolithography is also used and the deposited structures are heat treated for strengthening. As for LCD glass, chemical and thermomechanical stabilities have to be controlled carefully.

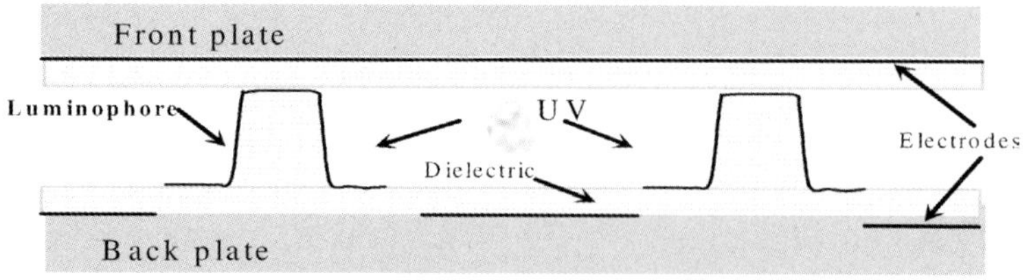

Fig. 12.4 Diagram of plasma display panel (PDP).

12.2.3
Glass Stability

Dedicated glasses have been fabricated with new compositions allowing for higher glass transition temperature (in order to improve the thermomechanical stability) and without or with reduced alkali ions for reduced chemical leaching. For instance, aluminosilicate glasses have been designed by Corning (codes 1737 and 7059) and Saint Gobain–Corning joint venture (CS25) as detailed by Caillot et al. 1997. As mentioned above, the back and front plates undergo a series of heat treatments until electronics are fully patterned. The treatment (the thermal history) depends on the type of materials used to fabricate the pixel-sized structure. Typical treatment temperature ranges between 500 and 650 °C. These elevated temperatures explain why the transition temperature of the substrates for such an application is critical. In fact, as discussed in detail in this book, glass is out of equilibrium and structural relaxation may happen when the glass approaches its transition temperature, this phenomenon being amplified as thermal treatment temperature and time increase. Notably, even though plates treatments are carried out at temperatures below T_g, structural relaxation may still happen. Virtually perfect alignment between front and back panels is required at pixel resolution. Whenever glass relaxation happens, structures fabricated at the front and back substrates shift since the supporting substrates are not dimensionally stable during the heat treatments. While assembling both substrates, misalignment may occur because of lateral shifts between pixels on the front plate and those on the back plate (Fig. 12.5).

Figure 12.6 shows the changes in the fictive temperature (structural relaxation) when a glass plate is treated at an elevated temperature ($T < T_g$) and cooled back. It is expected that structural relaxation amplitude would increase when higher temperatures and longer annealing periods are employed for the treatments of ceramic frits. Let us consider that different thermomechanical routes are used for the two panels

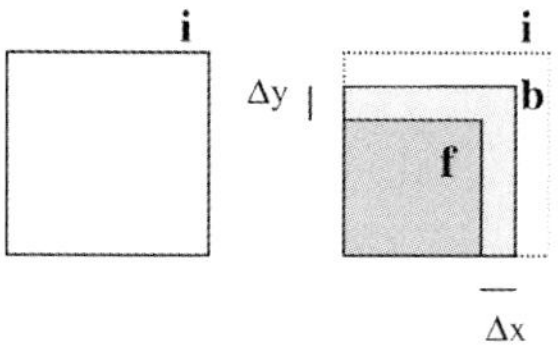

Fig. 12.5 Schematic representation of differential shrinkage. Shrinkage is exaggerated for clarity (f, front; b, back; I, initial).

(front and back). This is the case in practice since the ceramics deposited on both substrates are different and hence have to be treated differently. Both panels experience structural relaxation. Let us call the change in fictive temperature ΔT_f^f and ΔT_f^b for the front and back plates respectively. Since the thermomechanical routes are different we have $\Delta T_f^f \neq \Delta T_f^b$. This will lead to difference in structure between the front and back plates (supposing both substrates had initially the same structure):

$$\Delta T_f^f - \Delta T_f^b \tag{12.2}$$

Shrinkage between front and back plate can be estimated from

$$\frac{\Delta x}{x} = \frac{\Delta y}{y} = \alpha_s(\Delta T_f^f - \Delta T_f^b) \tag{12.3}$$

assuming isotropic shrinkage in the plane of the glass substrate (Fig. 12.5; Buehl and Ryszyticoskyj, 1991).

Since the performance in use requires the most perfect adjustment between the front and the backside electronics (remember that pixels are to be placed in front of each other), either shrinkage can be predicted or the glass needs to be stabilized before being heat treated. In the first case, masks used for ceramics deposition are designed in order to compensate for later shrinkage. In the second case, a pretreatment of the glass is designed in order to stabilize the structure so that shrinkage reduces in amplitude and becomes acceptable for the manufacturer.

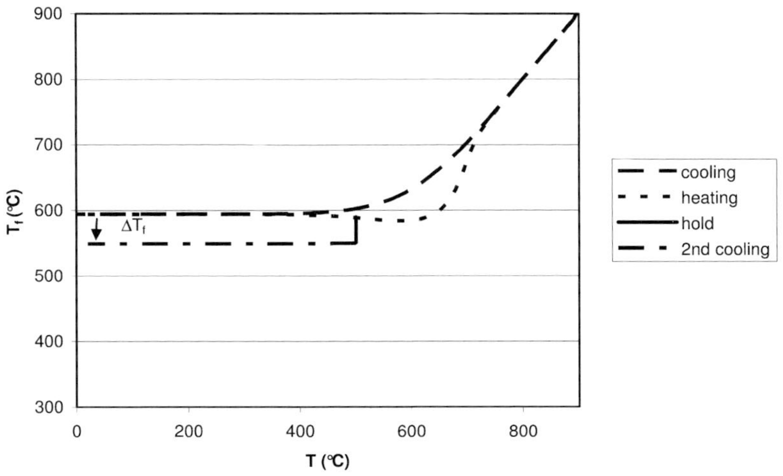

Fig. 12.6 Structural relaxation after a hold period at elevated temperature (here 500 °C). For comparison and clarity one complete cooling and heating curve is drawn.

12.3 Thin-Film Technology

Thin-film technology offers ways to improve glass strength and it is currently used for protecting containers (so-called hot-end coatings (HECs) and cold-end coatings (CECs); Chapter 10) and fibres (so called sizing) in production lines. Coatings also allow modification of functional properties of a glazing without having to change the substrate composition, which is time and yield consuming. In fact, it is difficult and expensive to change a float glass composition, since viscosity regimes have to be adapted accordingly and fabrication during the transition period has to be rejected. Therefore, developing thin-film technology has become of utmost importance for glass manufacturers since it allows minimization of industrial costs (see Section 12.1). Most common methods to produce films on glass substrates are chemical (chemical vapour deposition, CVD), physical (physical vapour deposition, PVD) and sol–gel methods. Because of the flexibility of these techniques, the efficiency of the industrial process and the performance of the related coatings, these functional glazing are becoming widespread.

Using these techniques, surfaces can be tailored in such a way that they bring a new function to the glazing (colour, water repellence, anti-solar properties, self-cleaning). Most commonly encountered functions are listed in Table 12.1 and associated to product utilization.

While transparency is important in the visible spectrum, IR transparency is a problem in buildings and in vehicles since it leads to excessive temperatures in summer and to thermal loss in winter. Therefore, new glass compositions have been developed to obtain a good compromise between optical transparency and IR insulation (Chapter 4). For the same purpose it is of interest to reflect IR radiation instead of absorbing it. This is achieved by treating the glazing surface with coatings. Multilayered coatings offer great efficiency in radiation selection. This speaks in favour of PVD that allows for the production of high-reliability coatings having 20 different sub-layers. These multilayers are composed of several silver, oxide and nitride layers (Lehmann, 2005). Unfortunately, PVD has to be carried out off-line while CVD is done on-line. Moreover, sputtering methods suppose the glass is flat.

Tab. 12.1 Current products that use coatings and their related functions.

Function	Product
Anti-abrasion	Fibre, containers
Low emissivity	Energy-conserving windows
Solar control + low-E	Windows in large buildings
FDP	TV, computers
Anti-reflection	TV, picture frames
Defogging	Supermarket freezers, windows in vehicles
Anti-reflection	TV, picture frames
Electrochromics	Automotive rear-view mirrors
Water repellence	Aeroplane windscreen
Self-cleaning	Self-cleaning windows

This is generally the case for buildings. When curved glazings are considered as in the case of vehicles this supposes that the coatings resist thermomechanically and keep their functionality when heat treated (during forming and/or tempering operations). However, there is an alternative followed by some glassmakers who deposit the PVD coating on the polyvinylbutyral (PVB) films used to laminate windscreens. The coating is then not submitted to the glass forming and/or tempering operations but instead only to the laminating operations (below 150 °C). Thermomechanical treatment resistance speaks in favour of CVD on-line methods since they offer more thermally resistant films as they are deposited at elevated temperatures while they present also minimized costs (no extra off-line operations like cleaning, annealing, etc.). R&D focuses today on the stability of PVD sputtered coatings during forming and tempering processes while alternative routes have also been suggested using sol–gel deposition techniques. These routes become very attractive when complex forms are to be produced as for containers.

More recently, glazings have been functionalized with titanium oxide (TiO_2) layers that allow for the production of self-cleaning glazing. These glazings were introduced in 2001 (Arbad et al., 2005) and are of much interest for buildings that present more and more glass surfaces. The photocatalytic property of TiO_2 causes the breakdown of organic material on the thin-film surface. The electron–hole pair formation in TiO_2 is induced by photon capture. This is followed by oxidation reactions that crack organic materials in a series of steps to clean the TiO_2 surface (Heller, 1995). This property when combined with hydrophilic behaviour allows the generation of the self-cleaning function at the glazing surface. Water (or falling rain) spreads all over the glazing removing dust while drying is uniform and clean (Saint Gobain Bioclean™, Pilkington Activ™). This phenomenon allows the glass to keep its transparency for over a year while an untreated glazing has to be cleaned (Chapter 4; Lehmann, 2005). In contrast, hydrophobic functions allow for water repellence (Fig. 12.7). When a glazing is inclined water drops downward and the glass surface remains dry (Fig. 12.8). Water repellence is of utmost importance when aeroplane safety is concerned. In fact, windshield wipers can be avoided while taking off and landing, improving the visibility for the pilots. It could be also of much interest when ageing of glass is concerned since developing impermeable and water-repellent films allow protection of the glass from water and thereby glass corrosion (Chapter 9). Both functions (hydrophilic or hydrophobic) are based on changing the glass surface

Fig. 12.7 Water drops on a superhydrophobic (non wetting) coated glass. (Courtesy of CEA.)

Fig. 12.8 Water control (hydrophobic) glass. (Copyright Saint-Gobain.)

energy ζ_{sl}. Figure 12.9 shows with a similar diagram to Fig. 5.12b the relevant parameters where the liquid l is generally water. Here again, the Young Dupré equation applies and allows for wetting angle to be determined and/or predicted ($\zeta_{sa} = \zeta_{sl} + \zeta_{la}\cos\theta$). This equation can be rewritten as $\cos\,\theta = (\zeta_{sa} - \zeta_{sl})/\zeta_{la}$ with extreme cases (i) for $\theta \to 0^\circ$ ($\cos\,\theta = 1$) corresponding to good wetting and (ii) $\theta \to 180^\circ$ ($\cos\,\theta = -1$) corresponding to poor wetting. The extreme cases are illustrated in Fig. 12.9, i.e. good wetting with a hydrophilic surface ($\theta \to 0$) and poor wetting with a hydrophobic surface ($\theta \to \pi$). Unfortunately, these coating treatments generally suffer poor mechanical resistance to contact stresses (Chapter 8) and more generally to mechanical fatigue. Therefore, R&D focuses on improving coating strength and adhesion to glass articles in order to keep the induced functionality over time. In this field nanoindentation and scratch tests are very promising tools (Chapter 8 and Appendix 8). The micrographs in Fig. 12.10 show for instance sol–gel films with scratches and delamination induced by severe conditions (Taber test). The hybrid coatings that are obtained as nanocomposites made of poly(methyl methacrylate) (PMMA) and silica show higher strength than sol–gel PMMA coatings

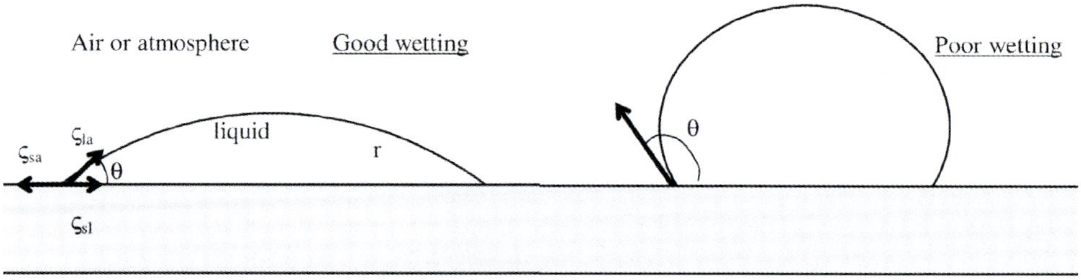

Fig. 12.9 Water drop formation on a glass substrate. Functionalizing the surface allows one to change drop geometry and induce water repellence.

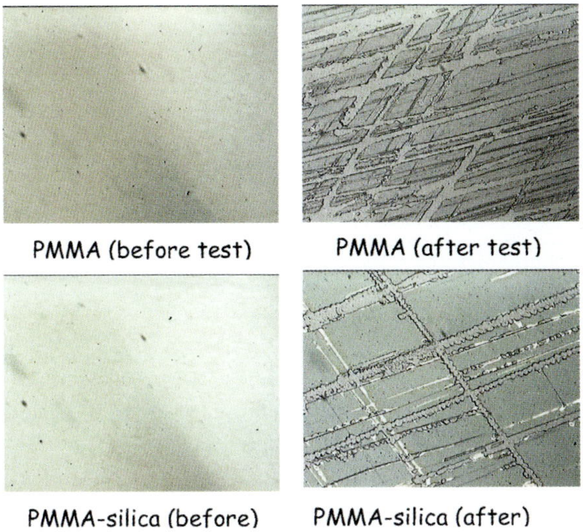

Fig. 12.10 Sol–gel spin-coated films before and after Taber testing (Mammeri, 2003). Notably, hybrid PMMA–silica (28% wt of PMMA) coating show better resistance than pure PMMA.

(see also Section 12.3.3). Delamination induces the loss of film functionalization since the film is removed from the glass surface.

12.3.1 Chemical Vapour Deposition

In chemical vapour deposition (CVD), precursor compounds (both gases and liquids) are vaporized in a reactor that spreads the resulting gas mixture uniformly over the substrate. Reaction in the gas phase happens at the proximity of a hot surface. One such reaction to form silica can be written for a $SiCl_4$ precursor as

$$SiCl_4 + O_2 \rightarrow SiO_2 + 2Cl_2 \quad (12.4)$$

In the flat glass industry, CVD happens on-line at the hot surface of the glass ribbon that moves continuously. Chemical reactions occur in the gas above the substrate and on the growing surface of the deposited film as shown schematically in Fig. 12.11

The high temperature of the ribbon in the forming float chamber and in the annealing lehr and the controlled atmosphere required for the tin bath provide possibilities for coating float glass. This supposes the coating is made uniformly over the whole width of the ribbon. The first manufactured coatings were made of tin oxide (SnO_2) developed for low-E glazing which was deposited from an organotin compound ($(CH_3)_2SnCl_2$) and oxygen. This coating phase is integrated nowadays to

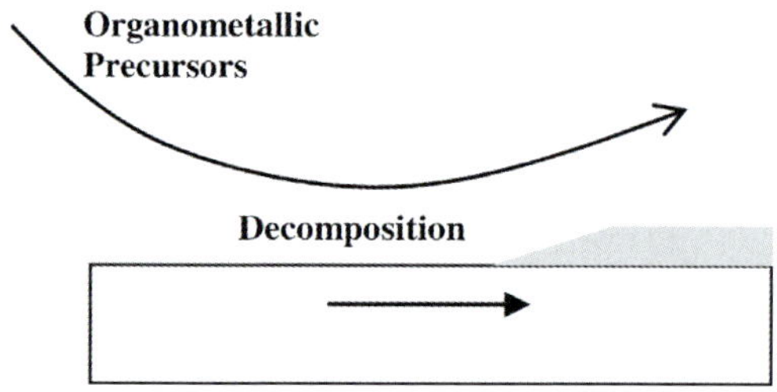

Fig. 12.11 CVD principles.

the float line saving off-line costs (Fig. 12.12; see Chapter 10). To prevent contamination of the float chamber atmosphere, a robust CVD reactor has to be well isolated from the rest of the forming chamber. The temperature range used for CVD coating on glass is 600–700 °C and is located between the float end and the lehr. Because of the continuously moving ribbon, high growing rates are required. As compared to PVD, CVD coatings resist better high temperatures but offer less flexibility in compositions and materials. The optical properties of CVD coatings, while very good for most architectural applications, are at present not sufficient for the most demanding applications. In particular lower solar transmissions are achieved by PVD coatings. As a result, automotive solar control is dominated by sputter-deposited coatings. Since economics drives technology in the glass industry, the push will be towards faster and better on-line coating processes. This is in favour of CVD techniques allowing shorter residence times (higher deposition efficiency).

12.3.2 Physical Vapour Deposition

In physical vapour deposition (PVD), a target of the deposited material is sputtered in a vacuum chamber by accelerated ions. Generally argon gas is used as the working

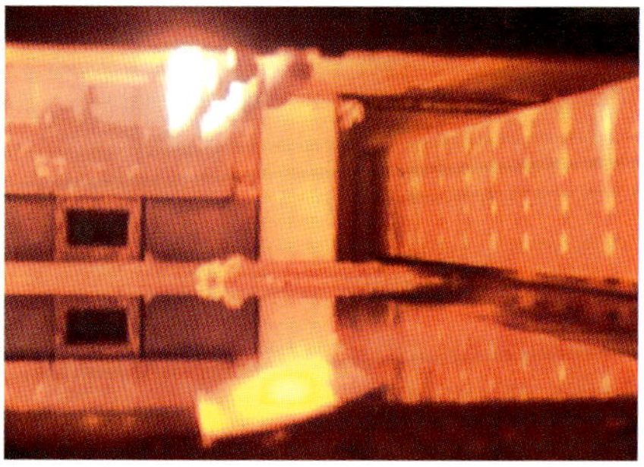

Fig. 12.12 Interior view of a float-glass bath showing a CVD coating unit (on the right-hand side. From Van Mol et al., 2005, with permission from the American Ceramics Society.)

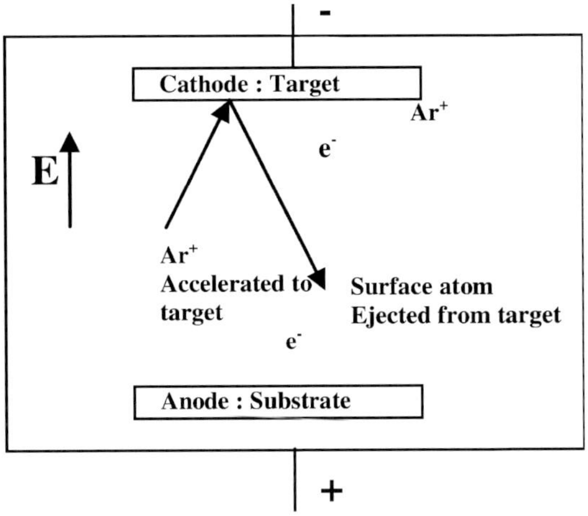

Fig. 12.13 PVD diode system principles.

gas (Figs. 12.13 and 12.14). These positive ions are accelerated through an electric field towards the cathode, ejecting target atoms by momentum transfer. The target and substrate are set parallel, the ejected atoms from the target impacting the substrate and progressively forming the coating. Metallic targets are preferred, when the target is a ceramic; alternate polarization is used to avoid charging the target electrically.

The sputtered atoms reach the substrate at an energy ranging between 1 and 100 eV depending on the machine (diode or magnetron) and operating conditions (voltage and pressure in the chamber). Collision between the deposited atoms and the film is referred as atom peening and this phenomenon yields dense films (Fig. 12.15), but also elevated compressive residual stresses that may eventually cause delamination of the coating. For a given voltage, important parameters of the process are then the substrate temperature and the pressure in the chamber that determines the activation energy of the species being deposited. According to the operating conditions, the growing structure will show different microstructures, as reviewed by Thornston and Hoffman (1989). At low activation energy coatings present columnar structure with small column size and high porosity (called zone 1 in the Thornston and Hoffman scheme). Increasing the substrate temperature or decreasing the chamber pressure allows formation of a denser columnar structure (called zone 2). At even higher temperature (called zone 3) diffusion is much enhanced allowing relaxation and recrystallization. Coatings are compact and formed by large grains. At a given temperature, using lower chamber pressure allows for fewer collisions between the extracted atoms and the working gas and yields higher compressive stresses as shown in Fig. 12.16 for tungsten-sputtered films. Such a phenomenon also happens to be dependent on the sputtering gas nature.

Injection of a reactive gas like O_2, N_2, CF_4 and H_2 into the chamber allows production of complex oxides, nitrides, oxynitrides, fluorides and hydride. This

(a)

(b)

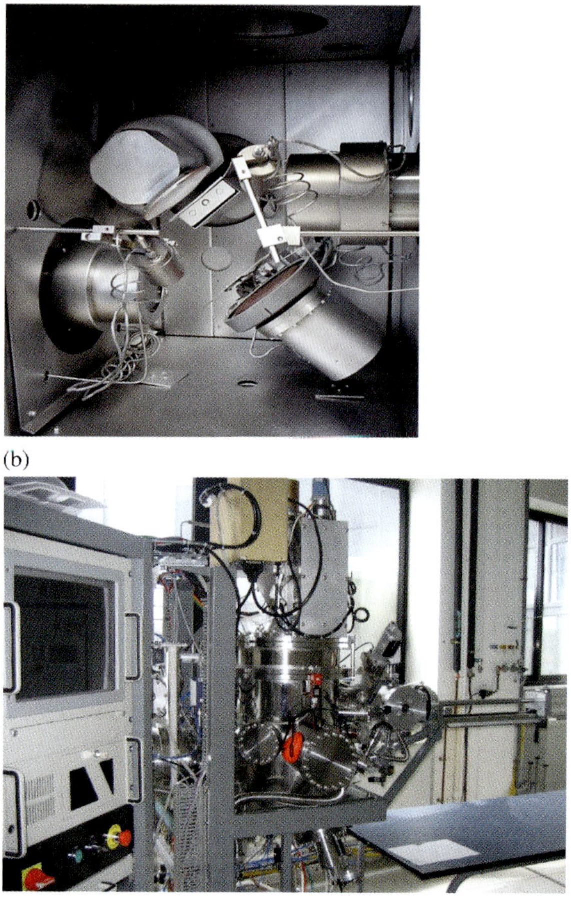

Fig. 12.14 (a) Inside of a PVD chamber designed for research purposes showing the argon gun (top right), target (top left), substrate holder (bottom right), assistance gun (bottom left). (b) Magnetron system chamber with its electronics.

injection has to be carefully managed since poisoning of the target induces low sputtering yields. Target bias or its annealing prevents such a poisoning by reactive gas. The advantages of sputtering techniques are the ability to deposit pure metals as well as metal compounds (oxide, nitride) and the availability of precursors that are the targets, manufactured from high-purity metals and the various reactant gases. The rate of sputtering is greatly increased by using magnetron cathodes. These have an array of magnets behind the target surface that, in combination with the electric field, trap the electrons near the target. The magnetic field is induced so that it is perpendicular to the electric field. The electrons are forced to drift at the target

Fig. 12.15 PVD sputtered gold film showing columnar grains (TEM cross-sectional view, white bar is 200 nm long).

surface causing a greater degree of working gas ionization. Higher growing rates are achieved so that magnetron sputtering has become widespread in industry (Fig. 12.17).

Sputtering is used off-line in glass industries but in a continuous process, where washed (in contrast, on-line CVD is employed on pristine glass surfaces) cut-size pieces of glass up to 3 m wide are continuously cycled in and out of the coater through its entry and exit locks (Fig. 12.17). Carrying out this process at an industrial scale calls

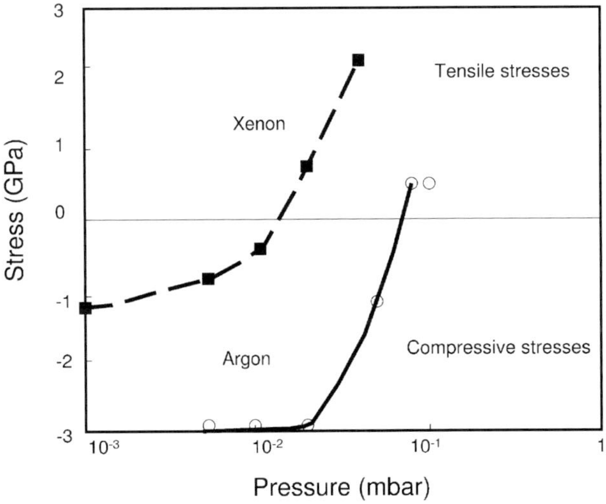

Fig. 12.16 Stress versus working gas pressure for 500 nm thick tungsten films sputtered in argon and xenon. (Reprinted from Mamor et al., 1995, with permission from Elsevier.)

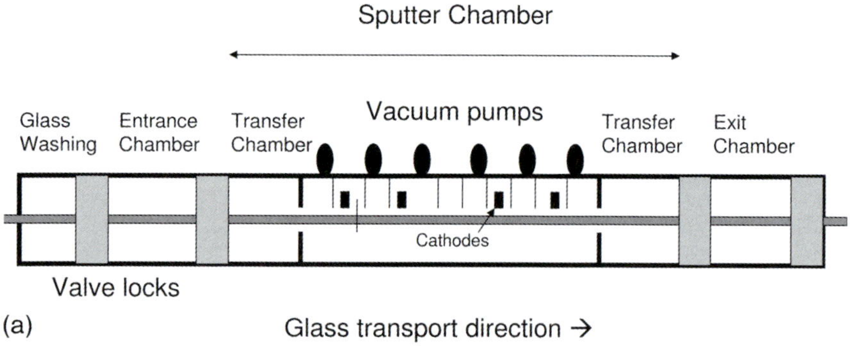

(a)

(b)

Fig. 12.17 (a) Magnetron off-line reactor schematic; (b) magnetron off-line reactor. (Copyright Saint-Gobain.)

for rather large and complex manufacturing plants. The coater consists of a multi-chamber vacuum system, where each chamber contains up to three sputter deposition targets. The glass moves at a constant speed so that the coatings build up sequentially. The number of chambers depends on the coating design (nature and number of individual layers). Indeed, multilayers can be produced alternating the deposited material. Direct current magnetron sputtering in the reactive mode for compound materials has been used for industrial large-area coating deposition. Recent developments in alternating current technology provide for increased deposition rates as well as decreased coating defect density and improved thickness uniformity across the width of the ribbon. This is a very important issue for building applications since thickness variations may induce colour variations which are not acceptable. Sputtered coatings are better reflectors of unwanted UV and IR radiations than CVD coatings, making them desirable for automotive applications where reduction in heat is critical. Unfortunately, sputtered coatings present relatively poor thermomechanical performance and cannot be bent at elevated temperatures. Moreover, they cannot be used in any exterior application such as the exterior surface of a glass wall in a building. As mentioned above, sputtering adds steps to the manufacturing process and increases costs, which are major drawbacks.

12.3.3 Sol–Gel Routes

Sol–gel science offers mild synthetic routes that allows organic and oxide layers to be formed at a glass surface using moderate temperatures (Brinker and Scherrer, 1990). As mentioned in Chapter 3, natural glass may also form in this way. These types of coatings allow addition, with a great flexibility, new functions to glazing such as water repellence, degradation of organics (self-cleaning with TiO_2 oxide) and protection (e.g. CLEANOSIL®, ABRASIL®). The process comprises the solution (sol) preparation with organometallic precursors $R'M(OR)_n$, M being Si or metals (Ti, Zr, etc.) and OR and R′ being hydrolyzed and non-hydrolyzed groups respectively. The precursors are first hydrolyzed and then progressively interact to form a gel because of the molecular condensation that yields a three-dimensional network (Fig. 12.18). The respective reactions are written

Hydrolysis:

$$M{-}OR + H_2O \rightarrow M{-}OH \qquad (12.5a)$$

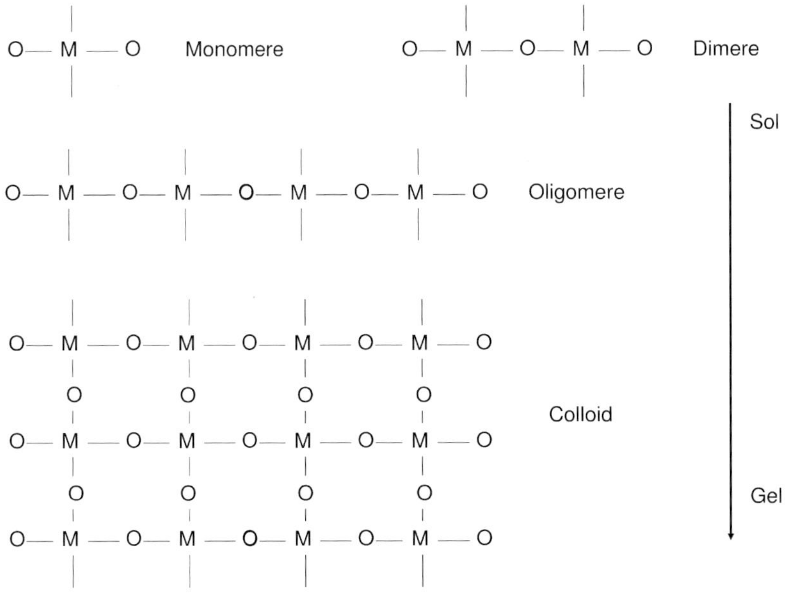

Fig. 12.18 Sol–gel path from the monomer to the colloid with intermediates as polymerization increases.

Condensation:

$$M{-}OR + HO{-}M \rightarrow M{-}O{-}M + ROH \qquad (12.5b)$$

$$M{-}OH + HO{-}M \rightarrow M{-}O{-}M + H_2O \qquad (12.5c)$$

To produce a film on a substrate the sol is coated on the substrate by dipping it into the solution or spinning the substrates with drops of sol; spray can also be used (Fig. 12.19). As discussed in detail for glass processing in Chapter 6, also here the combination of viscosity and velocity will control the sol–gel process (Section 6.1).

Let us first consider dip coating and define ν the velocity of the substrate extraction and η the viscosity of the sol–gel solution. The product $\eta\nu$ can be compared to surface energy ζ_{la} (subscript l and a indicate liquid and atmosphere respectively). When $\eta\nu \ll \zeta_{la}$ then the layer thickness is (Landau and Levich, 1942)

$$t_f = \frac{(\eta\nu)^{2/3}}{\zeta_{la}^{1/6}(\rho g)^{1/2}} \qquad (12.6)$$

where ρ is the film density and g the gravitational acceleration. This shows a thickness dependence with $\nu^{2/3}$, thicker films being obtained with a more rapid extraction from the sol. Spin coating yields a thickness dependence of $1/w^{2/3}$, where w is the constant spinning velocity (Meyerhofer, 1978).

Then drying happens with continuation of sol–gel reactions (condensation and reticulation). The last step is the thermal treatment to complete the sol–gel reactions. These steps allow for the solvent to evaporate and for strengthening the films. As a matter of fact, large shrinkage is induced during drying and annealing and residual stresses build in.

The sol–gel process allows for treatment of much more complex shapes than sputtering PVD techniques which are mostly dedicated to flat glass. Nonetheless, even for flat glass the sol–gel process becomes attractive, the main issue being the mechanical resistance to contact abrasion. Note that silanol-functionalized precursors allow for achieving good adhesion between the layer and the glass substrate since strong siloxane bonds can be formed at the interface as shown in Fig. 12.20. As

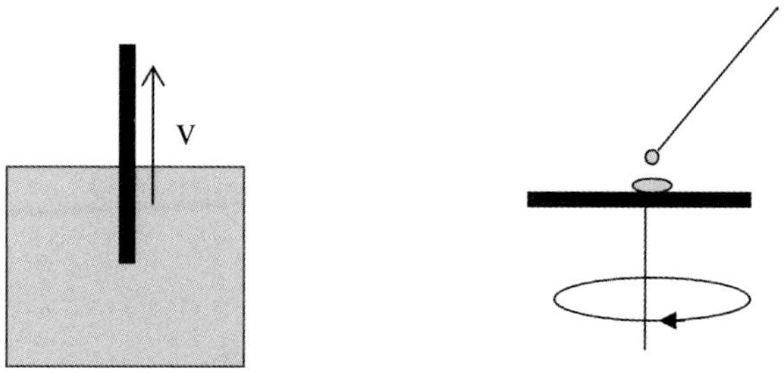

Fig. 12.19 Schematic representations of dip and spin coating.

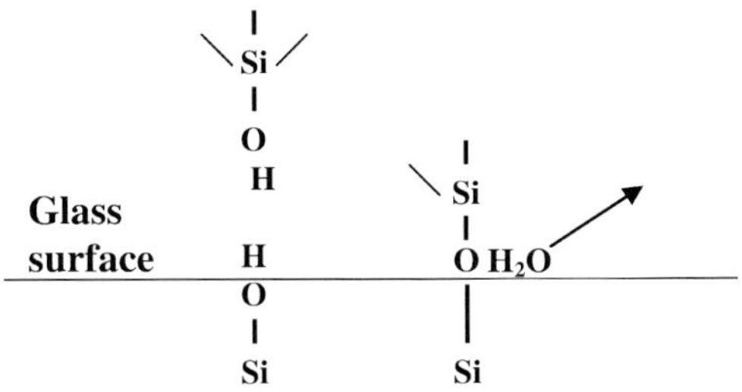

Fig. 12.20 Functionalizing a glass surface using Si–O–Si bonds.

discussed in Chapter 5, the surface has to be cleaned first of organic contaminants. The reaction is written

$$\mathrm{Si{-}OH + HO{-}Si \rightarrow Si{-}O{-}Si + H_2O} \tag{12.7}$$

Another promising potential of sol–gel routes is that both organic and inorganic functions can be tailored (so-called hybrid organic–inorganic materials). Strong hybrid films can be formed once polymer precursors are functionalized with Si–OH groups allowing for reaction with glass surfaces and condensation with the inorganic precursors. One such route is illustrated in Fig. 12.21 where the functionalized precursor is triethoxysilane-functionalized PMMA while the inorganic precursor is tetraethoxysilane (TEOS). Also, hydrolysis, condensation of precursors, drying and thermal treatments allow for fabricating and strengthening the coatings. The resulting coatings are shown in Fig. 12.9. Interestingly one is able to change the ratio between the concentrations of the polymer and silica and hence to tailor the elastic and plastic properties of the film as shown in Table 12.2 (Mammeri, 2003). For instance, the elastic–plastic ratio of such coatings varies in a decade range (H^3/E^2; Johnson 1985; Musil, 2000; Appendix 8) suggesting great changes in terms of tribological performance.

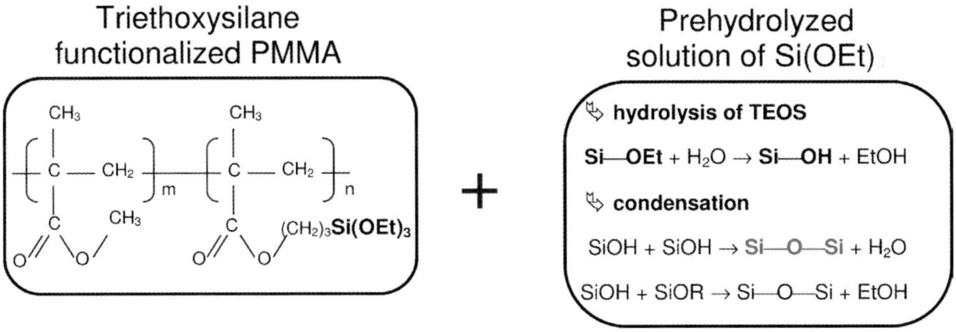

Fig. 12.21 Example of sol–gel reaction synopsis for $PMMA–SiO_2$ hybrid coating fabrication (after Mammeri, 2003).

Tab. 12.2 Mechanical characteristics of PMMA–SiO_2 hybrid materials with PMMA concentration decreasing from top to bottom. Value of PMMA in wt%. (After Mammeri, 2003.)

Sample	E (GPa)	H (GPa)	H^3/E^2 (MPa)
PMMA	4.1	0.25	1
PMMA 75%	7.6	0.49	2
PMMA 50%	6.6	0.53	4
PMMA 25%	9.5	0.84	7
Sol–gel silica	17.5	1.48	11

12.4 Residual Stresses in Thin Films

An important parameter to consider when fabricating thin coatings is the residual stresses amplitude. Residual stresses result from several phenomena: (i) the difference in thermal expansion coefficient between the thin film and the substrate on cooling from process temperature and (ii) the process itself (atomic peening in PVD deposition, drying and shrinking in sol–gel processing for instance; Fig. 12.22). Such stresses may be detrimental. When highly compressive, films may delaminate; when highly tensile, films may crack. As a matter of fact, process parameters have to be adjusted to achieve acceptable stress levels.

Let us consider a substrate coated on its two sides (a dip-coated glass for instance). Mechanical equilibrium requires (Hand et al., 2003)

$$\sigma_g t_H = 2\sigma_f t_f \tag{12.8}$$

where σ_g and σ_f are the stresses in the substrate and the film and t_H and t_f the thicknesses of the substrate and the film respectively. The stress in the film is

$$\sigma_f = \frac{E_f}{1-\nu_f}\varepsilon \tag{12.9}$$

while the stress in the substrate is obtained by combining Eqs. (12.8) and (12.9):

$$\sigma_g = \frac{E_f}{1-\nu_f}\frac{2t_f}{t_H}\varepsilon \tag{12.10}$$

where E_f and ν_f are Young's modulus and Poisson's ratio of the film.

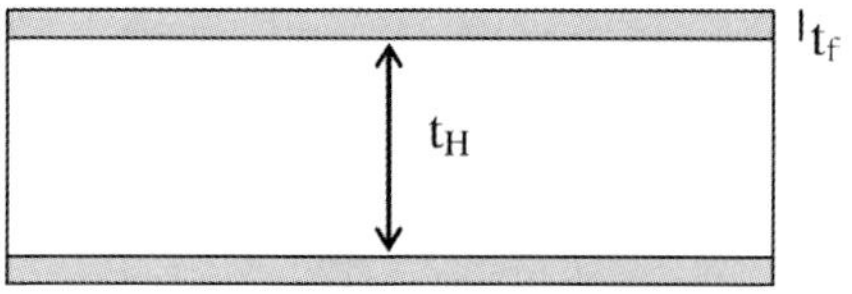

Fig. 12.22 Schematic representation of thin-coated substrate.

Supposing strains result mainly from the expansion coefficient mismatch between the film and substrate, we then have

$$\varepsilon = (\alpha_f - \alpha_g)\Delta T \tag{12.11}$$

where α_f and α_s are the thermal expansion coefficients of the film and the substrate respectively and ΔT the temperature change. According to the sign of the strain ε, the film is either under compressive or tensile stresses. When under tensile stresses, fracture of the film may appear and this is to be prevented as said before.

Thin-film residual stresses can be determined mainly by two different methods: (i) the Stoney 1909 curvature method and (ii) X-ray diffraction if films are crystalline. We emphasize the curvature method, which is the most popular and widely used. The X-ray diffraction method is developed in Appendix 12.

The method requires the coating of a thin substrate that will deform when the coated film is deposited. Let us imagine a three-step formation of the coated substrate (Fig. 12.23). The stress-free film (a) is adjusted to the substrate (b), while curvature results from mechanical equilibrium (c). Stoney's equation is written

$$\sigma_f = \frac{E_s}{1 - \nu_s} \frac{t_H^2}{6t_f} \frac{1}{R} \tag{12.12}$$

where subscripts s and f indicate substrate and film respectively, E_s and ν_s are Young's modulus and Poisson's ratio of the substrate and R is the curvature radius. When the substrate is thin enough the curvature induced by the film can be

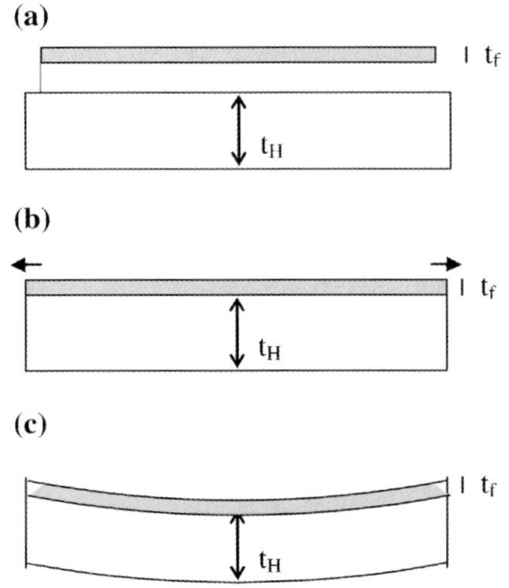

Fig. 12.23 Schematic of film stress–substrate curvature relations: (a) stress-free film and stress-free substrate; (b) film adjusted to substrate; (c) substrate curvature at mechanical equilibrium.

Fig. 12.24 Sol–gel film applications with decorative interest. Glassware demonstrating high quality optical appearance of crystal glass coated with dye coloured hybrid coatings. (Reprinted from Schottner, 2001, with permission from the American Chemical Society.)

measured using a contact probe or optical interferometry. Notably, the former equation involves only the mechanical characteristics of the substrate. Hence, such measurements require only knowledge of film thickness. Then, the procedure encompasses the measurement of the curvature of the substrate before deposition to check (or correct later) its flatness. The measurement is carried out again after deposition and R being determined, Eq. (12.12) is used to extract the residual stresses present in the coatings. This method applies to any type of films (PVD, CVD or sol–gel).

12.5 Applications of Coatings and Summary

Coatings are widely used in the glass industry. They were developed mostly to protect glass surfaces from erosion and abrasion. So called HEC and CEC are CVD on-line coating operations for containers (Chapter 10). Also, sizing is used on drawing lines to protect glass fibres from erosion and to improve adhesion when employed in composites (Chapter 10). As regards flat glass and despite relatively poor mechanical resistance achieved to date, CVD, PVD and sol–gel films allow new functions like UV and IR reflection and water repellence (Chapters 4 and 12) to be added to the glass without changing the composition of the glass matrix.[1] New opportunities have also

[1] Note that when lamininated glazings are used, films can be deposited on inside surfaces (namely 2 or 3) that are protected from contact damage.

appeared in the display market with the emergence of macroelectronics. Hence, thin films allow dedicated and attractive applications while mechanical performances as well as strength improvements are of growing interest. To develop further glass functionality, new coating designs are being achieved (Fig. 12.24). In this field the sol–gel technique is being employed and studied as a potential technique complementary to conventional CVD and PVD techniques in particular when complex shapes are to be treated. Recent works have also shown that surface mechanical performance can be improved when a designed film is deposited onto the damage surface (Chapter 8; Briard, 2004). The coating field is very active, in fact, and drives innovation in the glass industry.

13
Conclusion

This book has reviewed scientific and technological backgrounds required for understanding glass mechanics and technology approaches. This is a very important field since glass mechanics is involved at all stages of the industrial process. The field is very large since industrial processes are complex and result from a long-standing know-how. Glass fabrication involves complex thermomechanical routes to produce glazing while new functions are continuously invented and added. Glass manufacturing tools have been considerably transformed over the last century. The production yields for conventional soda-lime-silica glass allow worldwide production of glazing and containers with excellent performance. While glass is facing competition from other materials like polymers and metals, the combination of its strength, transparency and chemical inertness is still a great advantage. Glass can also be associated with other materials like polymers to form composites. As a matter of fact, it was realized some 50 years ago that the association of glass fibres with a polymer matrix yielded composites presenting high strength and lower density. The designed mechanical performances allow for load gains which are essential for transportation (airplanes in particular). Moreover, new functions are added to glazing while glass enters into nanotechnology developments. Industrial evolution is driven by the search of in-use performances comprising quantified indicators (strength, optical properties, industrial cost) and qualitative parameters (value, aesthetics, etc.). Also, the constant improvement of tools allows for guaranteeing quality and hence reducing industrial costs. Then, like in other industries, interdisciplinarity is greatly demanded to develop and produce new products facing commercial competition. Using a step-by-step approach we have illustrated how glass science and technology are applied. R&D stands at the meeting point where understanding of both materials science and processes is required. As illustrated in the book, both fields require deep insight and understanding of background knowledge. Innovation and progress are challenges that demand a large overview of glass science, industrial processes and the market. The intention of the book was to propose to the reader background, routes and perspectives into glass science and technology. Glass is a state of matter that is out of equilibrium and this has very important consequences for the process and the manufacturing of glass (strengthening, high-precision dimensioning). Flow, fracture and plasticity of glass are very important properties from the process point of view

Glass; Mechanics and Technology. Eric Le Bourhis

ISBN: 978-3-527-31549-9

and for in-use performance. As discussed in the book, much progress has been gained in the understanding of glass flow and fracture and these properties are carefully considered when designing new manufacturing tools. As extensively illustrated in the book, contact mechanics controls the density of superficial defects generated while transferring the products (including elevated temperatures). The distribution and importance of the superficial defects control the strength of the manufactured object. Glass quality depends on its homogeneity and this is directly related to the viscous properties of the glass melt. Quenching the glass allows for the production of residual stresses that will strengthen the glass and protect it from subcritical crack growth. The dimensional adjustment of displays at the pixel resolution requires control of the thermal history and knowledge of structural relaxation of glass substrates. Glass mechanics is one of the most important issues for the glassmaker. Numerical tools (using finite element modelling) are developed to approach the mechanics of complex industrial shapes following complex industrial thermomechanical routes. Such procedures allow for great savings in development costs.

There are still important questions that are debated and would allow for much further insight in the future. Non-Newtonian flow and contact resistance are important issues from both the scientific and technological points of view. Although changes in the composition of glass allow modification of its properties (infrared absorption, colour), glassmakers prefer nowadays using coatings to change surface properties. This means that surface properties are becoming of great importance. This opens new fields to glass R&D comprising nanomaterials. These progresses suppose that industrial tools can be developed at production scales. This is the case for chemical and physical vapour deposition coatings technologies that are becoming extensively used on-line and off-line. Coating techniques and properties are improved while the mechanical stability will allow the new functions to be retained over time. Nanomaterials developments suppose that investigation can be carried out at the relevant length scales. In this domain glass technology will benefit from developments obtained in the micro-electro-mechanical system (MEMS) domain (coating, photolithography, characterizing tools). As observed in opto- and micro-electronics and communications, the R&D time scales shorten requiring always more and more rapid adaptation to new achievements and technologies. The same is expected for the glass industry, and deeper understanding of border-line science and technology will allow great progress for leading industries.

Appendix 1
Light Absorption, Dispersion and Polarization

A1.1
Electromagnetic Spectrum

The electromagnetic spectrum (Fig. A1.1) is divided into seven main domains ranged according to their wavelength λ. We have

$$\lambda = cT = c/\nu$$

where c is the speed of light, T the period and ν the frequency. Another important equation is that relating photon energy E and frequency ν:

$$E = h\nu$$

where h is Planck's constant.

A1.2
Light Absorption

The light absorption of materials is related to the electronic and atomic (or ionic) polarizations under the electromagnetic field (see also Chapter 11). Absorption can be modelled using a simple approach considering the movement of the particles constituting the matter (e.g. electrons, ions). Their movement can be obtained from classical mechanics solving

$$m\ddot{r} + f\dot{r} + k'r = q(E + v^\wedge B) \approx qE$$

where m is the particle mass, q its charge and $-f\dot{r}$ and $k'r$ are the friction and attraction forces respectively. The electromagnetic force reduces in good approximation to its electrical component ($vB \ll E$).

Let us take the conventional academic notations for polarization vector p and wave vector k:

$$p = qr$$

Glass; Mechanics and Technology. Eric Le Bourhis

ISBN: 978-3-527-31549-9

γ-rays	X-rays	Ultraviolet rays UV	Visible Light	Infrared rays IR	Radar	Radio

Wavelength λ: 0.01nm, 10nm, 0.4μm, 0.7μm, 100 μm, 1 cm, 100 cm

Fig. A1.1 Electromagnetic spectrum.

$$E = E_o \exp(-i\omega\tau + i\kappa\rho)$$

$$k = \frac{\omega}{c} n$$

then we get (Feynmann et al., 1979; Bertin et al., 1984; Pérez et al., 1997)

$$p = \left[\frac{q^2}{\varepsilon_o m} \frac{1}{\omega_o^2 - \omega^2 - i\omega\gamma} \right] \varepsilon_o E.$$

where $\omega_o^2 = k'/m$ and $\gamma = f/m$.

Considering that the concentration of particles is N, then polarization P is

$$P = Np = \varepsilon_o \chi E = \varepsilon_o (n^2 \quad 1) E$$

where ε_o, χ and n are the vacuum permittivity, electric susceptibility and refractive index respectively. Refractive index n is determined to be

$$n^2 = 1 + N \left[\frac{q^2}{\varepsilon_o m} \frac{1}{\omega_o^2 - \omega^2 - i\omega\gamma} \right] \varepsilon_o$$

This relation can be generalized taking into account all types of polarization:

$$n^2 = 1 + \sum_i N_i \alpha_i$$

where subscript i refers to polarisation type, e.g. electronic, atomic, ionic, orientation. Note that for dense materials (fluids, glasses and crystals), the local field E_l is to be considered instead of the macroscopic field E. Then we get

$$\frac{n^2 - 1}{n^2 + 2} = \frac{\sum_i N_i \alpha_i}{3}$$

This last relation is called the Lorenz–Lorentz relation.

A metal is characterized by free electrons so that w_o is zero and n is complex. In fact metals are not transparent but reflect light and this is used in mirrors and low-E glazing:

$$n^2 = 1 + N \left[\frac{q^2}{\varepsilon_o m} \frac{1}{-\omega^2 - i\omega\gamma} \right] \varepsilon_o$$

Likewise glass absorbs UV radiation because of electronic polarization. In such case $w \sim w_o$, the 'friction term' γ is not negligible close to resonance and refractive index becomes a complex number since

$$n^2 = 1 + N\left[\frac{q^2}{\varepsilon_o m}\frac{1}{-i\omega\gamma}\right]\varepsilon_o$$

and the glass is not transparent.

A1.3 Light Dispersion

In contrast, away from the absorption bands ($\omega \neq \omega_o$) we have

$$n^2 = 1 + N\left[\frac{q^2}{\varepsilon_o m}\frac{1}{\omega_o^2 - \omega^2}\right]\varepsilon_o$$

The refractive index increases with the frequency (decreases with the wavelength). This corresponds to normal dispersion as opposed to anomal dispersion (n decreasing with w) at absorption bands. This phenomenon is difficult to investigate because of absorption. Dispersion is widely used in spectrometers, a prism being used to analyse the light intensity as a function of wavelength. Dispersion equation can be approximated in the 0.4–0.8 μm region ($\omega^2 \ll {\omega_o}^2$) by

$$n^2 = 1 + N\left[\frac{q^2}{\varepsilon_o m}\frac{1}{\omega_o^2(1-\omega^2/\omega_o^2)}\right]\varepsilon_o \approx 1 + N\left[\frac{q^2}{\varepsilon_o m}\frac{(1+\omega^2/\omega_o^2)}{\omega_o^2}\right]\varepsilon_o$$

$$\approx 1 + N\left[\frac{q^2}{\varepsilon_o m\omega_o^2}(1 + \lambda_o^2/\lambda^2)\right]\varepsilon_o$$

$$n^2 \approx A + B/\lambda^2$$

which is known as the empirical Cauchy equation.

Dispersion is generally defined as

$$n_F - n_C$$

where n_C and n_F are the refractive indices at $\lambda_C = 0.6563$ nm and $\lambda_F = 0.4861$ (H_α and H_β hydrogen lines). Reciprocal dispersion ν is

$$\nu = \frac{n_D - 1}{n_F - n_C}$$

where n_D is the refractive index at $\lambda_D = 0.5893$ nm (sodium D line).

A1.4 Light Polarization

Polarizers select light with a polarization along a given direction (in fact within a small angle aperture around the polarizer axis). In practice polymeric materials are used as polarizers (see also Appendix 7). Light polarization changes when refractive indices vary with the direction. Consider the three principal directions, namely 1, 2 and 3, and suppose light propagates along direction 3. Phase shift is written (see also Chapter 11)

$$\beta = \frac{2\pi}{\lambda}\delta = \frac{2\pi}{\lambda}d(n_1 - n_2)$$

where n_1, n_2 are the principal indices, λ the wavelength and d the distance propagated by the beam.

Suppose initially, light is polarized linearly and inclined by 45° to 1 and 2 principal axis (Fig. A1.2). We write the polarization vector as

$$a\cos\omega t\begin{vmatrix} 1 \\ 1 \end{vmatrix} = a\begin{vmatrix} \cos\omega t \\ \cos\omega t \end{vmatrix}$$

that transforms into

$$a\begin{vmatrix} \cos\omega t \\ \cos\omega t + \beta \end{vmatrix}$$

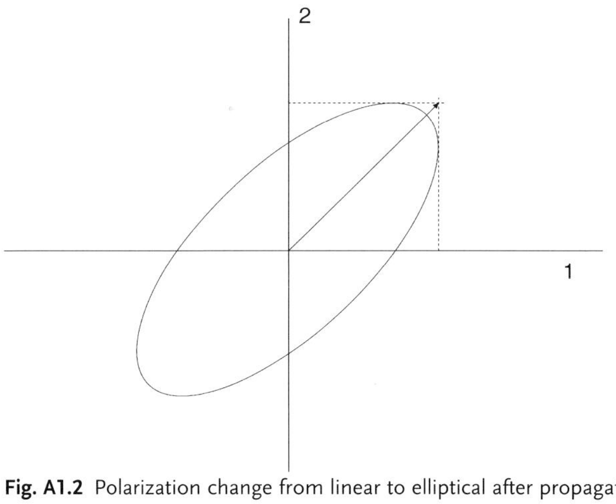

Fig. A1.2 Polarization change from linear to elliptical after propagation and phase shift β.

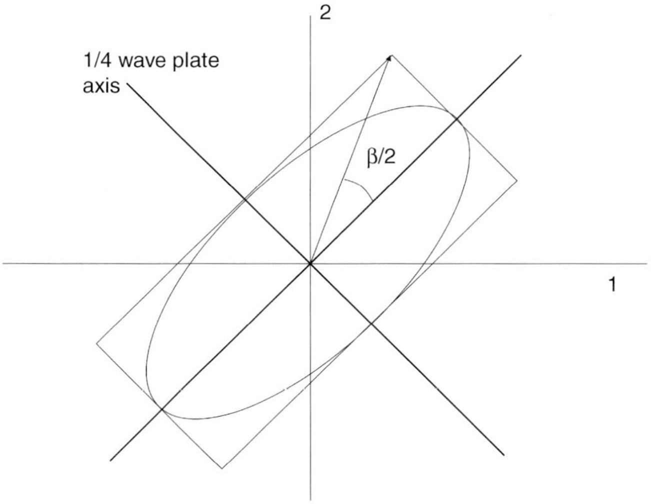

Fig. A1.3 Polarization change from elliptical to linear polarization after propagation through a $\lambda/4$ plate.

For $\beta = \pi/2$ a circularly polarized light emerges from the specimen. In the general case light is elliptically polarized (Fig. A1.2).

Next consider a 1/4 wave plate (as used for a Senarmont compensator; Chapter 11) inclined by 45° from principal directions 1 and 2 of a specimen, say, parallel to initial linear polarization (Fig. A1.3). Polarization vector is written in the 1/4 wave plate frame as

$$\left| \begin{array}{l} a\dfrac{\sqrt{2}}{2}(\cos\omega t + \cos\omega t + \beta) \\[2ex] a\dfrac{\sqrt{2}}{2}(-\cos\omega t + \cos\omega t + \beta) \end{array} \right. = \left| \begin{array}{l} a\sqrt{2}\cos\dfrac{\beta}{2}\cos\omega t + \dfrac{\beta}{2} \\[2ex] a\sqrt{2}\sin\dfrac{\beta}{2}\sin\omega t + \dfrac{\beta}{2} \end{array} \right.$$

which transforms after crossing the 1/4 wave plate into

$$\left| \begin{array}{l} a\sqrt{2}\cos\dfrac{\beta}{2}\cos\omega t + \dfrac{\beta}{2} \\[2ex] a\sqrt{2}\sin\dfrac{\beta}{2}\sin\omega t + \dfrac{\beta}{2} + \dfrac{\pi}{2} \end{array} \right. = \left| \begin{array}{l} a\sqrt{2}\cos\dfrac{\beta}{2}\cos\omega t + \dfrac{\beta}{2} \\[2ex] a\sqrt{2}\sin\dfrac{\beta}{2}\cos\omega t + \dfrac{\beta}{2} \end{array} \right. = a\sqrt{2}\cos\left(\omega t + \dfrac{\beta}{2}\right) \left| \begin{array}{l} \cos\dfrac{\beta}{2} \\[2ex] \sin\dfrac{\beta}{2} \end{array} \right.$$

That is, a linear polarized light with polarization vector forming an angle of $\beta/2$ with initial polarization vector. Hence, an analyser allows determination of $\beta/2$ after rotating to darkness as done when using a Senarmont compensator.

Appendix 2
Atomic Structure and Bond Formation

A2.1
Atomic Structure

Atoms are formed by a nucleus surrounded by electrons (Fig. A2.1). The nucleus contains neutrons and positively charged protons. The negatively charged electrons are held to the nucleus by an electrostatic attraction. The electrical charge carried by each electron and proton is $-/+e = -/\ 1.6 \times 10^{19}$ C respectively. The atom as a whole is electrically neutral. The number Z of protons is identical to the number of electrons and is called the atomic number. The number of neutrons is N, the total number of nucleons (protons + neutrons) is then $A = Z + N$. We usually use the notation ${}^{A}_{Z}\mathrm{At}$ (for instance ${}^{28}_{14}\mathrm{Si}$) to define atom structure.

Most of the mass of the atom is contained within nucleons. Then the atomic mass is about A g mol^{-1}. Atoms of the same element that contain a different number of neutrons in the nucleus are called isotopes and thus have a different atomic mass. The atomic mass used for such an element is an average value of those of the different isotopes and thus the atomic mass may not be a whole number. For instance the element Si has three isotopes: 92.2% ${}^{28}_{14}\mathrm{Si}$, 4.7% ${}^{29}_{14}\mathrm{Si}$, 3.09% ${}^{30}_{14}\mathrm{Si}$, its mass being $M = (0.922 \times 28) + (0.047 \times 29) + (0.0309 \times 30) = 28.1$ g mol^{-1}. Their concentration may change with time when the isotopes are unstable following exponential decay and allowing for dating objects containing such elements.

Electrons occupy discrete energy levels within the atoms. Each electron possesses a particular state with no possibility for another electron to have the same state; electrons occupy different quantum states following the Pauli exclusion principle (electrons being fermions) which are defined by four quantum numbers:

i. n the principal quantum number ($n \geq 1$) being an integral number ($n = 1, 2, 3\ldots$) that refers to the quantum shell to which the electrons belong (Fig. A2.1); these shells are often assigned a letter rather than a number, i.e. $K, L, M\ldots$ instead of $1, 2, 3\ldots$

ii. l the azimuthal quantum number also an integral number ($0 \leq l \leq n-1$)

Glass; Mechanics and Technology. Eric Le Bourhis

ISBN: 978-3-527-31549-9

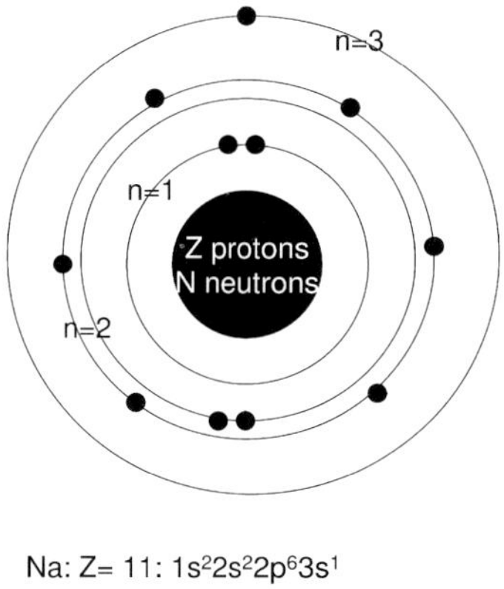

Fig. A2.1 Atomic structure (example of Na) using the Bohr model. Quantum mechanics suggests instead orbitals showing only probabilities.

iii. m_l the magnetic quantum number ($-l \leq m_l \leq l$)
iv. m_s the spin quantum number ±½ reflecting two different spin orientations.

For shorthand notations, only *n*, *l* numbers are used and a maximum of $2(2l+1)$ different electrons have these two same *n*, *l* quantum numbers. We denote electrons s for $l=0$, p for $l=1$, d for $l=2$, f for $l=3$.

The different quantum states ordered by *n* are then denoted as in Table A2.1.

The atomic structure is defined in its fundamental state (lowest energy) knowing the quantum states occupied by *Z* electrons. Oxygen with $Z=8$ has the following configuration: $1s^2 2s^2 2p^4$. For larger *Z*, the following quantum states are progressively used: 1s, 2s, 2p, 3s, 3p, 4s, 3d, 4p, 5s, 4d, 5p, 6s, 4f, 5d, 6p, 7s, 5f. This allows formation of the Mendeleev table (Fig. A2.2). Extensive information on every element is available at http://www.webelements.com/webelements/.

Tab. A2.1 Quantum numbers and states.

n	*l*	Quantum state	Maximum number of electrons
1	0	1s	2
2	0	2s	2
2	1	2p	6
3	0	3s	2
3	1	3p	6
3	2	3d	10
4	0	4s	2
4	1	4p	6
4	2	4d	10
4	3	4f	14

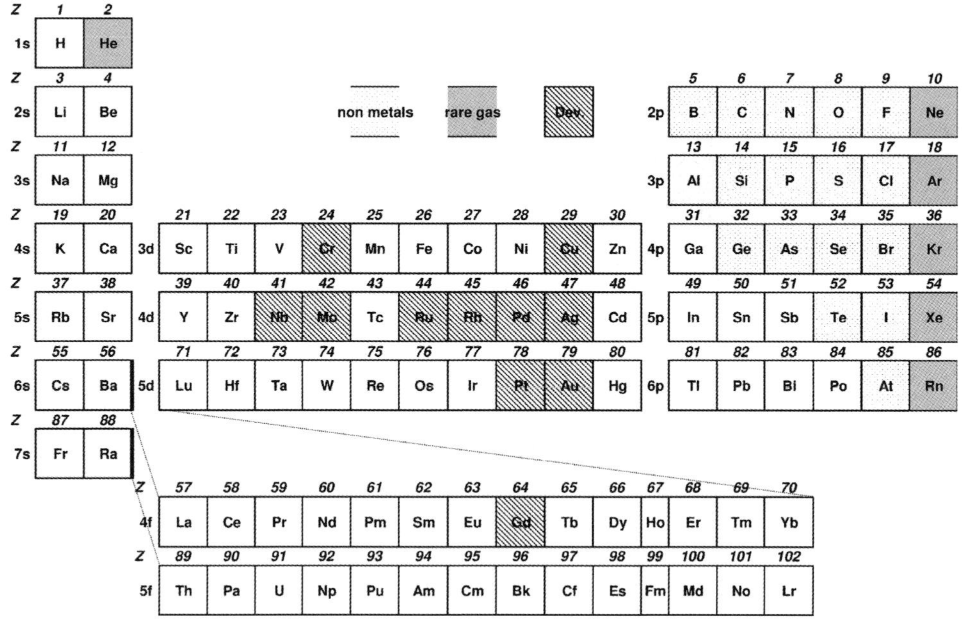

Fig. A2.2 Mendeleev table.

A2.2 Mendeleev Table

Deviations happen from expected electronic structures when d and f levels begin to fill (dashed in Fig. A2.2). For instance, one would expect Cr structure to be $1s^2 2s^2 2p^6 3s^2 3p^6 4s^2 3d^4$, it is instead $1s^2 2s^2 2p^6 3s^2 3p^6 4s^1 3d^5$.

The valence of an atom is related to its ability to enter into chemical combinations with other elements and is often determined by the number of electrons in the outermost s and p levels. Oxygen for instance with an atomic structure $1s^2 2s^2 2p^4$ shows a valence of 6. If an atom has a valence of zero (or 8), no electrons enter into chemical reactions and this element is inert. Examples are rare gases (grey colour in Fig. A2.2). Electronegativity describes the tendency of an atom to gain electrons and increases from the left- to right-hand side of the Mendeleev table (electropositivity describes the tendency to lose electrons and increases from the right- to left-hand side). Distributions of metals and non-metals can be displayed in the Mendeleev table. The dotted area shows non-metals on the right-hand side while the remaining elements (on the left-hand side) generally display a metallic nature.

A2.3 Bond Formation

Let us consider first an ionic bond that may form between a metal and a non-metal (for instance Na and Cl elements). The metal (Na) is electropositive while the

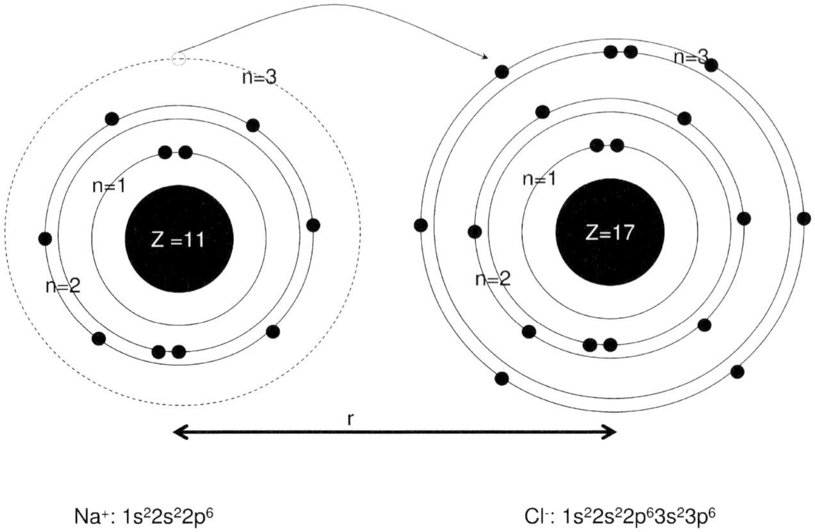

Fig. A2.3 Ionic bond structure (example of Na and Cl) using the Bohr model and ionic bond formation.

non-metal is electronegative; they will transform into Na^+ and Cl^- ions presenting full shells ($1s^22s^22p^6$ and $1s^22s^22p^63s^23p^6$ respectively). The electrostatic force between ions of opposite charges is attractive and is written (Fig. A2.3)

$$F_{\text{attr}} = -\frac{1}{4\pi\varepsilon_o}\frac{qq'}{r^2}$$

In terms of energy we have ($F_{\text{attr}} = -\nabla E_{\text{attr}}$)

$$E_{\text{attr}} = -\frac{1}{4\pi\varepsilon_o}\frac{qq'}{r}$$

Attraction is dominant at large distances; however at short distances, electron repulsion becomes predominant. Repulsion energy is written

$$E_{\text{rep}} = \frac{\alpha}{r^m}$$

with $m > 1$.

Potential energy is the sum of attraction and repulsion contributions:

$$E_{\text{p}} = E_{\text{rep}} + E_{\text{attr}} = \frac{\alpha}{r^m} - \frac{1}{4\pi\varepsilon_o}\frac{qq'}{r}$$

Potential energy is plotted in Fig. A2.4; r_o is the distance corresponding to the energy minimum (force is then zero). This corresponds to the equilibrium distance

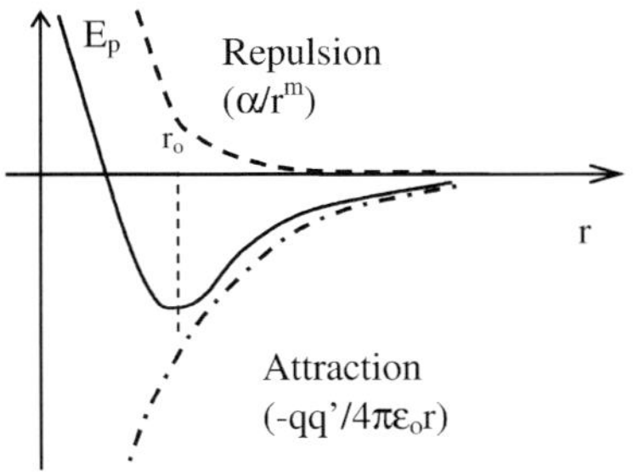

Fig. A2.4 Potential energy as a function of distance resulting from electron repulsion predominant at short distance and ion attraction predominant at long distance.

and is called the bond distance. Bond distance allows for ion size to be determined. In the present case $r_o = r_{Na^+} + r_{Cl^-}$.

Ionic crystals are formed by ions that all interact electrostatically with r^{-1} being a decreasing function. A simple calculation of the attractive energy can be made for a one-dimensional chain of N ions of opposite charge ($q = q'$; Kittel, 1983; Fig. A2.5):

$$\ln(1+x) = \left(x - \frac{x^2}{2} + \frac{x^3}{3} - \frac{x^4}{4} \cdots\right)$$

$$NE_{\text{attr}} = -2N\frac{qq'}{4\pi\varepsilon_o}\left(\frac{1}{r} - \frac{1}{2r} + \frac{1}{3r} - \frac{1}{4r}\cdots\right)$$

$$= -2N\frac{1}{4\pi\varepsilon_o}\frac{qq'}{r}\left(1 - \frac{1}{2} + \frac{1}{3} - \frac{1}{4}\cdots\right) = -2N\frac{1}{4\pi\varepsilon_o}\frac{qq'}{r}\ln 2$$

$$E_{\text{attr}} = -2\frac{1}{4\pi\varepsilon_o}\frac{1}{r}\ln 2 = -\frac{2\ln 2}{4\pi\varepsilon_o}\frac{1}{r} = -\frac{\beta}{r}$$

where 2 ln 2 (= 1.38) is the Madelung constant for a one-dimensional chain. For a three-dimensional solid, the calculation is more difficult and yields a Madelung constant of 1.74 for NaCl.

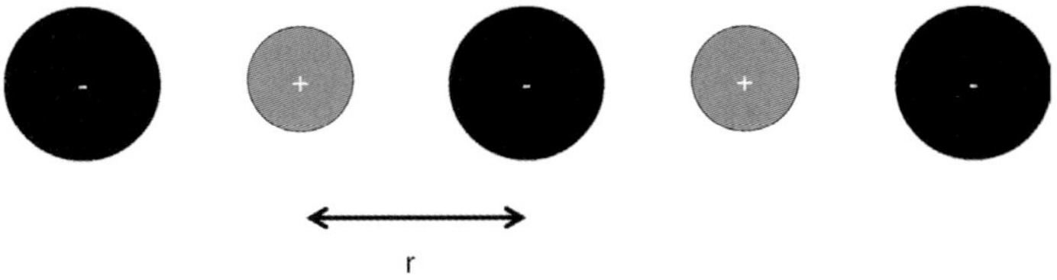

Fig. A2.5 One-dimensional and periodical ion chain.

The variation of E_p can be generalized to any bonds and solids in the following form:

$$E_p = E_{rep} + E_{attr} = \frac{\alpha}{r^m} - \frac{\beta}{r^n}$$

with $m > n$ since repulsion is dominant over attraction at low r. The Lennard-Jones potential is written for $m = 12$ and $n = 6$.

Appendix 3
Thermal Expansion and Elasticity

A3.1
The α–E Trend

Materials' thermal expansion coefficient α and stiffness (elastic modulus E) are related. In fact a ceramic (stiff material) will expand less than a polymer. This general trend is observed in the α–E chart (Fig. 2.8) where all classes of materials are considered. This trend is mathematically formulated as

$$\alpha = \frac{\gamma_G \rho c_v}{3E}$$

where the Grüneisen 'constant' γ_G is not a constant but varies within one order of magnitude ($0.4 < \gamma_G < 4$), ρ is the density and c_v the specific heat $c_v \approx c_p$.

A simple approach to thermal expansion is proposed by Zarzycki 1982 and Kittel 1983. We shall review briefly here their approaches.

A3.2
Qualitative Approach

Figure A3.1 shows schematically the potential energy of bonded atoms. The continuous line is for a more rigid material since it shows a deeper potential. Parameter r_o corresponds to the equilibrium distance at $T = 0$ K. This temperature cannot be produced, so for clarity we take this temperature as a reference point. At $T > 0$ K, atoms oscillate around their equilibrium position. It is important to note that the potential is not symmetric. For small oscillations, one generally neglects this phenomenon and takes a parabolic function in the form $E_p = (1/2)k(r - r_o)^2$ also called the harmonic potential. Force is then $F = -k(r - r_o)$. Under such an assumption, no thermal expansion is found (Fig. A3.2).

Now considering the fact that the potential is not symmetric, the average distance between atoms shifts towards the right-hand side of Fig. A3.1, or the materials are expected to expand when they are heated. Here, we do not consider any phase

Glass; Mechanics and Technology. Eric Le Bourhis

ISBN: 978-3-527-31549-9

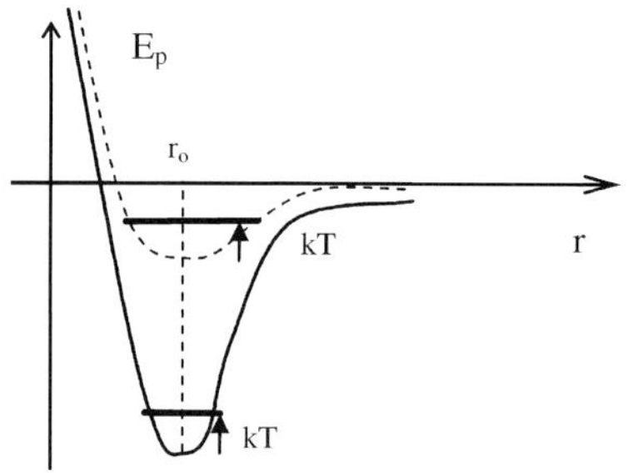

Fig. A3.1 Potential energy as a function of distance for weak (dashed line) and strong (continuous line) bonds.

changes which either yield contraction or expansion. Materials showing a deeper potential (stiffer materials) show less potential asymmetry (for the same additional thermal energy kT). Hence, these materials present a lower thermal expansion coefficient than materials having a shallow potential (soft materials).

Considering now silicate glasses, when modifiers are introduced in the silicate glass, NBOs are formed and yield Si–O potential asymmetry and hence lead to large thermal expansion. In contrast, in borosilicate glasses, B atoms participate in glass network formation and no NBOs are formed maintaining the potential symmetry and hence less thermal expansion is observed for these glasses.

A3.3
Expansion Modelling

In a first approximation we may add one asymmetric term to the harmonic potential that we have defined above:

$$E_p = (1/2)k(r - r_o)^2 + g(r - r_o)^3 = cx^2 + gx^3$$

with $c = ½k$, $x = r - r_o$, $g < 0$ and we assume $gx^3 \ll kT$. When $g = 0$, we get again a harmonic or parabolic potential.

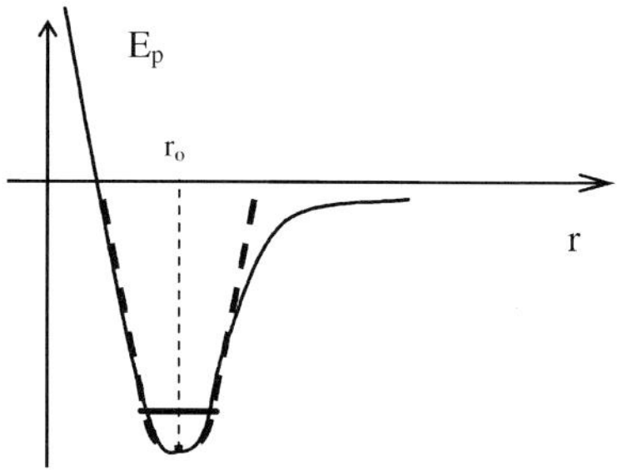

Fig. A3.2 Potential energy as a function of distance curve showing the harmonic approximation.

Following Kittel 1983, we consider a Boltzmann distribution and the average distance shift at temperature T is

$$\langle x \rangle = \frac{\int_{-\infty}^{\infty} x \exp(-E_p/kT)}{\int_{-\infty}^{\infty} \exp(-E_p/kT)}$$

With the following approximation:

$$\exp\left(-\frac{E_p}{kT}\right) = \exp\left(-\frac{cx^2 + gx^3}{kT}\right) = \exp\left(-\frac{cx^2}{kT}\right)\left(1 - \frac{gx^3}{kT}\right)$$

a series of calculations yields

$$\langle x \rangle = \frac{-3g}{4c^2} kT$$

This simple approach then yields a constant thermal expansion coefficient proportional to g (asymmetric term):

$$\alpha = \frac{-3g}{4c^2} \frac{k}{r_0}$$

We find that stronger binding (larger value of c, stiffer materials) yields indeed lower thermal expansion while larger asymmetry (larger value of g) yields larger thermal expansion.

A3.4
Differential Expansion Measurement

A dilatometer is used to determine the thermal expansion coefficient of a specimen submitted to a thermal ramp (constant heating rate). Not only does the specimen expand but also the dilatometer (Fig. A3.3). Hence, differential expansion is

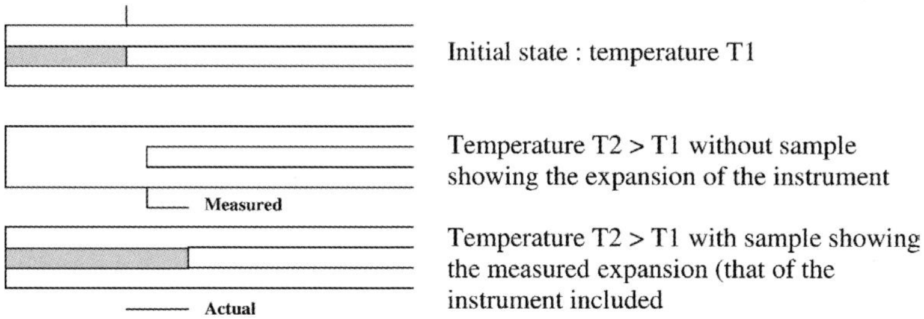

Fig. A3.3 Schematic representation of thermal expansion of both sample and dilatometer.

measured and correction is necessary to include the dilatometer expansion in the experimental data. Consider for instance an alumina specimen in an alumina cylindrical support: the apparent expansion would be zero because all expand the same relative length. A diagram can be constructed showing the measured and actual (and superior) expansion (Fig. A3.3).

Appendix 4
Falling Sphere Viscometer and Fining of Glass

A4.1
Falling Sphere

A4.1.1
Asymptotic Regime

Let us consider a sphere of radius r, volume V and mass $M = \rho_S V$ (ρ_S being the sphere density) falling in a viscous fluid of density ρ (Fig. A4.1). It is subjected to three different forces: (i) gravitational force, (ii) Archimedes reaction and (iii) viscous friction. Newton's principle is written

$$Mg + A + f = Ma$$

where g is the gravitational acceleration and a is the acceleration of the ball. We have

$$Mg = \rho_s Vg$$

$$A = -\rho Vg$$

$$f = -kv$$

where v is the velocity and k is taken as the Stokes coefficient:

$$k = 6\pi\eta r$$

The ball reaches a constant velocity after a transient period ($v = v_l$, $a = 0$) that is given by

$$Mg + A + f = 0$$

$$(\rho_s - \rho)Vg = kv_l$$

Glass; Mechanics and Technology. Eric Le Bourhis

ISBN: 978-3-527-31549-9

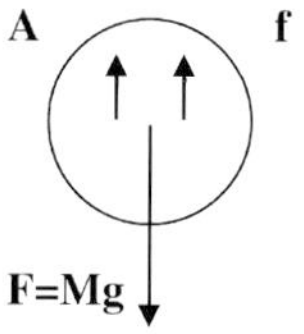

Fig. A4.1 Falling sphere in a viscous liquid.

We finally find

$$v_l = \frac{(\rho_s - \rho)Vg}{k} = \frac{2gr^2(\rho_s - \rho)}{9\eta}$$

A4.1.2
Transient Regime

When a falling sphere viscometer is used, one supposes that the asymptotic regime is reached or the transient regime (from velocity $v = 0$ to v_l) is achieved. This regime is obtained solving

$$Mg + A + f = Ma$$

This equation is rewritten as

$$\rho_s V \frac{dv}{dt} + kv = Vg(\rho_s - \rho)$$

with initially

$$v(t = 0) = 0$$

The solution is

$$v(t) = v_l[1 - \exp(-t/T)]$$

with

$$T = \rho_s V/k$$

The solution is qualitatively plotted (Fig. A4.2) for two different values of k (or η). Notably when viscosity increases T and v_l decrease. Hence, the asymptotic regime is

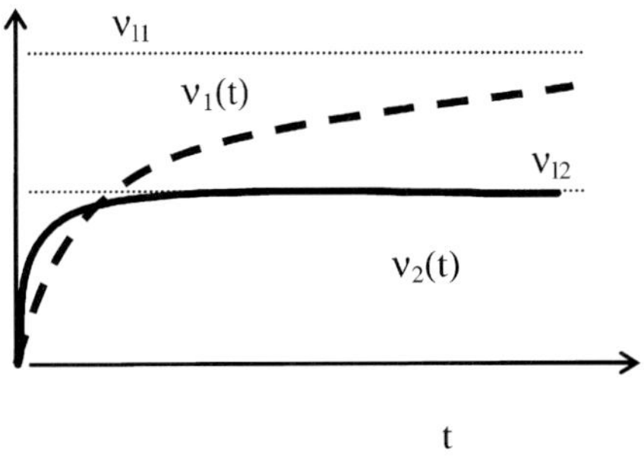

Fig. A4.2 Velocity evolution with time of a falling sphere in a viscous liquid ($k_1 < k_2$).

reached more rapidly and after a shorter falling distance. Therefore such conditions are more favourable for the method to be employed.

A4.2 Fining of Glass

Let us consider the limit velocity solution and consider a bubble formed in a glass melt. The asymptotic solution becomes

$$v_l = -\frac{2gr^2\rho}{9\eta}$$

It is of utmost importance to eliminate bubbles in the glass (Fig. A4.3). Considering a period of time t, the distance d that a bubble travels is

$$d = v_l t = \frac{2gr^2\rho}{9\eta}t$$

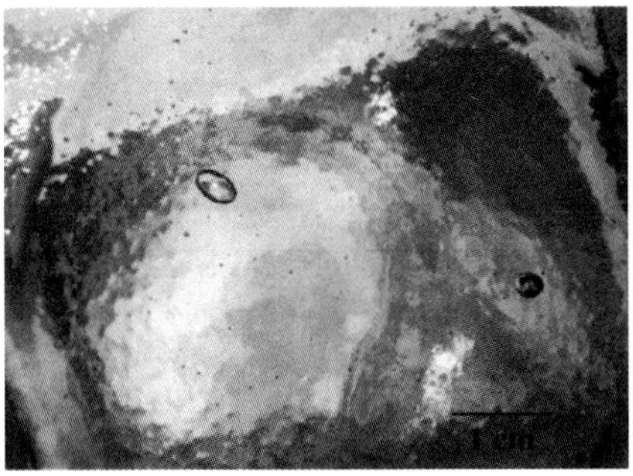

Fig. A4.3 Handmade and old-fashioned container showing a large amount of bubbles with different sizes.

showing a dependence on the square of the bubble size and an inverse dependence on the viscosity:

$$d \propto \frac{r^2}{\eta} t$$

At 1330 °C in a standard melt, it takes 1 minute for a 1 mm diameter bubble to move a distance of 1 μm, while this time becomes 5 h and 20 days for 100 μm and 10 μm diameter bubbles respectively.

There are two main ways to accelerate the elimination of bubbles (Chapter 10): (i) decreasing the viscosity, which is done by imposing a higher temperature at the surface; and (ii) increasing the bubbles size, which is done using fining agents.

Appendix 5
Theoretical Strength of a Solid

Let us consider like Orowan (1949) the potential energy of bonded atoms as shown schematically in Fig. A5.1. Parameter r_o corresponds to the equilibrium distance (rigorously, only at $T = 0$ K; see Appendix 3). Let us apply a force to separate the atoms:

$$F = \mathrm{d}E_p/\mathrm{d}r \text{ (positive sign since it is the applied force)}$$

Figure A5.1 transforms into Fig. A5.2 when the applied force is plotted as a function of the distance, equilibrium distance r_o being obtained for $F = 0$. The force curve maximizes at F_m. Let us consider a specimen loaded under applied force F with unit section surface made of n atoms (Fig. A5.3). The applied stress is

$$\sigma = nF$$

Note that we only consider the interactions between atoms in front of each other. This is the first assumption. Theoretical strength is written

$$\sigma_{th} = nF_m$$

We have in the elastic domain (for $r \sim r_o$)

$$\sigma = E\varepsilon = E\frac{r - r_o}{r_o}$$

While in this domain we have also

$$\sigma = n\left.\frac{\mathrm{d}F}{\mathrm{d}r}\right|_{r=r_o}(r - r_o)$$

The work W necessary to cause the fracture of the specimen corresponds to the area shown in Fig. A5.2 and is

$$W = \int_{r_o}^{\infty} nF(r')\mathrm{d}r' = 2\zeta$$

where ζ is the surface energy (two surfaces are created on fracture).

Glass; Mechanics and Technology. Eric Le Bourhis

ISBN: 978-3-527-31549-9

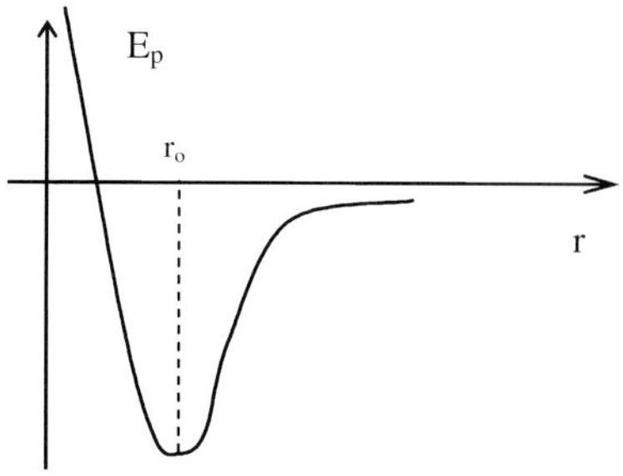

Fig. A5.1 Potential energy as a function of distance.

Next, we shall approximate the $F(r)$ curve by a sinusoidal function to facilitate the calculations as suggested by Gilman (1949). This second approximation leads to the underestimation of the work W; the calculation shows that the theoretical strength is much greater than the observed one (see Section 7.1). In fact the model suggested here leads to the multiple fracture of the specimen since the situation for a given plane is to be considered for all planes. Thereafter, the specimen would be sliced into monatomic planes after such a test.

$$F(r) = F_m \sin\left(2\pi \frac{r - r_o}{\lambda}\right)$$

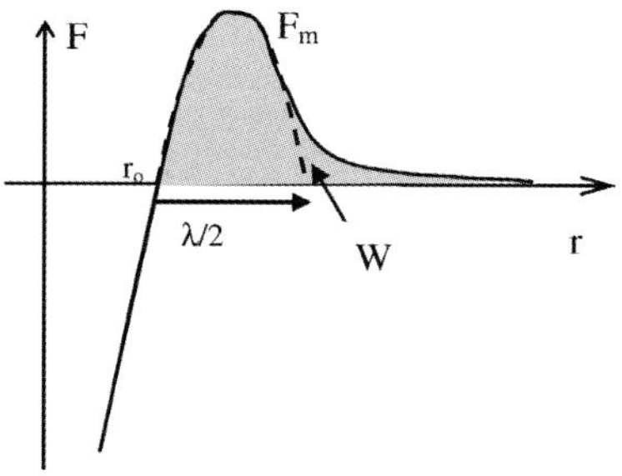

Fig. A5.2 Applied force as a function of distance.

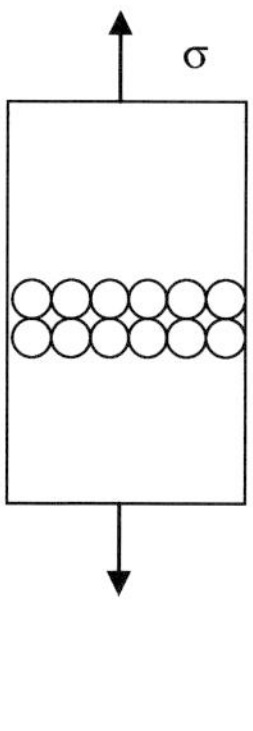

Fig. A5.3 Tensile test on a unit surface (n atoms).

Then in the elastic domain (for $r \sim r_o$) we have

$$F(r) = F_m\left(2\pi\frac{r - r_o}{\lambda}\right)$$

so that

$$E = nF_m\left(2\pi\frac{r_o}{\lambda}\right) = \sigma_{th}\left(2\pi\frac{r_o}{\lambda}\right)$$

while

$$W = 2\zeta > \int_{r_o}^{r_o+\lambda/2} nF(r')\mathrm{d}r' = nF_m\frac{\lambda}{\pi} = \sigma_{th}\frac{\lambda}{\pi}$$

Then multiplying the latter two equations allows one to obtain

$$\sigma_{th}^2 r_o = E\zeta$$

that is rewritten

$$\sigma_{th} = \sqrt{\frac{E\zeta}{r_o}}$$

Appendix 6
Weibull Analysis

The investigation of the strength of brittle solids requires numerous experiments since fracture is statistical. This results from the flaw size distribution of the material. For a first experimental approach, one can consider a population of $N = 50$ fibre specimens of the same length sectioned in flexure (Fig. A6.1). Care is taken in order not to damage the tested length (contact damage, finger moisture). We suggest the fibre segments be manipulated only by their extremities.

The procedure consists in testing under tensile stress optical fibres up to fracture. The loading curve $F - \Delta l$ is checked as to its linearity up to fracture. The N stress values at fracture are then ranked in ascending order. Defining x as the number of unfractured specimens at stress σ, the probability estimator is taken as

$$P_s(V) = (x - 0.5)/N \qquad \text{(A6.1)}$$

where V is the volume of the considered fibres. We use here this simplest expression for the estimator. The precise form of the estimator is still the subject of ongoing research. Several expression for $P_s(V)$ have been proposed in the literature (Bergman, 1984; Wu et al., 2004, 2006): $P_s(V) = x/N$ (simplest one but not generally used because it is not quite statistically correct), $P_s(V) = x/(N+1)$, $P_s(V) = (x - 0.3)/(N + 0.4)$, $P_s(V) = (x - 0.25)/(N + 0.25)$, $P_s(V) = (x - 0.375)/(N + 0.25)$ and more generally $P_s(V) = (x - \alpha)/(N + \beta)$. The step-by-step procedure is then shown in Table A6.1. Figure A6.2 shows $P_s(V)$ as a function of stress for a studied population.

The next step consists in obtaining a Weibull diagram plotting $\ln\{\ln[1/P_s(V)]\}$ as a function of $\ln(\sigma)$. We use here the simplest method, i.e. the linear regression (LR) method. For the weighted linear regression, maximum likelihood and moments methods see Wu et al. 2006. For a single defect population (bimodal populations are not uncommon and lead to strong deviations from a linear plot) the Weibull diagram shows a linear evolution which allows determination of the Weibull modulus, that is, the slope of the linear fit and σ_o from the y-axis intercept

Glass; Mechanics and Technology. Eric Le Bourhis

ISBN: 978-3-527-31549-9

Tab. A6.1 Glass fibre strength and step procedure to determine Weibull parameters.

σ	$\ln(\sigma)$	x	$P_s(V) = (x - 0.5)/N$	$\ln\{\ln[1/P_s(V)]\}$	$P_s(V)$, Weibull
2.10	0.74	49	0.99	−4.58	0.98
2.11	0.75	48	0.97	−3.47	0.97
2.11	0.75	48	0.97	−3.47	0.97
2.26	0.82	46	0.93	−2.60	0.89
2.26	0.82	46	0.93	−2.60	0.89
2.30	0.83	44	0.89	−2.13	0.84
2.30	0.83	44	0.89	−2.13	0.84
2.30	0.83	44	0.89	−2.13	0.84
2.30	0.83	44	0.89	−2.13	0.84
2.31	0.84	40	0.81	−1.53	0.83
2.32	0.84	39	0.79	−1.42	0.81
2.32	0.84	39	0.79	−1.42	0.81
2.32	0.84	39	0.79	−1.42	0.81
2.32	0.84	39	0.79	−1.42	0.81
2.34	0.85	35	0.70	−1.05	0.78
2.34	0.85	35	0.70	−1.05	0.78
2.34	0.85	35	0.70	−1.05	0.78
2.36	0.86	32	0.64	−0.82	0.74
2.37	0.86	31	0.62	−0.75	0.72
2.38	0.87	30	0.60	−0.68	0.70
2.42	0.88	29	0.58	−0.61	0.60
2.42	0.88	29	0.58	−0.61	0.60
2.42	0.88	29	0.58	−0.61	0.60
2.44	0.89	26	0.52	−0.43	0.54
2.46	0.90	25	0.50	−0.37	0.48
2.46	0.90	25	0.50	−0.37	0.48
2.47	0.90	23	0.46	−0.25	0.45
2.47	0.90	23	0.46	−0.25	0.45
2.47	0.90	23	0.46	−0.25	0.45
2.47	0.90	23	0.46	−0.25	0.45
2.47	0.90	23	0.46	−0.25	0.45
2.50	0.92	18	0.36	0.03	0.36
2.50	0.92	18	0.36	0.03	0.36
2.50	0.92	18	0.36	0.03	0.36
2.52	0.92	15	0.30	0.20	0.29
2.52	0.92	15	0.30	0.20	0.29
2.54	0.93	13	0.26	0.31	0.24
2.54	0.93	13	0.26	0.31	0.24
2.55	0.94	11	0.21	0.43	0.21
2.55	0.94	11	0.21	0.43	0.21
2.58	0.95	9	0.17	0.56	0.13
2.58	0.95	9	0.17	0.56	0.13
2.58	0.95	9	0.17	0.56	0.13
2.58	0.95	9	0.17	0.56	0.13
2.58	0.95	9	0.17	0.56	0.13
2.62	0.96	4	0.07	0.97	0.07
2.62	0.96	4	0.07	0.97	0.07
2.63	0.97	2	0.03	1.25	0.05
2.65	0.97	1	0.01	1.52	0.03

Fig. A6.1 Rolled glass fibre.

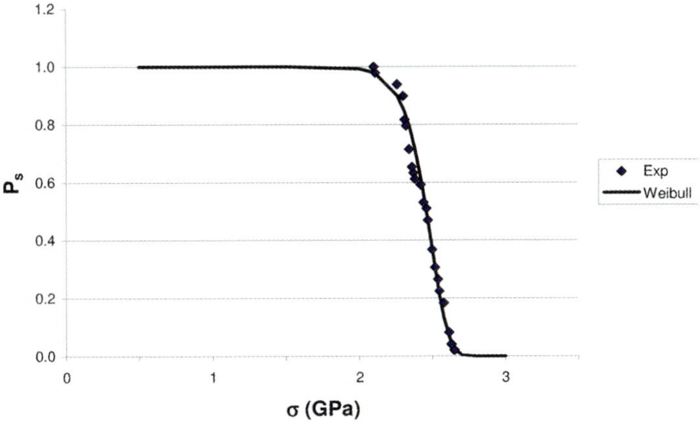

Fig. A6.2 $P_s(V)$ as a function of stress σ and Weibull modelled curve with the parameters extracted from least squares fit of $\ln\{\ln[1/P_s(V)]\}$ versus $\ln(\sigma)$ as illustrated in Fig. A6.3.

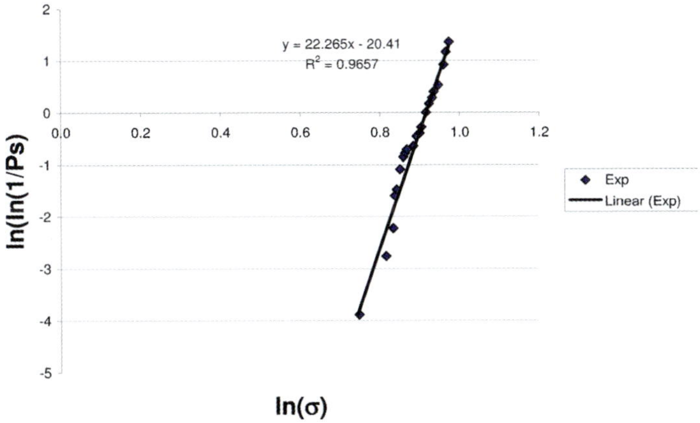

Fig. A6.3 Procedure to determine Weibull parameters from a least squares fit of $\ln\{\ln[1/P_s(V)]\}$ versus $\ln(\sigma)$.

since

$$\ln\left[\ln\left(\frac{1}{P_s(V)}\right)\right] = m\ln(\sigma) - m\ln(\sigma_o) \qquad \text{(A6.2)}$$

This is illustrated in Table A6.1 and Fig. A6.3. The Weibull adjustment is then shown on the $P_s(\sigma)$ plot (Fig. A6.3). Two defect populations may sometimes be observed, due to damage induced while making the experiments. In such a case, the analysis can be made separately for the two populations.

Appendix 7
Photoelastic Set-Up for Lectures

A7.1
Set-Up for Photoelastic Projection

The proposed set-up comprises a projector used for the projection of slides (Fig. A7.1). This is available in most classrooms and allows the demonstration of transient (Fig. A7.2) and residual stresses (Fig. A7.3). A polarizer or analyser can be obtained as $1 \times 1\,m^2$ panel (HN32 from Sarelec, 91560 Crosne, France, the panels being sold with one edge parallel to the polarization axis). It has to be cut into two squares with area covering the projector plate. For protection these square polarizers can be placed and sealed between two square and annealed glass sheets. One of the panels serves as the polarizer and is set on the projector. The other one serves as the analyser and is rotated by 90° from the polarizer. The glass specimens are then set in between. The image can be projected for a classroom focusing on the glass specimen, and examples are discussed below. Before setting a specimen one checks that polarizer and analyser are shifted by 90° when darkness appears on the screen.

A7.2
Example of a Beam under Flexion (Transient Stresses)

A $3.5 \times 20 \times 400\,mm^3$ glass beam is cut from an annealed glass sheet and the edges abraded. It is to be set under flexion taking the beam by its extremities. The beam is then placed at 45° from the polarization axis. Again analyser and polarizer are rotated preliminarily by 90° to get darkness on the screen, the contrast being maximum under such conditions while vanishing is observed when aligning the beam along polarization axis (isoclinics along principal stress). The photoelastic contrast increases as applied stresses increases. One observes in Fig. A7.2 under white contrast both edges, one being under compressive stress, the other being under tensile stress (top and bottom respectively) while the beam centre shows minimum

Glass; Mechanics and Technology. Eric Le Bourhis

ISBN: 978-3-527-31549-9

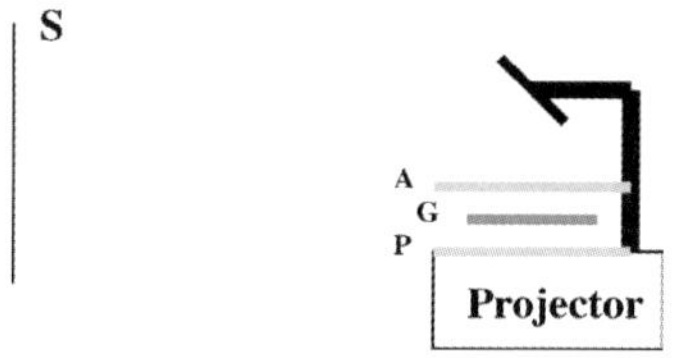

Fig. A7.1 Set-up for photoelasticity experiments in a classroom (A, analyser; P, polarizer; G, glass specimen; S, screen).

intensity at the neutral line. Note that one does not distinguish here tensile and compressive stresses. This can be done adding a wavelength sheet to the set-up. When flexure is released, photoelastic contrast vanishes, showing that these applied stresses are transient.

A7.3 Example of Tempered Specimens (Residual Stresses)

Two glass specimens $3.5 \times 100 \times 100$ and $6 \times 100 \times 100\,\text{mm}^3$ were thermally tempered under the same conditions. The two shown in Fig. A7.3 correspond to these two specimens and were obtained under the same conditions with the analyser axis perpendicular to the polarizer axis. Again darkness is to be observed before starting the observation of glass specimens. The degree of temper is expected to be more important for the thicker glass since a larger thermal gradient is generated (Chapter 6). One indeed observes more contrast in the photoelastic image for the thicker specimen. This is particularly obvious at the four corners of the specimens where coloured contrasts are even obtained for the thicker specimens. Notably, not only

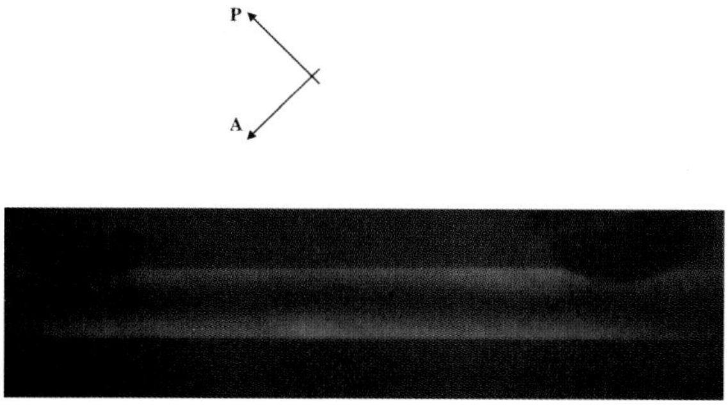

Fig. A7.2 Photoelasticity image of a beam under flexion (P, polarizer; A, analyser).

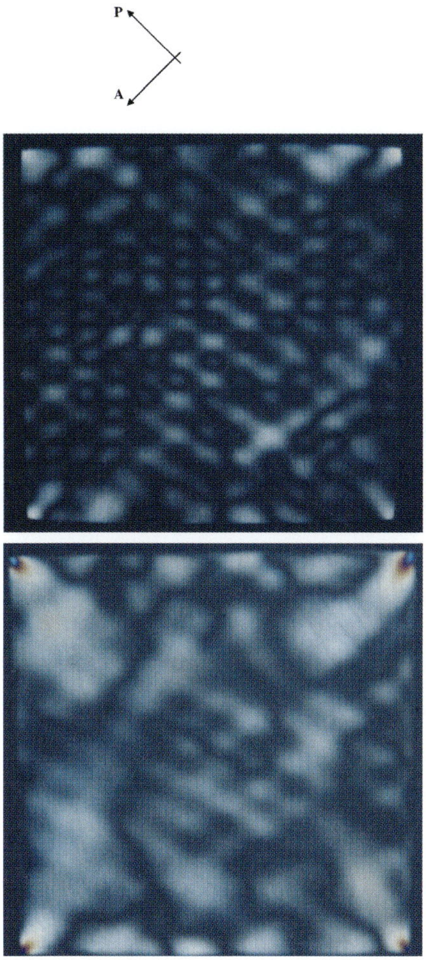

Fig. A7.3 Photoelasticity images of two specimens, $3.5 \times 100 \times 100$ and $6 \times 100 \times 100\,mm^3$, thermally tempered and observed under the same conditions. The thicker one shows more contrast (P, polarizer; A, analyser).

does the stress intensity play a role in the photoelastic contrast here, but also the glass thickness t_H since phase shift $\beta = (2\pi/\lambda)\delta$ is proportional to glass thickness and principal stresses difference $\sigma_1 - \sigma_2$ (Chapter 11). One further observes the signature left by the nozzles as seen at dawn on car backlights (Chapters 3 and 6). These contrasts appear without applying any forces on the specimens showing that stresses are residual.

Appendix 8
Instrumented Nanoindentation Applied to Thin Films

A8.1
Instrumented Nanoindentation

Instrumented indentation allows for monitoring the penetration of a tip into a specimen as a function of load. This technique has become very popular in thin-film technology (Fischer-Cripps, 2004a). In fact, when low loads are used, the observation of indents becomes tedious, requiring dedicated instruments like atomic force and/or transmission electron microscopes, and is time consuming (Le Bourhis and Patriarche, 2003). Therefore, it has become of importance to extract the mechanical properties from the loading–unloading curves (Fig. A8.1). The test consists of loading up to a maximum value F_m (a hold period may be programmed) and unloading to zero. A typical curve is shown in Fig. A8.1. The loading and unloading curves separate when irreversible deformation is generated in the specimen (only the elastic deformation recovers on unloading).

Most indentation machines are load controlled. The tip is fixed to a coil merged into a magnetic field. Controlling the intensity of the current through the coil allows the force to be controlled. Because of the Joule effect one has to consider carefully the inherent thermal drift of the machine. Displacement is determined using a capacitive sensor (Fig. A8.2). The tips are categorized into sharp and blunt (see also Chapter 8). Sharp tips are generally three-faced Berkovich tips allowing for a better definition of the contact area A_c than four-sided Vickers tips. However, the aspect ratio is the same for both tips ($A_c = 24.5 h_c^2$, where h_c is the contact depth). Spherical 'blunt' tips are used to determine the overall stress–strain $\sigma(\varepsilon)$ curves (Le Bourhis and Patriarche, 2003; Section A8.2).

The profile of a deformed surface under load is shown in Fig. A8.3. The cone loaded by force F_m penetrates the specimen down to h_m. The specimen and cone are in contact to depth h_c, that is, up to point B. The specimen is observed to deform elastically around the contact area as discussed also in Chapter 8. Assuming unloading is purely elastic, the surrounding surface recovers to its original position shown as A in Fig. A8.3.

The mean pressure under load can be extracted from the load–penetration curves (Fig. A8.1) employing the well-known Oliver and Pharr and Field and Swain

Glass; Mechanics and Technology. Eric Le Bourhis

ISBN: 978-3-527-31549-9

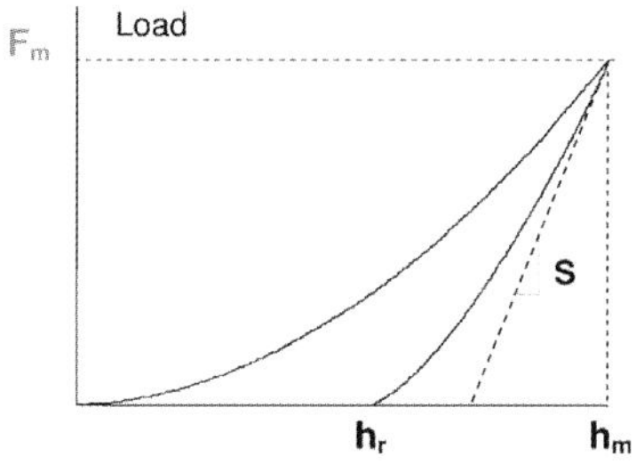

Fig. A8.1 Typical loading–unloading curve. These curves separate because the specimen has deformed 'plastically' on loading while unloading is mainly elastic.

methods, named after their respective authors (Oliver and Pharr, 1992; Field and Swain, 1993). These methods have several limitations. Firstly, one major assumption of both methods is that only elastic recovery happens while unloading; this might not be strictly the case as revealed for very slight indents and when films show viscoelastic response (Le Bourhis and Patriarche, 2003). Secondly, the methodology was derived from the solution to elastic contact problems, while prior to unloading the contact conditions involve plastic deformation. Thirdly, the material around the contact may exhibit substantial pile-up or sink-in, the projected contact area being then either underestimated or overestimated, yielding errors in the hardness values (Oliver and Pharr, 1992; Field and Swain, 1993).

Considering a sharp contact, the unloading curve can be fitted by (Oliver and Pharr, 1992)

$$F = k(h - h_r)^m, \; 1 < m < 2 \tag{A8.1}$$

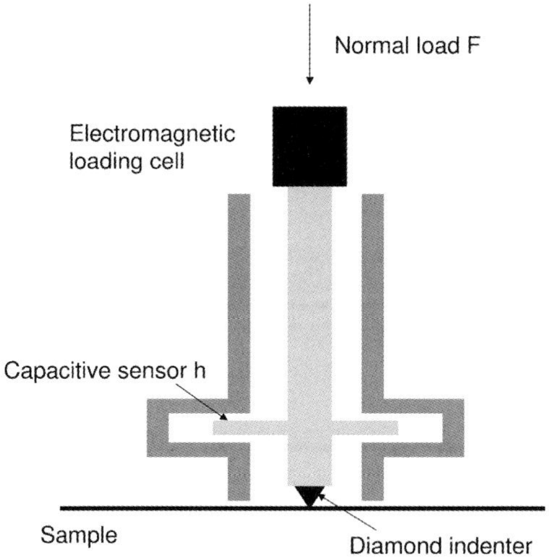

Fig. A8.2 Instrumented indentation principle. Loading is electromagnetic while the displacement sensor is capacitive.

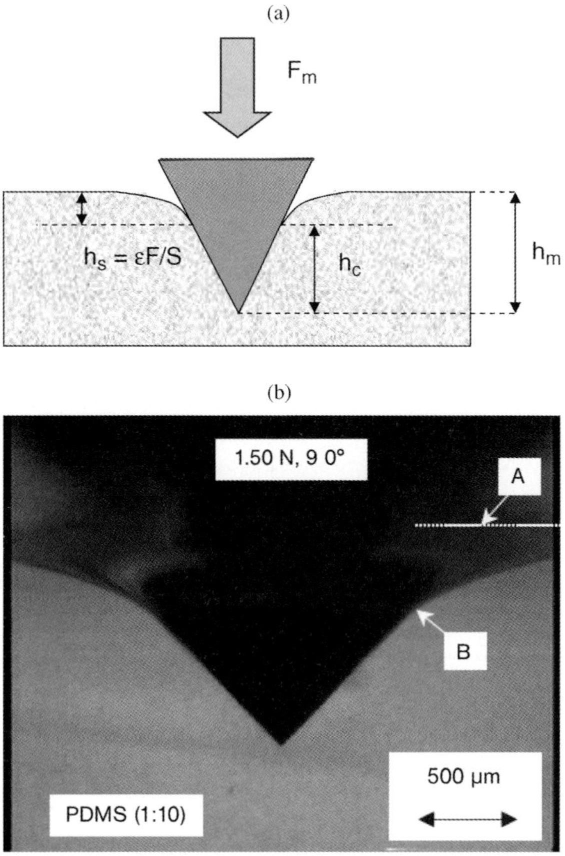

Fig. A8.3 (a) Schematic indentation profile; (b) Experimental indentation profile (Reprinted from Lim and Chaudhri, 2004, with permission from Taylor & Francis Ltd http://www.informa world.com). Original surface is shown as A; contact between specimen and indenter is up to B. Elastic deformation of the surface is observed all around the indenter.

where F is the load applied to the indenter, h its penetration and h_r the residual depth (after complete unloading). Contact stiffness ($S = dF/dh$) at maximum load is then determined as

$$S = km(h_m - h_r)^{m-1} \tag{A8.2}$$

where h_m is the maximum penetration of the indenter. The contact depth h_c (Fig. A8.3) is then given by

$$h_c = h_m - \varepsilon F_m/S \tag{A8.3}$$

where h_m, F_m and S are the penetration, force and stiffness at maximum load and ε is obtained from the following equations (Woirgard and Dargenton, 1997; Fig. A8.4). We adopt the modified Sneddon (1965) solution in the following, although it is still debated in the literature (Lim and Chaudhri, 2004, 2006):

$$\varepsilon = m(1 - \gamma) \tag{A8.4}$$

$$\gamma = \frac{1}{\sqrt{\pi}} \frac{\Gamma[(1/2)n + 1/2]}{\Gamma[(1/2)n + 1]} \tag{A8.5}$$

where Γ is the gamma function, $\Gamma(x) = \int_0^{\infty} e^{-t} t^{x-1} dt$, and

$$n = 1/(m - 1) \tag{A8.6}$$

where m is the exponent of the unloading curve defined above.

Then, from the contact depth knowledge, contact area is determined by

$$A_c = 24.5 h_c^2 \tag{A8.7}$$

and hardness is calculated as

$$H = F_m / A_c \tag{A8.8}$$

It should be noted that hardness values obtained from instrumented nanoindentation and microindentation may differ significantly and this phenomenon is invoked as the indentation size effect (ISE).

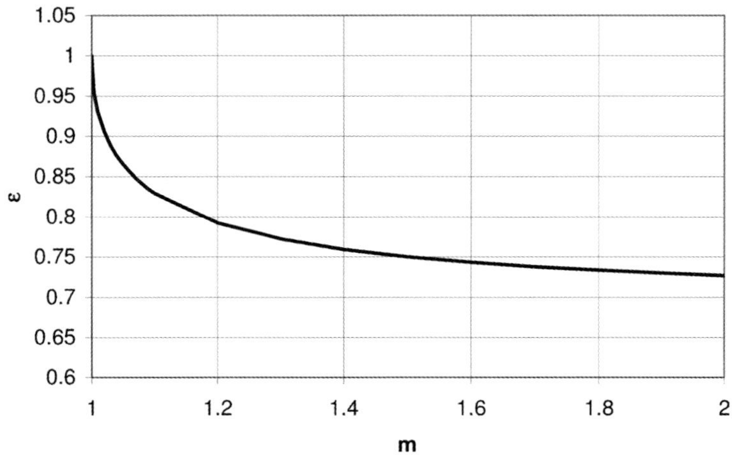

Fig. A8.4 $\varepsilon(m)$ curve.

Another advantage of instrumented nanoindentation is that the elastic properties of materials can be determined. The composite modulus (specimen–tip) E^* is extracted from

$$S = \frac{\partial F}{\partial h} = 2\beta E^* \sqrt{\frac{A_c}{\pi}} \tag{A8.9}$$

where S is the stiffness at maximum penetration, A_c the contact depth and β is a correcting factor depending on the tip geometry ($\beta = 1.034$ for a Berkovich tip). The composite modulus (specimen–tip) E^* is

$$\frac{1}{E^*} = \frac{1 - \nu_D^2}{E_D} + \frac{1 - \nu_S^2}{E_S} \tag{A8.10}$$

where E and ν are Young's modulus and Poisson's ratio, and subscripts D and S indicate diamond and specimen respectively ($E_D = 1141$ GPa, $\nu_D = 0.07$). The reduced modulus is often referred to as

$$\frac{1}{E_r} = \frac{1 - \nu_S^2}{E_S} \tag{A8.11}$$

Note that when viscoelastic materials are considered, it is important to control carefully the loading–unloading procedure (Fischer-Cripps, 2004a; Mammeri et al., 2004, 2005). In fact, for such kinds of materials indentation response is time dependent with a characteristic relaxation time $T_R = \eta/G$, where η and G are the viscosity and shear modulus of the material (Chapter 6; see also Section A8.5.1). Interestingly, some instrumented machines allow for adding a force modulation to the applied force at a given frequency and hence dynamic mechanical analysis can be performed (Chapter 6; Fischer-Cripps, 2004a).

A8.2 Indentation Strain Field

When indenting a sample a plastic zone is generated and surrounded by the undeformed solid. Hence, a plastic strain gradient remains in the specimens after unloading. The representative indentation strain ε_R is defined as (Tabor, 1951; Johnson, 1985)

$$\varepsilon_R = 0.2d/D \text{ for a spherical indenter} \tag{A8.12a}$$

and

$$\varepsilon_R = 0.2\tan\beta \text{ for a conical indenter} \tag{A8.12b}$$

where d and D are the residual indent diameter (or contact area diameter) and the sphere diameter respectively, while β is the angle between the original surface and the conical indenter flank. ε_R increases continuously as load is increased under a spherical tip (Eq. A8.12a). This point has motivated the use of a spherical tip to determine the overall $\sigma(\varepsilon)$ stress–strain curve (Field and Swain, 1993). Under a conical indenter the representative strain remains unchanged (assuming a perfect sharpness of the tip), a property that is often referred to as self-similarity or auto-similarity. The self-similarity of the strain field under a conical indenter is attributed to the geometric similarity of the tip ($A_c \propto h_c^2$, where h_c is the contact depth). Self-similarity is also assumed for non-axisymmetric tips like Berkovich and Vickers tips. Parameter β is then the angle between the original surface and the flank of a conical indenter with equivalent aspect ratio. For instance, for the Vickers and Berkovich tips which are most commonly used, the contact area is $A_c = 24.5\, h_c^2$ so that β is to be taken as 19°. A question has risen as to the significance of ε_R in view of strain gradient (Chaudhri, 1998). Furthermore, Eqs (A8.12) were established originally for metals and applied regardless of the type of material. Indentation residual strain can be determined experimentally in brittle materials (AlAs/GaAs superlattice) using a focussed ion beam (FIB) milled cross-section through the centre of a Vickers indentation. The sub-layers in the superlattice allow then mapping of the deformation, the interfaces between the AlAs (or GaAs) sub-layers being used as markers (Largeau et al., 2005). The distance e between two GaAs layers at different points of the sample was compared to the period e_o of the superlattice (Fig. A8.5) and the ε_z strain component along the loading axis can be calculated using equation $\varepsilon_z = (e - e_o)/e_o$.

Figure A8.5 shows the ε_z contours, and as reported for metals very high strains and strain gradients are generated just beneath the indent site. The highest ε_z value is determined to be about 28 % in good agreement with Chaudhri (1998) who estimated the strain ε to be as large as 20 % under a Vickers indenter in heavily work-hardened mild steel. These strain values are considerably higher than the representative strain value ε_R that is commonly accepted for a Vickers geometry (about 7.2 % (Eq. (A8.12b) with $\beta = 19°$ for a Vickers tip). The representative strain ε_R is hence to be considered as an averaged strain. Nonetheless, it is still unclear which deformed volume should be considered when averaging the strain (Chaudhri, 1998; Largeau et al., 2002).

A8.3
Hardness, Yield Stress and Representative Flow Stress

Hardness measures the resistance of a surface to contact loading (Tabor, 1951, 1996; Fischer-Cripps, 2004a). Traditionally, a spherical or a Vickers diamond (four-sided) pyramid is loaded on the investigated surface. The hardness (or mean pressure) p is defined by the ratio between the applied load and the projected surface of the permanent deformation left by the indenter:

$$p = F_m / A_p \tag{A8.13}$$

We denote it p in this appendix (H in the main text) in order to compare microhardness p and nanohardness H as extracted from nanoindentation curves. As discussed in detail, p and H may differ. Macro-, micro- and nanoindentation

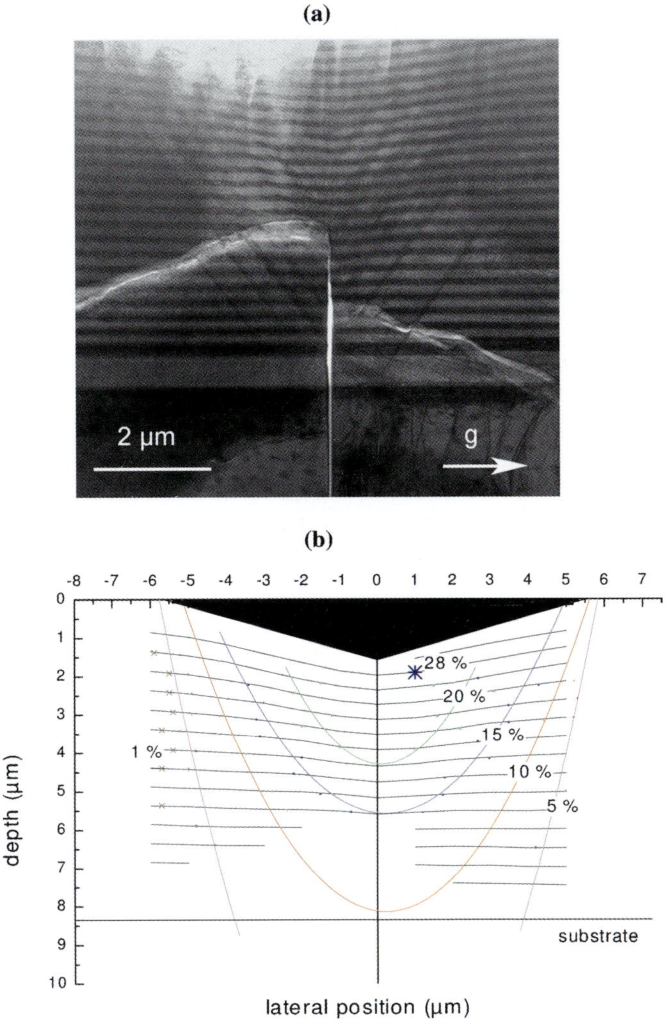

Fig. A8.5 (a) FIB-TEM cross-sectional view of an indented brittle semiconductor (AlAs/GaAs superlattice); (b) ε_z strain contours as obtained from (a). Strain values are given in percent. The dark area drawn on the graph represents the material punched into the sample (residual indentation).

domains have been defined and hardness values are to be compared in the respective domain since an ISE is reported for many types of materials (Gao et al., 1999; Sangwal et al., 2000; Elmustafa and Stone, 2003). In the macro- and microindentation domains the surface is generally determined after unloading by optical microscopy. We shall refer to this area as A_p (projected area defined above). Going to the nanoindentation domain, observation of the indent becomes tedious and requires scanning electron microscopy (SEM), atomic force microscopy (AFM) or eventually transmission electron microscopy (TEM) resolutions. Hence, hardness is determined under load and is defined as the ratio of the force to the projected contact area A_c (Section A8.1):

$$H = F_m / A_c \tag{A8.14}$$

It should be noted also that hardness values p and H may differ significantly. Let us consider for instance the elastic domain (before plastic onset). The indent recovers entirely so that $A_p = 0$ and p is not defined. In contrast, H can be determined advantageously by mean of an instrumented indentation machine. Also, plastic relaxation of the indents may yield differences between A_p and A_c and this phenomenon has been invoked sometimes to interpret the ISE.

It should be noted that for elastic–perfectly plastic materials, the mean pressure under a conical indenter p is generally assumed to be three times the yield stress Y (Tabor, 1951; Johnson, 1985):

$$p = 3Y \tag{A8.15a}$$

For the particular case of a Vickers tip, Eq. (A8.15a) is rewritten

$$H_v = 2.8Y \tag{A8.15b}$$

These relationships can be extended to strain-hardening materials with good approximation if Y is replaced by the representative flow stress Y_R measured in simple compression at the representative strain ε_R (Johnson, 1985):

$$p = 3Y_R = 3Y(\varepsilon_R) \tag{A8.15c}$$

Strain hardening is often taken as $Y_R = K\varepsilon_R^{1/n}$ (Hollomon, 1945). In the literature, it must be said that there has been some confusion between the different Eqs (A8.15) which have been used regardless of their domain of validity. However, the aperture of the tips commonly used is generally large so that the representative strain is below 8 %, which for low strain-hardening materials leads to $Y \sim Y_R$. Nonetheless, sharp cube-corner tips are now being employed to lower the plastic onset and, hence, to deform the specimens under smaller loads. For such a tip, Eqs. (A8.15) should be used more carefully. In terms of resolved yield stress Eqs (A8.15) and (A8.15c) are rewritten $p = 6\tau_c$ and $p = 6\tau_R$ (where τ_R is the representative resolved yield stress and using the Tresca factor). In fact, according to the class of materials, the relationship between and H and Y (or H and τ) is to be reconsidered with care since ratio H/Y varies (Gilman, 1973; Johnson, 1985). Hence we shall introduce a constraint factor C into Eqs (A8.15) which is expected to vary according to the type of material:

$$p = CY_R \tag{A8.15d}$$

A general relationship is written (Marsh, 1964; Johnson, 1985)

$$\frac{p}{Y_R} = \frac{2}{3}\left[1 + \ln\left(\frac{1}{0.6}\frac{E\varepsilon_R}{Y_R}\right)\right] \tag{A8.15e}$$

Indentation with a conical tip allows determination of the representative flow stress Y_R at the representative indentation strain ε_R. Following Johnson's model, Harvey et al. (1993) and Kramer et al. (1999) suggested that the representative flow stress Y_R of a material could be determined measuring the plastic zone radius c as a function of the maximum load F applied to the indenter using Eq. (A8.16a) (Le Bourhis and Patriarche, 2003):

$$Y_R = \frac{3}{2\pi}\frac{F}{c^2} \tag{A8.16a}$$

This relation is established balancing the radial forces at the elastic–plastic boundary (Kramer et al., 1999; Fig. A8.6):

$$F = \pi c^2 \sigma_r(c) \tag{A8.16b}$$

where the radial stress σ_r at the elastic–plastic boundary is given by (Johnson, 1985)

$$\sigma_r(c) = -\frac{2}{3} Y_R \tag{A8.16c}$$

The proper use of Eq. (A8.16a) supposes that small indenting loads are applied to the specimens in order to avoid fracture and remain in the plastic regime. This is possible using the nanoindentation technique. Equation (A8.16a) indicates that the plastic zone size varies linearly with the square root of the applied load. Figure A8.7b plots the central plastic zone size c as well as the rosette arm length (LA and SA, Fig. A8.7a) as a function of the square root of the maximum load, $\sqrt{F_{max}}$, measured in an indented brittle material (single-crystal (1 1 1) GaAs). All the sizes can be measured precisely on TEM plan views (Fig. A8.7a). Very interestingly it can be observed that all sizes (central zone radius c as well as the SA and LA sizes) are observed to increase linearly with $\sqrt{F_{max}}$ as expected from Eq. (A8.16a) (greater discrepancy is obtained for the LA sizes however).

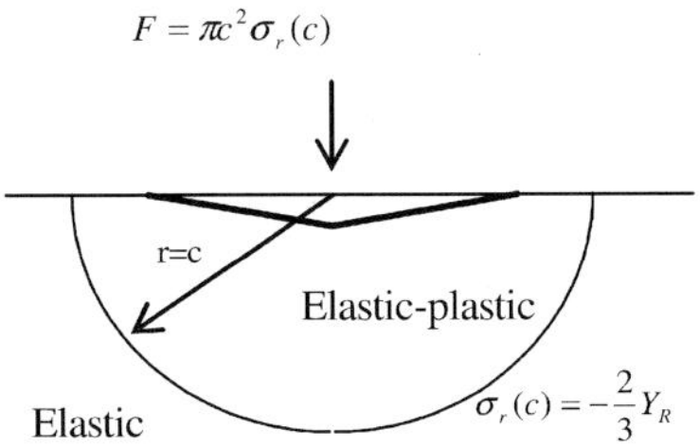

Fig. A8.6 Indentation stress balance.

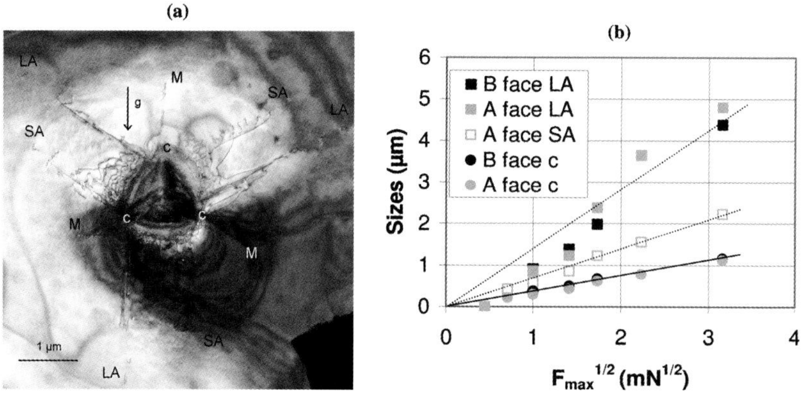

Fig. A8.7 (a) TEM plan view image of an indent site made under 5 mN at room temperature in a brittle semiconductor ((1 1 1)A face of GaAs, $g = [2\bar{2}0]$ diffraction condition). The central plastic zone (c), long arms (LA), short arms (SA) and microtwins (M) are shown. (b) Central plastic zone radius c and lengths of the rosette arms (LA and SA) as a function of the square root of the maximum applied load ($\sqrt{F_m}$) ((1 1 1)A and $(\bar{1}\bar{1}\bar{1})$B indented surfaces of GaAs crystal; Le Bourhis et al., 2004).

It should be noted that AFM can also be employed to determine the plastic zone size measuring the topography of the indent proximity (Harvey et al., 1993; Kramer et al., 1999). In that case, the plastic zone size is estimated from the upset region or pile-up. Except for very small indents where plastic recovery is observed, the upset zone radius determined by AFM shows also a square root dependence on load (Lilleodden et al., 1995). Interestingly, microindentation at elevated temperature also yields a square root dependence on load of the rosette arm length in a wide variety of materials as reviewed by Chaudhri (2004).

Using the linear variation of the central plastic zone with $\sqrt{F_{max}}$ plotted in Fig. A8.7, the flow stress is estimated to be about 3.4 GPa yielding an H/Y_R ratio about 2.5 close to the value expected from the expanding cavity model ($H/Y_R \sim 3$; Johnson, 1985).

A8.4
Coating–Substrate Composite Response

Using low load and increasing progressively the load, the instrumented nanoindentation technique allows for investigation of the elastic–plastic response of coatings as well as their adhesion to the substrate (Fig. A8.8). Indeed, the probe size can be adjusted to the film thickness using appropriate loads. For dedicated study, in particular when surface roughness is important, cross-sectional nanoindentation can be carried out on smoothly prepared cross-sections and a mechanical profile is extracted (Marot et al., 2002).

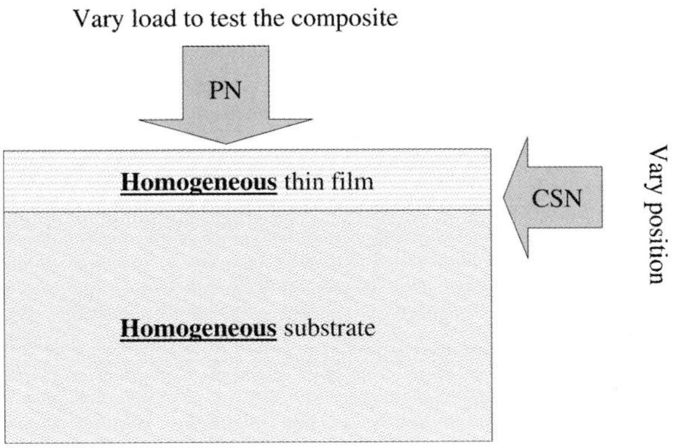

Fig. A8.8 Two different ways to characterize the indentation response of thin films either testing a prepared cross-section under constant load (CSN) or testing the film surface under decreasing loads (PN).

Mencik et al. 1997 reviewed models to predict the indentation modulus of homogeneous thin films on a homogeneous substrate as a function of the penetration. All of them consider the substrate and film properties and suppose the films are homogeneous:

$$\frac{1}{E_r} = \frac{1}{E_r^s}\phi\left(\frac{h_c}{e}\right) + \frac{1}{E_r^f}\left[1 - \phi\left(\frac{h_c}{e}\right)\right] \tag{A8.17}$$

where superscripts s and f indicate the substrate and film respectively. E_r is the measured substrate–film composite property. Assuming a given function ϕ (for $h_c \to 0$, $\phi \to 0$ and for $h_c \to \infty$, $\phi \to 1$), allows determination of the unknowns (E_r^f) once E_r^s is characterized and E_r measured as a function of contact penetration h_c (Mencik et al., 1997).

A similar procedure can be applied to determine the hardness of the films. In fact, the combined contributions of the coating and the substrate are measured. Notably, at equal indentation depth, the influence of the substrate on the measured hardness is less than on the indentation modulus. Indeed, the plastic strain field is much less extended spatially than the elastic strain field (Fig. A8.9). From Fig. A8.9, one further observes that the 10% rule of thumb (in other words, film response is determined with relative penetration depth less than 0.1) has to be used very carefully.

Beegan and Laugier (2005) reviewed models to predict the indentation hardness of homogeneous thin films on a homogeneous substrate as a function of penetration. As for moduli, all of them consider the substrate and film properties and suppose the films are homogeneous (Bhattacharya and Nix, 1988; Puchi-Cabrera, 2002):

$$H^* = H_s + (H_f - H_s)\phi'\left(\frac{h_c}{e}\right) \tag{A8.18}$$

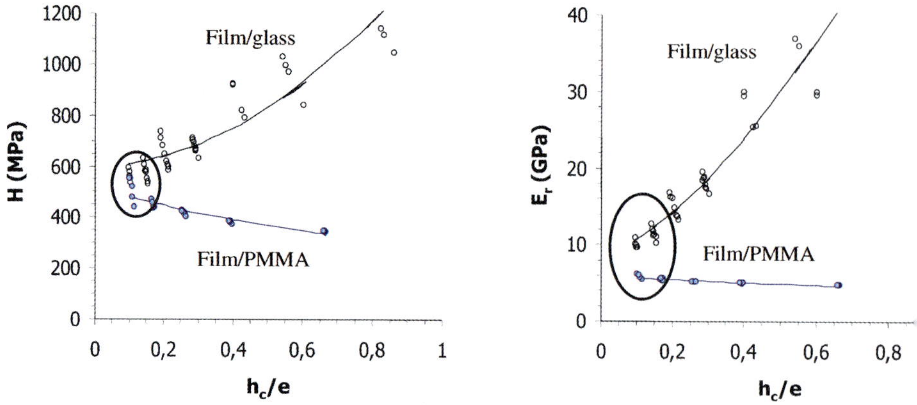

Fig. A8.9 Hardness and modulus evolution with relative penetration for the same film deposited on two different substrates. More rapid convergence at small depth is observed for hardness. The film is a hybrid TEOS–PMMA film deposited onto glass or PMMA substrates (Mammeri, 2003).

where subscripts s and f indicate the substrate and film respectively. H^* is the measured substrate–film composite property. Assuming a given function ϕ' (for $h_c \rightarrow 0$, $\phi' \rightarrow 1$ and for $h_c \rightarrow \infty$, $\phi' \rightarrow 0$), allows determination of the unknowns (H_f) once H_s is characterized and H^* measured as a function of contact penetration h_c.

Recently Veprek et al. (2005) emphasized that to measure correctly the hardness of hard coatings on softer substrates, the maximum indentation depth must not exceed 10 % of the thickness of the coating because, above this value, plastic deformation occurs in the softer substrate. Recent FEM simulations of indentation into superhard coatings with hardness $\geq$40 GPa revealed that because of the higher strength of these materials, the maximum indentation depth should not exceed about 5 % of the coating thickness in order to avoid plastic deformation in the substrate. Therefore, in order to avoid possible artefacts, Veprek et al. (2005) recommend to use relatively high loads of 30–150 mN, with indentation depths >0.3 μm not exceeding about 5 % of the thickness of the coatings. Their conclusion is that to meet these requirements, coatings $\geq$6 μm thick are needed. This requirement is not straightforwardly met (Tavares et al., 2005). Instead, as discussed above the survey of the mechanical response as a function of penetration allows extraction of the film contribution.

A8.5 Time-Dependent Response

As mentioned in Section A8.1, when viscoelastic materials are considered, the loading–unloading procedure has to be considered carefully since the mechanical response is time dependent (see also Chapter 6). Time dependence of the indentation response has been the subject of earlier work (Radok, 1957; Lee and Radok, 1960; Ting, 1966; Johnson, 1985). Because of the growing interest in polymeric and hybrid sol–gel coatings new interest has focused on this subject (Shimizu et al., 1999; Oyen and Cook, 2003; Fischer-Cripps, 2004b; Mammeri et al., 2004; Atanacio et al., 2005; Sakai et al., 2005). Moreover instrumented indentation allows for the monitoring of the overall $F(h)$ curves and allows creep to be studied when a load plateau is programmed. It is important to separate again sharp and blunt indenter geometries. In fact, under a blunt indenter, a general assumption is that plasticity can be neglected. In that case the viscoelastic solution can be obtained from the elastic–viscoelastic correspondence using the hereditary Boltzmann integral. For a sharp indenter, it has to be considered whether plasticity can or cannot be neglected. In the first case, the approach is the same as for a blunt contact. If plasticity is to be considered then the solution is complex since elastic, plastic and viscous behaviours are to be considered together. Under indentation, of course, the deformation and stress fields are complex with a combination of multiaxial compression and shear.

We review here phenomenological approaches allowing for $F(h)$ loading curves to be predicted and for creep to be analysed.

A8.5.1
Viscoelastic Indentation Curves

The solution for an elastic contact was given by Hertz (1981) and Sneddon (1965) (Chapter 8) for a blunt and sharp indenter respectively. The solutions can be written in the general form (Chapter 8; we suppose the indenter is much more rigid than the specimen)

$$F = A\frac{E}{1-\nu^2}h^n \tag{A8.19a}$$

where A is a geometrical factor and n an exponent equal to 1.5 and 2 for a spherical and a cone indenter respectively. Equation (A8.19a) is rewritten for an incompressible solid ($\nu = 0.5$) as

$$F = A\frac{2G}{1-\nu}h^n = 4AGh^n \tag{A8.19b}$$

Using the elastic–viscoelastic correspondence with hereditary Boltzmann integral allows for the general solution to be found as (Shimizu et al., 1999; Sakai, 2002; Fisher-Cripps, 2004b; Cheng and Cheng, 2005; Oyen, 2005)

$$F(t) = 4A\int_0^t G(t-t')\frac{dh(t')^n}{dt'}dt' \tag{A8.20}$$

$$h^n(t) = \frac{1}{4A}\int_0^t J(t-t')\frac{dF(t')}{dt'}dt' \tag{A8.21}$$

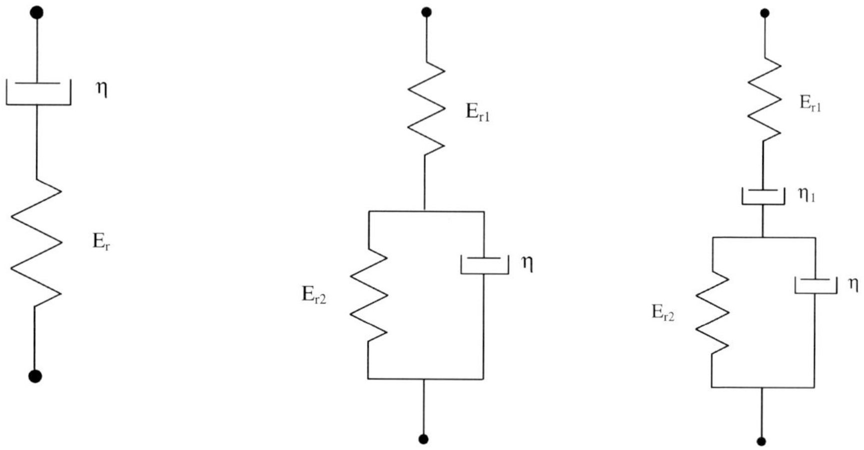

Fig. A8.10 Viscoelastic solids (Voigt, Voigt and spring in series, Burger) used to model the indentation creep of viscoelastic solids.

The solution of this integral is completely dependent on the choice of the material relaxation modulus $G(t)$ and compliance $J(t)$.

Fischer-Cripps (2004b) proposed empirical solutions for creep analysis during a load plateau (at F_m) based on Maxwell and Voigt elements. The solutions for a conical indenter (Fig. A8.10) are

$$\text{Maxwell element}: h^2(t) = \frac{\pi}{2} F_m \cot\alpha \left[\frac{1}{E_r} + \frac{1}{\eta} t\right] \tag{A8.22}$$

$$\text{Voigt and spring elements in series}: h^2(t) = \frac{\pi}{2} F_m \cot\alpha \left[\frac{1}{E_{r1}} + \frac{1}{E_{r2}}\left(1 - e^{-tE_{r2}/\eta}\right)\right] \tag{A8.23}$$

$$\text{Burger viscoelastic solid}: h^2(t) = \frac{\pi}{2} F_m \cot\alpha \left[\frac{1}{E_{r1}} + \frac{1}{E_{r2}}\left(1 - e^{-tE_{r2}/\eta_2}\right) + \frac{1}{\eta_1} t\right] \tag{A8.24}$$

where α is the semi-angle of the cone, E_{ri} the respective reduced moduli and η_i the respective viscosities. These solutions can also be obtained for a spherical contact (Fisher-Cripps, 2004b).

The models are applied to a HEMA–TEOS sol–gel film under indentation creep (Fig. A8.11). The Maxwell element shows the expected linear dependence of the square of the penetration. This does not fit the experimental data well. In fact, it is necessary to consider a more complex scheme of the solid in order to obtain a representative behaviour. A satisfactory fit is obtained using either a combination in series of Voigt and spring elements or a Burger solid.

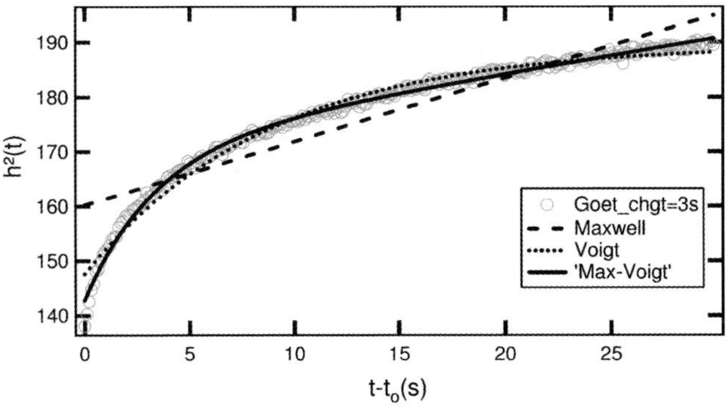

Fig. A8.11 Indentation creep curve of HEMA–TEOS thin film (Chemin, 2007).

A8.5.2 Viscous Elastic–Plastic Indentation *F*(*h*) Curves

Sharp indentation curves of time-dependent materials can be modelled using the Oyen and Cook (2003) viscous elastic–plastic (VEP) model which extends the Sakai (1999) representation of elastic–plastic (EP) materials (Chapter 8).

The total displacement of the indenter is taken as the sum of the elastic h_e, plastic h_p and viscous h_v displacements:

$$h = h_e + h_p + h_v \tag{A8.25}$$

Because of the geometric similarity of a sharp conical (or Berkovich) indenter, all constitutive equations show the square of the penetration and are assumed to be

$$F = AE_r h_e^2 \tag{A8.26a}$$

$$F = BHh_p^2 \tag{A8.26b}$$

$$F = C\eta(\mathrm{d}h_v/\mathrm{d}t)^2 \tag{A8.26c}$$

where A (4.4), B (24.6 for a Berkovich indenter) and C are geometrical factors, and E_r, H and η the reduced modulus, the true hardness and the quadratic viscosity respectively. Taking the derivation of the displacement equation and combining with constitutive equations yields

$$\bar{h} = \frac{\mathrm{d}h}{\mathrm{d}t} = \frac{1}{2\sqrt{AE_r}}\frac{1}{\sqrt{F}}\frac{\mathrm{d}F}{\mathrm{d}t} + \frac{1}{2\sqrt{BH}}\frac{1}{\sqrt{F}}\frac{\mathrm{d}F}{\mathrm{d}t} + \frac{\sqrt{F}}{C\eta} \tag{A8.27}$$

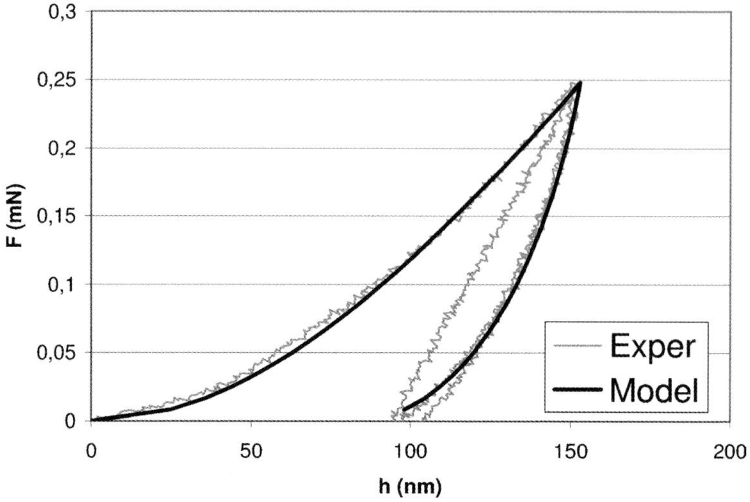

Fig. A8.12 Loading–unloading curves of a PMMA–TEOS specimen obtained under periodic load–reload sequence. The model curve is obtained for $E_r = 7.6$ GPa, $H = 3.9$ GPa, $C\eta = 1.0 \times 10^{15}$ Pa s^{-2}. (Reprinted from Mammeri et al., 2004, with permission from Elsevier.)

Once the loading (or unloading) rate is set constant or zero (hold period), Eq (A8.27) can be solved easily yielding $h(t)$. This is illustrated for PMMA–TEOS sol–gel coatings in Figs. A8.12 and A8.13 that show the $h(t)$ solutions under two different loading procedures (Mammeri et al. 2004). One of these shows a holding

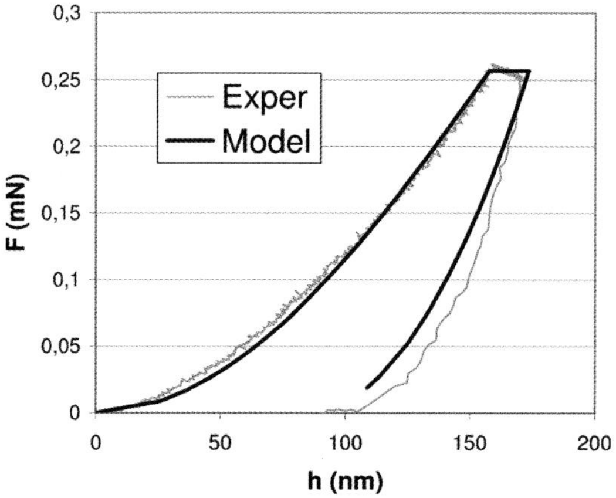

Fig. A8.13 Loading–unloading curves of a PMMA–TEOS specimen obtained under loading–hold–rapid unloading sequence. The model curve is obtained for $E_r = 7.2$ GPa, $H = 3.9$ GPa, $C\eta = 0.9 \times 10^{15}$ Pa s^{-2}. (Reprinted from Mammeri et al., 2004, with permission from Elsevier.)

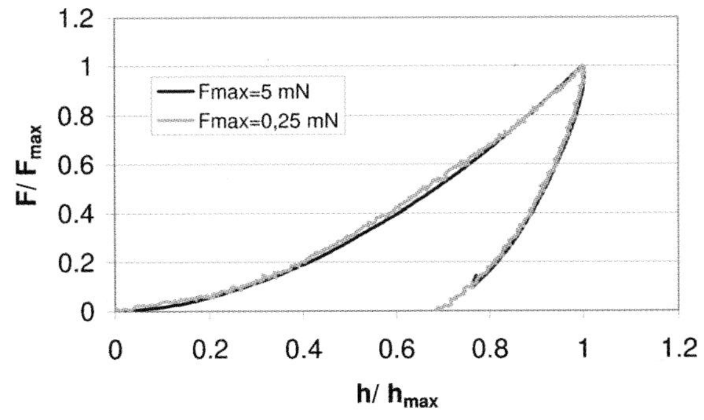

Fig. A8.14 Normalized loading–unloading curves obtained under two different loads but same loading and unloading times. Specimen is PMMA. Note the superposition of the curves.

period during which time-dependent penetration is observed. Without this holding period, the unloading curve shows unusual curvature: a 'nose' can be observed under low unloading rates and results from viscous flow. The presence of such a nose may be detrimental to extract the mechanical properties from the related unloading curves.

It is observed that the time-dependent part of the deformation (viscous deformation h_v) allows prediction of creep under a holding period (Fig. A8.13). It should be noted that viscous flow happens during the whole indentation procedure. When the unloading time is not negligible as compared to the relaxation time of the material, viscous flow cannot be neglected during unloading. The relaxation time τ of the VEP solid may be extracted from the fitting parameter (Oyen and Cook, 2003; Mammeri et al., 2004):

$$\tau = \sqrt{\frac{C\eta}{AE_r}} \tag{A8.28}$$

An interesting prediction of the VEP model is that the response is controlled by the total rise time and not by the actual loading rate, which is illustrated for PMMA in Fig. A8.14. The curves were obtained under two different loads but keeping the same loading and unloading time (about 30 s). It is observed using normalized load F/F_{max} and normalized depth h/h_{max} that the curves obtained under two different maximum loads (namely 0.25 and 5 mN) but with the same loading and unloading time superimpose quite well.

A8.6 Elastic–Plastic Ratios

We introduced in Chapter 9 the brittleness index as the hardness–toughness ratio to characterize the competition between the plastic and brittle deformations of glass.

Here again, a similar approach can be used instructively in order to balance the elastic and plastic deformations. Hence, we introduce hardness–elastic modulus ratios in order to quantify the competition between elastic and plastic responses. An instrumented indentation machine allows for extraction of hardness H and elastic modulus E; different ratios have been considered, the simplest one being H/E. The higher the ratio the higher the resistance to plastic deformation and hence the higher the elastic compliance of the film.

Some authors also use ratio H/E^* that can be related theoretically to residual depth–maximum depth ratio (Chen and Bull, 2006) and experimentally verified (Fouquet et al., 2004; Chen and Bull, 2006).

Taking again the unloading curve function (Section A8.1) we have

$$F = k(h - h_r)^m \tag{A8.29}$$

Contact stiffness at maximum load is

$$S = km(h_m - h_r)^{m-1} = mF_m/(h_m - h_r)$$

$$\text{Hence } h_m - h_r = mF_m/S \tag{A8.30}$$

or

$$1 - h_r/h_m = mF_m/(Sh_m) \tag{A8.31}$$

Moreover $h_m = (\pi/2)h_c$ (Sneddon, 1965), $F_m = HA_c$ and $S = \partial F/\partial h = 2\beta E^* \sqrt{A_c/\pi}$, then

$$1 - \frac{h_r}{h_m} = m\frac{HA_c}{2\beta E^* \sqrt{A_c/\pi}(\pi/2)h_c} = m\frac{H\sqrt{A_c}}{\beta E^* \sqrt{\pi}h_c} = \frac{H}{E^*}\frac{m\sqrt{24.5}}{\beta\sqrt{\pi}}$$

or

$$1 - \frac{h_r}{h_m} \approx \frac{H}{E^*} \tag{A8.32}$$

As regards indentation, Joslin and Oliver (1990) introduced the ratio H/E^{*2} that can be extracted without requiring any tip area calibration of the nanoindenter

Tab. A8.1 Elastic–plastic ratios and units.

Elastic–plastic ratio	Unit
H/E	–
H/E^*	–
H/E^{*2}	Pa^{-1}
H^3/E^2	Pa

since we have a simple relationship between this ratio and force-to-stiffness ratio F/S^2:

$$\frac{H}{E^{*2}} = \frac{F}{S^2}\frac{2\beta}{\pi} \tag{A8.33}$$

Another important ratio is H^3/E^2 that has units of pressure and is representative of the film resistance to plastic deformation (Johnson, 1985). This ratio is widely employed in the development of hard coatings for tribological applications (Musil, 2000). Various elastic–plastic ratios and units are summarized in Table A8.1.

Appendix 9
Strain and Stress

A9.1
Stress and Strain

There are two kinds of forces: surface forces or body forces. Surface forces are distributed over the surface of the specimen (for instance contact between objects), while body forces are distributed within the volume (for instance gravitational, electric, magnetic forces).

The mean stress σ_m resulting from a force F acting over a surface or section S is

$$\sigma_m = \frac{F}{S}$$

where F is expressed in N, S in m^2 and σ_m in Pa. Positive values are used for tensile stresses and negative for compressive stresses.

In general, the force is inclined to the surface and decomposes into its normal N and tangential T components (Fig. A9.1). The related normal and shear stress components are

$$\sigma = \frac{N}{S}$$

$$\tau = \frac{T}{S}$$

Strain refers to the relative displacement resulting from the applied stresses.

Consider a solid under normal stress σ. It deforms from size l_o to l (Fig. A9.2). Strain is defined as the relative deformation and is written

$$\varepsilon = \frac{l - l_o}{l_o}$$

The logarithmic form is also used:

$$\varepsilon = \ln \frac{l}{l_o}$$

Glass; Mechanics and Technology. Eric Le Bourhis

ISBN: 978-3-527-31549-9

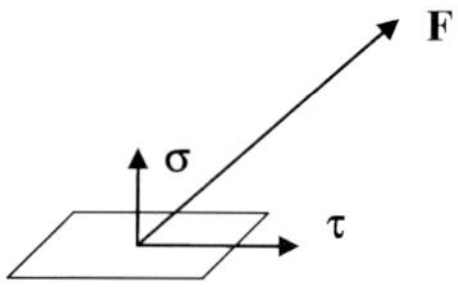

Fig. A9.1 Surface force, normal and shear stresses.

and reduces for small strains to (see also Appendix 12)

$$\varepsilon = \ln\frac{l}{l_o} = \ln\frac{l_o + l - l_o}{l_o} = \ln\left(1 + \frac{l - l_o}{l_o}\right) \approx \frac{l - l_o}{l_o}$$

Side deformation also occurs and lateral strain is $-\nu\epsilon$ where ν is Poisson's ratio ($\nu \leq 0.5$). $\nu = 0.5$ corresponds to volume conservative deformation. As detailed in the main part of this book, Poisson's ratio depends on glass type and increases as glass structure packing density increases. It is ~0.2 for oxide glasses ($C_g \sim 0.5$) and 0.4 for metallic glasses ($C_g > 0.7$).

Consider now a solid under shear stress τ (Fig. A9.2). Distortion appears and shear strain γ is written

$$\gamma = \tan\theta \approx \theta$$

A9.2
Stress and Strain Tensors

Considering a general loading of a specimen requires the stress and strain tensors to be considered (Fig. A9.3). Stress tensor $[\sigma]$ is

$$[\sigma] = \begin{pmatrix} \sigma_{11} & \sigma_{12} & \sigma_{13} \\ \sigma_{21} & \sigma_{22} & \sigma_{23} \\ \sigma_{31} & \sigma_{32} & \sigma_{33} \end{pmatrix}$$

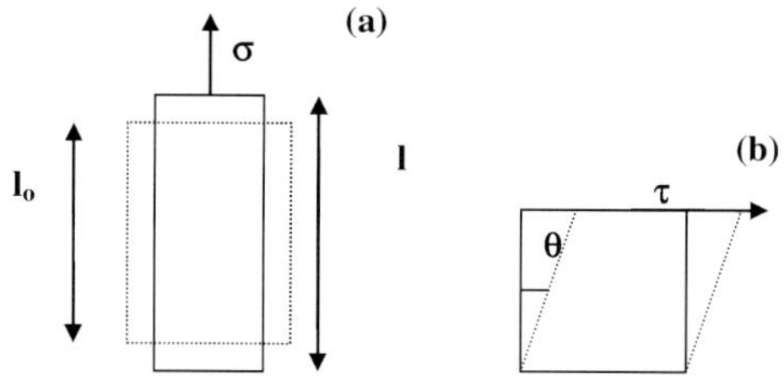

Fig. A9.2 Solid under (a) stress and (b) shear stress; elastic deformation and distortion are exaggerated for clarity.

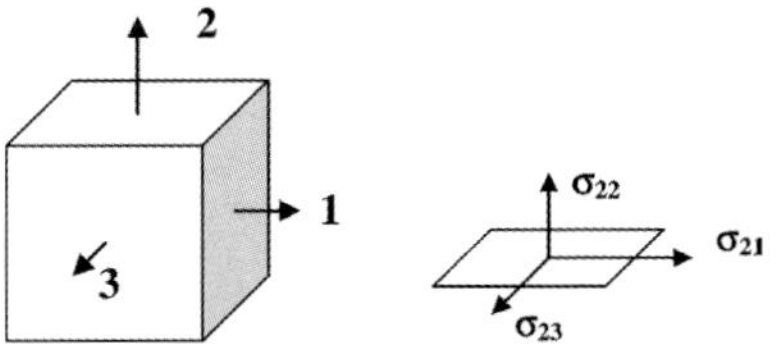

Fig. A9.3 Identification of components of stress tensor.

Or using stress σ and shear stress τ notations

$$[\sigma] = \begin{pmatrix} \sigma_{11} & \tau_{12} & \tau_{13} \\ \tau_{21} & \sigma_{22} & \tau_{23} \\ \tau_{31} & \tau_{32} & \sigma_{33} \end{pmatrix}$$

Let us define u, v, w as the displacement vector of any point x, y, z of the solid (Fig. A9.4). Normal strains are written

$$\varepsilon_{11} = \frac{\partial u}{\partial x}, \quad \varepsilon_{22} = \frac{\partial v}{\partial y}, \quad \varepsilon_{33} = \frac{\partial w}{\partial z}$$

while shear strains are written

$$\gamma_{12} = 2\varepsilon_{12} = \frac{\partial v}{\partial x} + \frac{\partial u}{\partial y}, \quad \gamma_{23} = 2\varepsilon_{23} = \frac{\partial w}{\partial y} + \frac{\partial v}{\partial z}, \quad \gamma_{31} = 2\varepsilon_{31} = \frac{\partial u}{\partial z} + \frac{\partial w}{\partial x}$$

Strain tensor [ε] is

$$[\varepsilon] = \begin{pmatrix} \varepsilon_{11} & \varepsilon_{12} & \varepsilon_{13} \\ \varepsilon_{21} & \varepsilon_{22} & \varepsilon_{23} \\ \varepsilon_{31} & \varepsilon_{32} & \varepsilon_{33} \end{pmatrix}$$

Or using strain ϵ and shear strain γ notations

$$[\varepsilon] = \begin{pmatrix} \varepsilon_{11} & \gamma_{12}/2 & \gamma_{13}/2 \\ \gamma_{21}/2 & \varepsilon_{22} & \gamma_{23}/2 \\ \gamma_{31}/2 & \gamma_{32}/2 & \varepsilon_{33} \end{pmatrix}$$

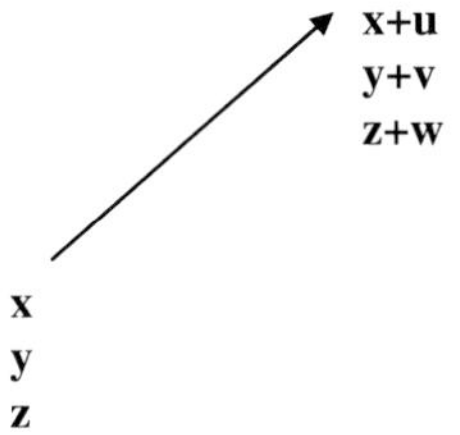

Fig. A9.4 Displacement vector.

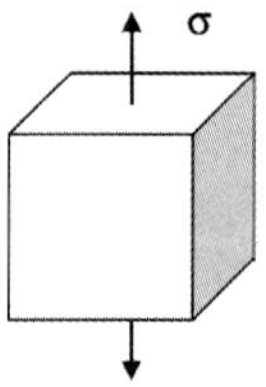

Fig. A9.5 Specimen under uniaxial stress.

A9.3 Uniaxial Tensile Test

Under uniaxial tensile testing (Fig. A9.5), the stress tensor is

$$\begin{pmatrix} \sigma & 0 & 0 \\ 0 & 0 & 0 \\ 0 & 0 & 0 \end{pmatrix}$$

The strain tensor is in the elastic domain

$$\begin{pmatrix} \sigma/E & 0 & 0 \\ 0 & -\nu\sigma/E & 0 \\ 0 & 0 & -\nu\sigma/E \end{pmatrix}$$

where E is Young's modulus and ν Poisson's ratio.

A9.4 Simple Shear

Under simple shear stress (Fig. A9.6), the stress tensor is

$$\begin{pmatrix} 0 & \tau & 0 \\ \tau & 0 & 0 \\ 0 & 0 & 0 \end{pmatrix}$$

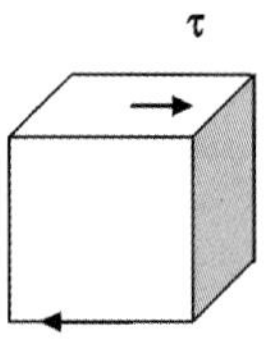

Fig. A9.6 Specimen under simple shear.

The strain tensor becomes in the elastic domain

$$\begin{pmatrix} 0 & \tau/2G & 0 \\ \tau/2G & 0 & 0 \\ 0 & 0 & 0 \end{pmatrix}$$

where G is the shear modulus ($E = 2G(1+\nu)$). For a viscous flow the strain tensor derivative is

$$\begin{pmatrix} 0 & \overline{\gamma}/2 & 0 \\ \overline{\gamma}/2 & 0 & 0 \\ 0 & 0 & 0 \end{pmatrix} = \begin{pmatrix} 0 & \tau/2\eta & 0 \\ \tau/2\eta & 0 & 0 \\ 0 & 0 & 0 \end{pmatrix}$$

A9.5 Plane Stress

Under plane stress or two-dimensional state of stress, the stress tensor becomes

$$\begin{pmatrix} \sigma_{11} & \tau_{12} & 0 \\ \tau_{21} & \sigma_{22} & 0 \\ 0 & 0 & 0 \end{pmatrix} = \begin{pmatrix} \sigma_x & \tau_{xy} & 0 \\ \tau_{xy} & \sigma_y & 0 \\ 0 & 0 & 0 \end{pmatrix}$$

Principal stresses are obtained after diagonalization of the stress tensor. Eigenvalues λ are obtained solving $[\sigma]\ [x] = \lambda[x]$ or $([\sigma] - \lambda[1])[x] = [0]$. $\lambda_\pm$ are the roots of the equation

$$det([\sigma] - \lambda[1]) = 0 = (\sigma_x - \lambda)(\sigma_y - l) - \tau_{xy}^2 = \lambda^2 - \lambda(\sigma_x + \sigma_y) + \sigma_x\sigma_y - \tau_{xy}^2$$

We get then

$$\lambda_\pm = \frac{\sigma_x + \sigma_y \pm \sqrt{(\sigma_x + \sigma_y)^2 - 4(\sigma_x\sigma_y - \tau_{xy}^2)}}{2}$$

$$= \frac{\sigma_x + \sigma_y \pm \sqrt{(\sigma_x - \sigma_y)^2 + 4\tau_{xy}^2}}{2} = \frac{\sigma_x + \sigma_y}{2} \pm \sqrt{\left(\frac{\sigma_x - \sigma_y}{2}\right)^2 + \tau_{xy}^2}$$

Principal stress tensor is written

$$\begin{pmatrix} \sigma_1 & 0 & 0 \\ 0 & \sigma_2 & 0 \\ 0 & 0 & 0 \end{pmatrix}$$

with

$$\sigma_1 = \frac{\sigma_x + \sigma_y}{2} + \sqrt{\left(\frac{\sigma_x - \sigma_y}{2}\right)^2 + \tau_{xy}^2}$$

$$\sigma_2 = \frac{\sigma_x + \sigma_y}{2} - \sqrt{\left(\frac{\sigma_x - \sigma_y}{2}\right)^2 + \tau_{xy}^2}$$

Under equibiaxial stresses, the stress tensor becomes

$$\begin{pmatrix} \sigma & 0 & 0 \\ 0 & \sigma & 0 \\ 0 & 0 & 0 \end{pmatrix}$$

Opposite principal stresses is also referred as pure shear

$$\begin{pmatrix} \sigma & 0 & 0 \\ 0 & -\sigma & 0 \\ 0 & 0 & 0 \end{pmatrix}$$

A9.6
Hydrostatic Pressure and Stress Deviator

The stress tensor can be separated into a deviator (trace is null) and a hydrostatic tensor:

$$\begin{pmatrix} \sigma_{11} & \sigma_{12} & \sigma_{13} \\ \sigma_{21} & \sigma_{22} & \sigma_{23} \\ \sigma_{31} & \sigma_{32} & \sigma_{33} \end{pmatrix}$$

$$= \begin{pmatrix} \sigma_{11} - \frac{1}{3}\mathrm{Tr}([\sigma]) & \sigma_{12} & \sigma_{13} \\ \sigma_{21} & \sigma_{22} - \frac{1}{3}\mathrm{Tr}([\sigma]) & \sigma_{23} \\ \sigma_{31} & \sigma_{32} & \sigma_{33} - \frac{1}{3}\mathrm{Tr}([\sigma]) \end{pmatrix}$$

$$+ \begin{pmatrix} \frac{1}{3}\mathrm{Tr}([\sigma]) & 0 & 0 \\ 0 & \frac{1}{3}\mathrm{Tr}([\sigma]) & 0 \\ 0 & 0 & \frac{1}{3}\mathrm{Tr}([\sigma]) \end{pmatrix}$$

$$= \begin{pmatrix} s_{11} & s_{12} & s_{13} \\ s_{21} & s_{22} & s_{23} \\ s_{31} & s_{32} & s_{33} \end{pmatrix} + \frac{1}{3}\mathrm{Tr}([\sigma])[I] = \begin{pmatrix} s_{11} & s_{12} & s_{13} \\ s_{21} & s_{22} & s_{23} \\ s_{31} & s_{32} & s_{33} \end{pmatrix} + \sigma[I]$$

with

$$\mathrm{Tr}([\sigma]) = \sigma_{11} + \sigma_{22} + \sigma_{33}$$

The hydrostatic pressure is

$$-\sigma = -\mathrm{Tr}([\sigma])/3 = -(\sigma_{11} + \sigma_{22} + \sigma_{33})/3$$

For simple shear stress, the stress tensor reduces to its deviator (hydrostatic tensor is zero).

A9.7 Generalized Hooke's Law

Under complex loading and in the elastic domain the stress–strain relationship is referred to as the generalized Hooke's law:

$$\sigma_{ij} = C_{ijkl}\varepsilon_{kl}$$

where C_{ijkl} are the stiffness tensor components. This equation can be inverted into

$$\varepsilon_{ij} = J_{ijkl}\sigma_{kl}$$

where J_{ijkl} are the compliance tensor components. For cubic materials, there are only three independent components in the C_{ijkl} matrix. Hooke's law is then written

$$\begin{pmatrix} \sigma_{11} \\ \sigma_{22} \\ \sigma_{33} \\ \sigma_{12} \\ \sigma_{13} \\ \sigma_{23} \end{pmatrix} = \begin{pmatrix} C_{11} & C_{12} & C_{12} & 0 & 0 & 0 \\ C_{12} & C_{11} & C_{12} & 0 & 0 & 0 \\ C_{12} & C_{12} & C_{11} & 0 & 0 & 0 \\ 0 & 0 & 0 & C_{44} & 0 & 0 \\ 0 & 0 & 0 & 0 & C_{44} & 0 \\ 0 & 0 & 0 & 0 & 0 & C_{44} \end{pmatrix} \begin{pmatrix} \varepsilon_{11} \\ \varepsilon_{22} \\ \varepsilon_{33} \\ 2\varepsilon_{12} \\ 2\varepsilon_{13} \\ 2\varepsilon_{23} \end{pmatrix}$$

For an isotropic material, the stiffness tensor can be further simplified and Hooke's law is written

$$\begin{pmatrix} \sigma_{11} & \sigma_{12} & \sigma_{13} \\ \sigma_{21} & \sigma_{22} & \sigma_{23} \\ \sigma_{31} & \sigma_{32} & \sigma_{33} \end{pmatrix} = \lambda \mathrm{Tr}([\varepsilon]) \begin{pmatrix} 1 & 0 & 0 \\ 0 & 1 & 0 \\ 0 & 0 & 1 \end{pmatrix} + 2\mu \begin{pmatrix} \varepsilon_{11} & \varepsilon_{12} & \varepsilon_{13} \\ \varepsilon_{21} & \varepsilon_{22} & \varepsilon_{23} \\ \varepsilon_{31} & \varepsilon_{32} & \varepsilon_{33} \end{pmatrix}$$

with $\mathrm{Tr}([\varepsilon]) = \varepsilon_{11} + \varepsilon_{22} + \varepsilon_{33}$, $C_{11} = \lambda + 2\mu$, $C_{12} = \lambda$ and $C_{44} = \mu$. Then $C_{44} = ½(C_{11} - C_{12})$, $A = C_{44}/½(C_{11} - C_{12}) = 1$; see also Appendix 12.

Parameters λ and μ are Lame coefficients related to Poisson's ratio ν and Young's modulus *E*:

$$\nu = \frac{\lambda}{2(\lambda + \mu)} = \frac{C_{12}}{C_{11} + C_{12}}$$

$$E = \frac{\mu(3\lambda + 2\mu)}{\lambda + \mu} = \frac{(C_{11} - C_{12})(C_{11} + 2C_{12})}{C_{11} + C_{12}}$$

The strain tensor is written

$$\begin{pmatrix} \varepsilon_{11} & \varepsilon_{12} & \varepsilon_{13} \\ \varepsilon_{21} & \varepsilon_{22} & \varepsilon_{23} \\ \varepsilon_{31} & \varepsilon_{32} & \varepsilon_{33} \end{pmatrix} = \frac{1+\nu}{E} \begin{pmatrix} \sigma_{11} & \sigma_{12} & \sigma_{13} \\ \sigma_{21} & \sigma_{22} & \sigma_{23} \\ \sigma_{31} & \sigma_{32} & \sigma_{33} \end{pmatrix} - \mathrm{Tr}([\sigma]) \frac{\nu}{E} \begin{pmatrix} 1 & 0 & 0 \\ 0 & 1 & 0 \\ 0 & 0 & 1 \end{pmatrix}$$

A9.8
Kelvin and Maxwell Models

Viscoelasticity is obtained combining elastic and viscous components, the Kelvin and Maxwell models being the most simple (Chapter 6).

The stress tensor for a Kelvin solid is

$$\begin{pmatrix} \sigma_{11} & \sigma_{12} & \sigma_{13} \\ \sigma_{21} & \sigma_{22} & \sigma_{23} \\ \sigma_{31} & \sigma_{32} & \sigma_{33} \end{pmatrix} = \lambda(\mathrm{Tr}([\varepsilon]) + \theta_\lambda \mathrm{Tr}([\bar{\varepsilon}])) \begin{pmatrix} 1 & 0 & 0 \\ 0 & 1 & 0 \\ 0 & 0 & 1 \end{pmatrix}$$
$$+ 2\mu \left[\begin{pmatrix} \varepsilon_{11} & \varepsilon_{12} & \varepsilon_{13} \\ \varepsilon_{21} & \varepsilon_{22} & \varepsilon_{23} \\ \varepsilon_{31} & \varepsilon_{32} & \varepsilon_{33} \end{pmatrix} + \theta_\mu \frac{\mathrm{d}}{\mathrm{d}t} \begin{pmatrix} \varepsilon_{11} & \varepsilon_{12} & \varepsilon_{13} \\ \varepsilon_{21} & \varepsilon_{22} & \varepsilon_{23} \\ \varepsilon_{31} & \varepsilon_{32} & \varepsilon_{33} \end{pmatrix} \right]$$

The strain tensor for a Maxwell solid is

$$\frac{\mathrm{d}}{\mathrm{d}t} \begin{pmatrix} \varepsilon_{11} & \varepsilon_{12} & \varepsilon_{13} \\ \varepsilon_{21} & \varepsilon_{22} & \varepsilon_{23} \\ \varepsilon_{31} & \varepsilon_{32} & \varepsilon_{33} \end{pmatrix}$$
$$= \frac{1+\nu}{E} \left[\frac{\mathrm{d}}{\mathrm{d}t} \begin{pmatrix} \sigma_{11} & \sigma_{12} & \sigma_{13} \\ \sigma_{21} & \sigma_{22} & \sigma_{23} \\ \sigma_{31} & \sigma_{32} & \sigma_{33} \end{pmatrix} + \frac{1}{\tau_1} \begin{pmatrix} \sigma_{11} & \sigma_{12} & \sigma_{13} \\ \sigma_{21} & \sigma_{22} & \sigma_{23} \\ \sigma_{31} & \sigma_{32} & \sigma_{33} \end{pmatrix} \right]$$
$$- \left(\mathrm{Tr}([\bar{\sigma}]) + \frac{\mathrm{Tr}([\sigma])}{\tau_2} \right) \frac{\nu}{E} \begin{pmatrix} 1 & 0 & 0 \\ 0 & 1 & 0 \\ 0 & 0 & 1 \end{pmatrix}$$

where θ_λ, θ_μ, τ_1 and τ_2 are coefficients that depend on the viscosity of the material.

A9.9 Generalized Maxwell Model

The stress and strain tensors are separated into their deviators (s_{ij}, e_{ij}) and volume ($\sigma[I]$, $\varepsilon[I]$ hydrostatic) tensors (Section A9.5). The stress deviator is written

$$s_{ij}(t) = \int_0^t \frac{de_{ij}}{dt''} G(t - t')dt'$$

while hydrostatic tensor is written

$$\sigma(t) = \int_0^t \frac{d\varepsilon}{dt'} K(t - t')dt'$$

with

$$\sigma_{ij}(t) = s_{ij}(t) + \sigma\delta_{ij}$$

$$\varepsilon_{ij}(t) = s_{ij}(t) + \varepsilon\delta_{ij}$$

$$\delta_{ij} = 1 \text{ when } i = j, \delta_{ij} = 0 \text{ when } i \neq j$$

$$G(t) = 2G\psi_1(t)$$

$$K(t) = 3K_e - (3K_e - 3K_g)\psi_2(t)$$

where G and K_g are the shear and bulk instantaneous elastic constants and K_e the bulk equilibrium elastic modulus.

Using a generalized Maxwell model, relaxation functions is written

$$\psi_1(t) = \sum_{k=1}^{n} w_{1k}\exp(-t/T_{1k}^R)$$

and

$$\psi_2(t) = \sum_{k=1}^{n} w_{2k}\exp(-t/T_{2k}^R)$$

Average shear relaxation time is given by

$$T_R = \frac{\eta_R}{G} = \sum_{k=1}^{n} w_{1k} T_{1k}^R$$

Hydrostatic relaxation time is often assumed to be ten times the shear relaxation time (Guillemet and Gy, 1996):

$$\psi_2(t) = \psi_1(t/10)$$

When stress is not pure shear or pure hydrostatic pressure, one can always consider any stress state be the sum of both former components. Under uniaxial loading, relaxation times should be multiplied by (Gy et al., 1994; Guillemet and Gy, 1996)

$$\frac{3}{2(1+\nu)} \approx 1.2$$

while for equibiaxial stresses (existing in flat glass) the factor is

$$\frac{3(1-\nu)}{2(1+\nu)} \approx 1.9$$

Appendix 10
Flow and Plasticity in Glass

A10.1
Introduction

Mechanical properties were presented in two separated parts in the main text:

i. Chapter 6 presented elevated temperature behaviour characterized by viscous flow.
ii. Chapters 7 and 8 presented low-temperature behaviour characterized by brittleness and plasticity.

This presentation tends to separate the phenomena. Plasticity and viscous flow might have similarities and Spaepen (1977) proposed a unified perspective, strain rate and temperature controlling the mechanical response (Spaepen, 1977; Steif et al., 1982).

Figure A10.1 shows the shear stress–temperature map for a glass as proposed originally by Frost and Ashby (1982) for crystalline materials. At elevated temperatures and low strain rates, homogeneous flow controls the glass behaviour. Going to lower temperatures and larger strain rates changes homogeneous flow into heterogeneous flow. As regards homogeneous flow a transition from Newtonian to non-Newtonian flow appears once strain rate increases. This is further analysed below.

Spaepen (1977) proposed a general strain–stress relationship as

$$\dot{\gamma} = \Delta f \exp\left(-a\frac{v^*}{v_\mathrm{f}}\right) 2\nu_\mathrm{E} \sinh\left(\frac{\tau\Omega}{2kT}\right) \exp\left(-\frac{\Delta g}{kT}\right)$$

The first term on the right-hand side refers either to the homogeneity ($\Delta f = 1$, deformed volume fraction being 100%) or to the heterogeneity ($\Delta f \rightarrow 0$) of the deformation. The second term follows free-volume theory and takes into account the glass fragility, that is, the viscosity–temperature deviation from the Arrhenius law (Chapter 6). This term is a constant at a given temperature (constant free volume) unless elevated stresses allow for increasing free volume within the material. This in fact may happen in a reduced volume of the deformed specimen, that is, within slip bands (or less viscous bands). This phenomenon can be taken into account by taking $\Delta f \rightarrow 0$ and reduced viscosity in related bands

Glass; Mechanics and Technology. Eric Le Bourhis

ISBN: 978-3-527-31549-9

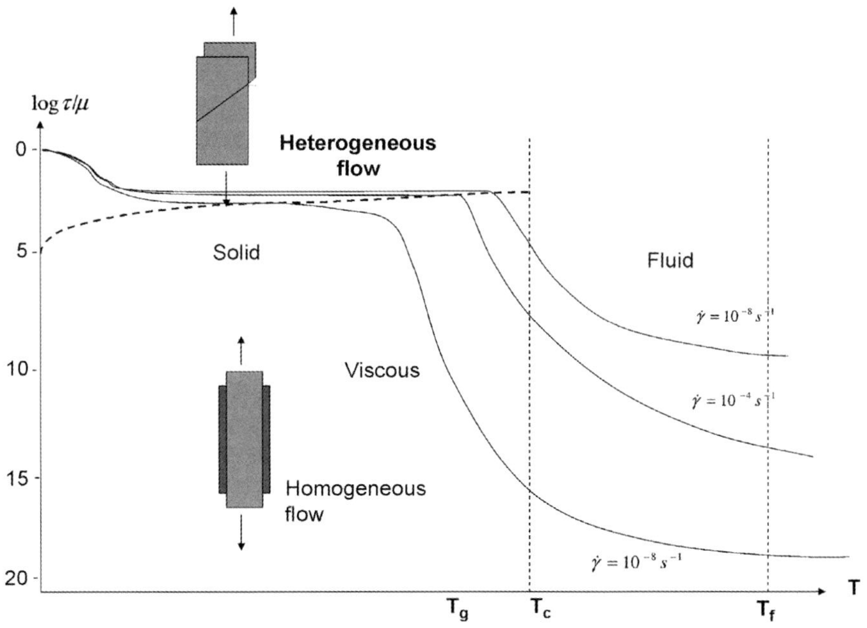

Fig. A10.1 Deformation map. (Reprinted from Spaepen, 1977, with permission from Elsevier.)

(heterogeneous flow). The third term reveals that flow is biased along the stress direction. For low stresses it reduces to

$$\sinh\left(\frac{\tau\Omega}{2kT}\right) \approx \frac{\tau\Omega}{2kT}$$

yielding Newtonian flow (strain rate being proportional to shear stress). The fourth term is the Eyring term (Macedo and Litovitz (1965) hybrid solution; see also Section 6.1.3.4).

A10.2
From Newtonian to Non-Newtonian Flow

Simmons et al. (1988) showed evidences of a transition from Newtonian to non-Newtonian flow when strain rate was increased or temperature lowered. The data plotted in Fig. A10.2 correspond to silico-soda-lime glass. As strain rate increase, apparent viscosity decreases.

The empirical relationship proposed for the apparent viscosity by those authors was

$$\eta = \frac{\eta_o}{1 + \xi\dot{\varepsilon}}$$

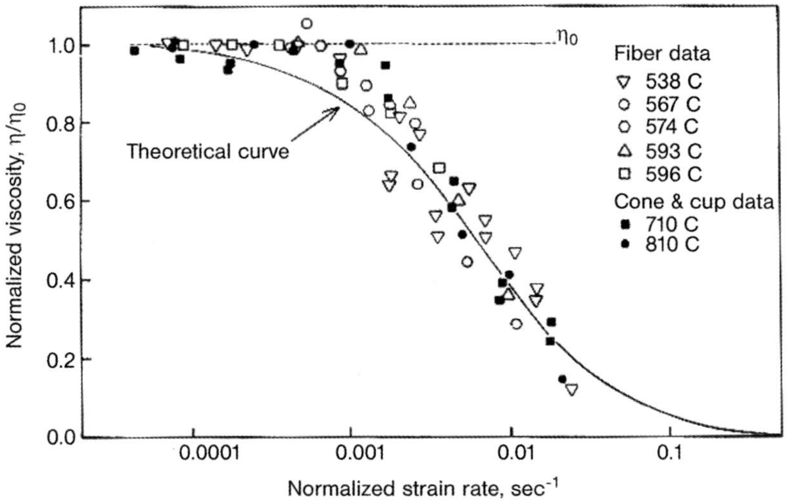

Fig. A10.2 Normalized viscosity versus normalized strain rate. (Reprinted from Simmons et al., 1988, with permission from Elsevier.)

where η_o is the Newtonian shear viscosity and ξ an adjustable parameter. Hence SLS glass shows a reduced apparent viscosity under elevated strain rate. Similar behaviour was reported more recently for metallic glasses in the supercooled domain (Kawamura et al., 1998; Lu et al., 2003). Moreover, homogeneous non-Newtonian flow was observed to yield superplasticity, which has important industrial applications.

At room temperature (below glass transition) non-Newtonian flow has attracted much attention in the past two decades thanks to instrumented indentation tools that allow for strain rate controlled experiments (Appendix 8). Hardness H is related to yield stress Y,[1)] the dependence of which on strain rate may be written in the form (Han and Tomozawa, 1990; Keulen, 1993)

$$Y = H/3 = b\bar{\varepsilon}^m$$

where b is a constant. Indentation strain rate ε is for a sharp indenter (Han and Tomozawa, 1990; Le Bourhis and Patriarche, 2003)

$$\bar{\varepsilon} = \frac{\bar{h}}{h}$$

[1)] We may neglect strain hardening in glass (Appendix 8; $Y_R = Y$). Notably deformation of glass is shown to be elastic–perfectly plastic (Bruck et al., 1994; Xing et al., 1998; Inoue, 2000). In crystalline solids, the motion and mutual interaction of dislocations gives rise to work hardening, that is, increasing flow stress with strain that yields stable plastic flow. Instead, in glasses there are no dislocations and plastic deformation is unstable.

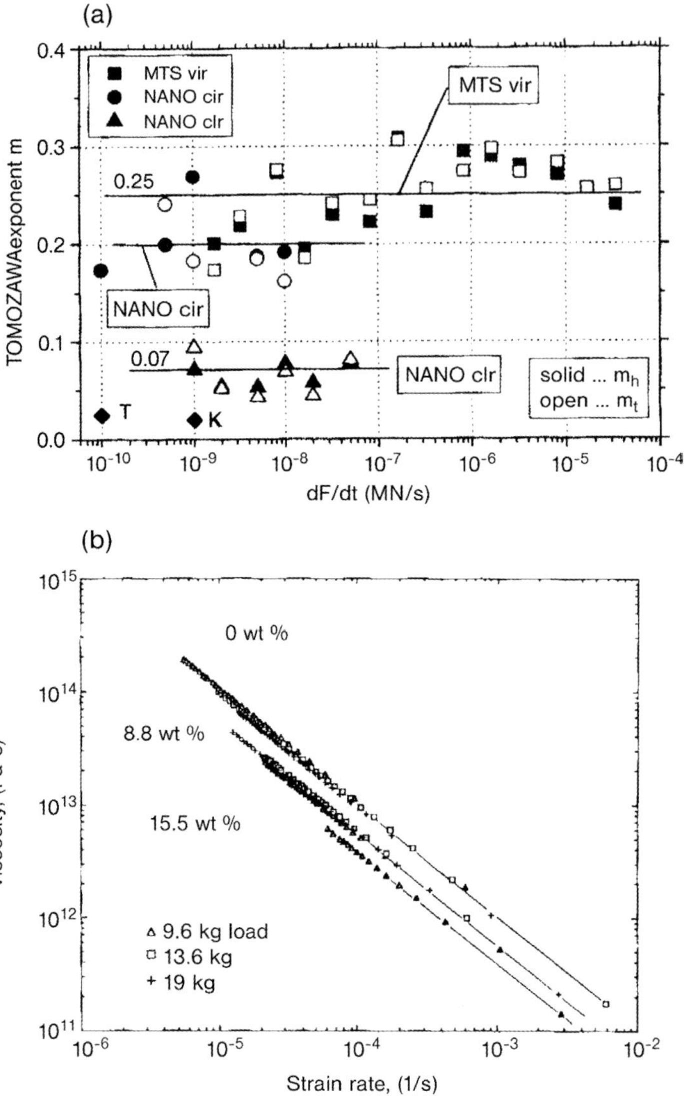

Fig. A10.3 (a) Strain rate exponent. T means Han and Tomozawa (1990), K means Keulen, 1993). (Reprinted from Grau et al., 1998, with permission from Blackwell Publishing.) (b) Viscosity as a function of strain rate. (Reprinted from Han and Tomozawa, 1990, with permission from Blackwell Publishing.)

When $m = 1$ flow is Newtonian, while non-Newtonian flow is obtained for $m \neq 1$. A great deviation from Newtonian flow for soda-lime-silica glass has been observed with m in the range 0.05–0.3 reported in Fig. A10.3 (Han and Tomozawa, 1990; Grau et al., 1998), that is, m is determined to be much below 1. This yields reduced viscous resistance as observed in Fig. A10.3 and suggested by Douglas (1958). Apparent viscosity can be extracted from indentation data dividing stress and strain rate (Chapter 6). Hence viscosity is written

$$\eta = c\bar{\varepsilon}^{m-1}$$

where c is a constant. For $m = 1$, viscosity is independent of strain rate. The reported experimental viscosity values are shown in Fig. A10.3 as a function of strain rate showing instead a dramatic drop of the apparent viscosity as the strain rate increases. The reported changes in apparent viscosity explain in part the large discrepancy reported for glass viscosity at room temperature in the literature as extensively discussed in Chapter 6.

Also, Han and Tomozawa (1990) insist on the influence of glass surface hydration. In fact, water is known to enter a glass surface and change its superficial property (Chapter 5, Appendix 13). Both molecular diffusion and chemical reaction with the glass network happen. Of much importance also is the intrinsic structure of the glass at such a temperature that is not at equilibrium and dependent on the thermal history (Chapter 6).

A10.3
From Homogeneous to Heterogeneous Flow

At temperatures below the glass transition, non-Newtonian flow is generally also associated with heterogeneous flow. Figures A10.4 and A10.5 show such slip bands in indented and uniaxially compressed glass specimens. The nature of these slip bands has been the subject of controversy and still attracts much scientific interest. Hagan (1979) showed that confined plasticity at an indent tip in silico-soda-lime glass yielded slip lines that could be observed on sectioning the sample through the indent site. It has been suggested that such slip bands could correspond to Na-rich zones allowing for enhanced flow in soda-lime-silica glass. As discussed in Chapter 9, Na ion diffusion is known to be enhanced under elevated stresses as revealed by Nghiem (1998) at crack tip vicinity. Interestingly, in metallic glasses, flow can be induced in large volumes and the resulting slip lines obtained under uniaxial compression look similar to those observed at the sides of crystalline samples deformed under compression (Pampillo and Chen, 1974; Fig. A10.5).

More recently, Schuh et al. (2004) using instrumented indentation showed that metallic glass loading response was dependent on strain rate with emphasized discontinuities in the curves observed under low strain rate. In fact, serrated flow has been reported extensively for many metallic glasses.[2)] Such serrations are observed in

[2)] Metallic glasses have attracted much attention since they can be deformed as bulk over a wide range of temperature including temperatures below T_g (Hufnagel, 1996) and have become available as bulk, the most well known being Vitreloy developed by Johnson's team at Caltech (Johnson, 2006). This allowed for great progress in the understanding of plasticity and flow of glass.

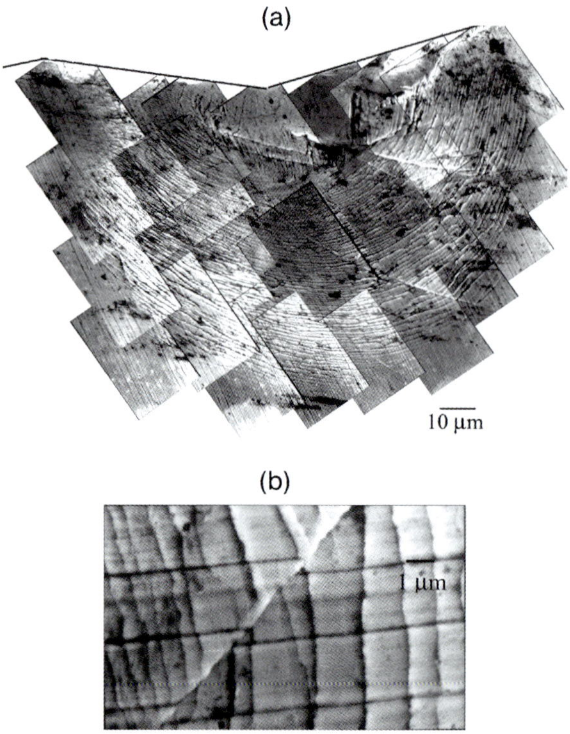

Fig. A10.4 Heterogeneous flow in indented metallic glass. (Reprinted from Ramamurty et al., 2005, with permission from Elsevier.) (b) The shear band morphology underneath the indenter shown in a higher magnification image obtained from within the deformed region of (a).

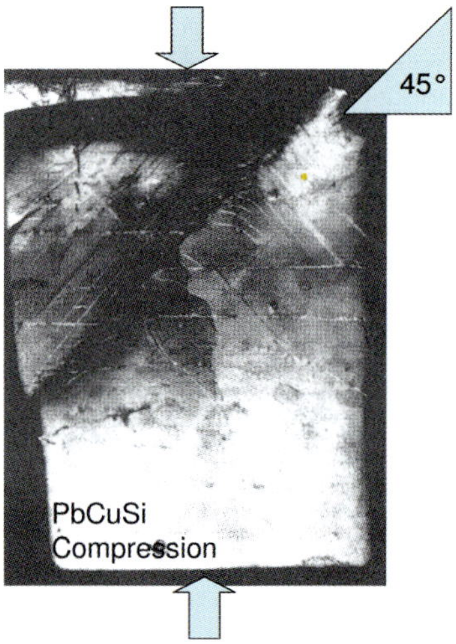

Fig. A10.5 Heterogeneous flow in uniaxially compressed metallic glasses. (Reprinted from Pampillo and Chen, 1974, with permission from Elsevier.)

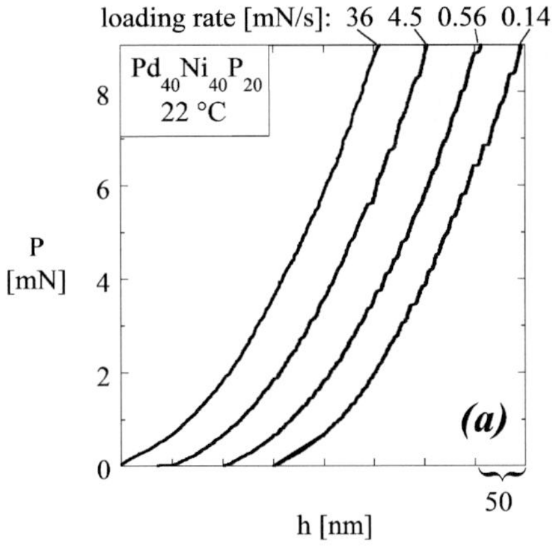

Fig. A10.6 Serration on loading curves of metallic glasses. (Reprinted from Schuh et al., 2004, with permission from Elsevier.)

Fig. A10.6 as small jumps on the curve. The correlation between serration and slip lines is however still debated.

All these observations led several authors to consider the dislocation concept in amorphous solids (G'Sell, 2001). Notably, phenomenologically plasticity and heterogeneous non-Newtonian flow are similar and both characterized by localized flow. Spaepen (1977) suggested that increased free volume (lower viscosity) in localized slip bands ($\Delta f \rightarrow 0$) was responsible for this behaviour in glass. Figure A10.7 shows how under elevated stress an atom forces its path even though initially the free volume at its surrounding is too small for a jump (see also Chapter 6). Supposing such a phenomenon can happen implies that the free volume increases under elevated stresses. This is suspected to happen within slip bands where free volume increase leads to viscosity decrease. Experiments could be conducted on metallic glass bulk under compression showing localized bands along the maximum shear stress direction (45°, in fact the angle is slightly less than 45° in compression (~42°)

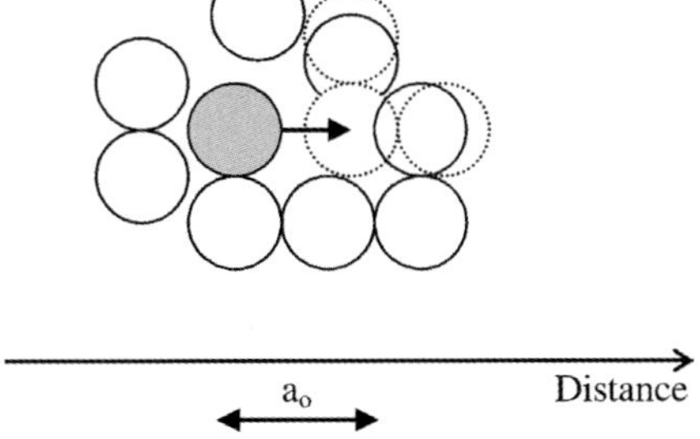

Fig. A10.7 Free volume creation.

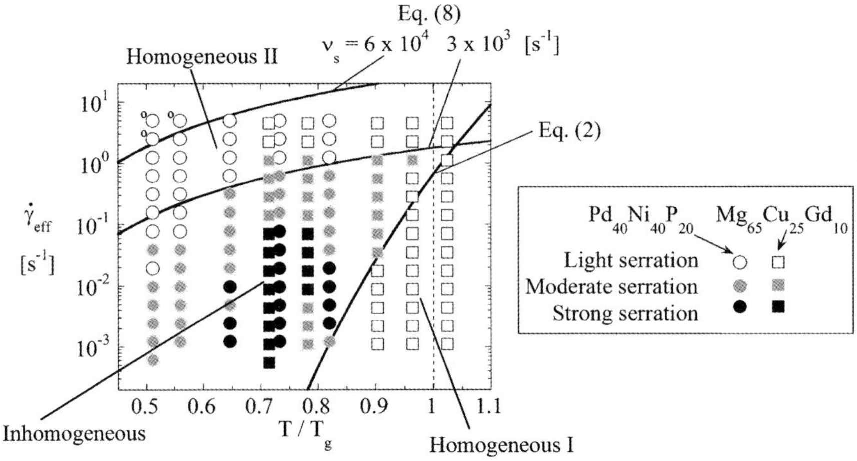

Fig. A10.8 Transition from homogeneous Newtonian to non-Newtonian heterogeneous and non-Newtonian homogeneous flow. (Reprinted from Schuh et al., 2004, with permission from Elsevier.)

and slightly above in tension (~50–57°); Fig. A10.5). It is of interest to mention that Pampillo and Chen (1974) showed preferential etching of these slip bands. More recently, slip bands were revealed under lighter contrast in TEM micrographs (Kim et al., 2002; Chang et al., 2006).

Based on serration events, Schuh et al. (2004) showed a new transition from heterogeneous non-Newtonian to homogeneous non-Newtonian flow in metallic glasses using indentation tools under different temperatures (Fig. A10.8). Those authors showed such transitions at very elevated loading strain rates for which only light serrations were observed similar to those observed for very small strain rates. Hence, they suggested that the inhomogeneous deformation domain was surrounded by homogeneous deformation domains; one at small strain rates, the other at large strain rates with frontiers depending on temperature. The question whether at high strain rates flow is homogeneous or inhomogeneous with finer slip bands and serration is still under debate (Jiang and Atzmon, 2003).

Appendix 11
Finite Element Analysis

Finite element analysis (FEA) or finite element modelling (FEM) decomposes a solid into a mesh made of discrete elements. For each element, the relevant parameters are supposed to be known (physical, chemical, mechanical properties) or locally determined (stress, temperature). Then after each increment of time the overall behaviour of the object can be determined. A step-by-step procedure is used to determine the behaviour of a glass object with time. Notably, when the shape varies the mesh has to be redefined continuously. This technique has become widely used since it allows complex loading and geometries to be considered that make it very attractive for industry and R&D.

A11.1
FEM of the Pressing of a Parison

We consider first the pressing of a parison (Chapter 10; press-blow and press-press techniques). The parison is under a stress gradient and a temperature gradient. Also at elevated temperatures the influence of gravity on glass flow comes into play. Figure A11.1 shows the mesh that is to be defined again after each penetration increment. Pressure field shows a gradient with maxima of intensity at the penetrator proximity. Such a pressure field in the glass can be calculated under realistic conditions considering the outside pressure from the mould. The numerical tool can be iterated to adjust viscosity regime, mould geometry and hence achieve optimized glass distribution over the articles. This issue is very important in order to reduce container weight.

A11.2
FEM of the Precision Moulding of a Glass Lens

The second process example considers high-precision lens moulding (Fig. A11.2). Here, great precision in the dimensional calculation is required. The glass blank is moulded between two parallel moulds (upper and lower ones) under pressure. Both

Glass; Mechanics and Technology. Eric Le Bourhis

ISBN: 978-3-527-31549-9

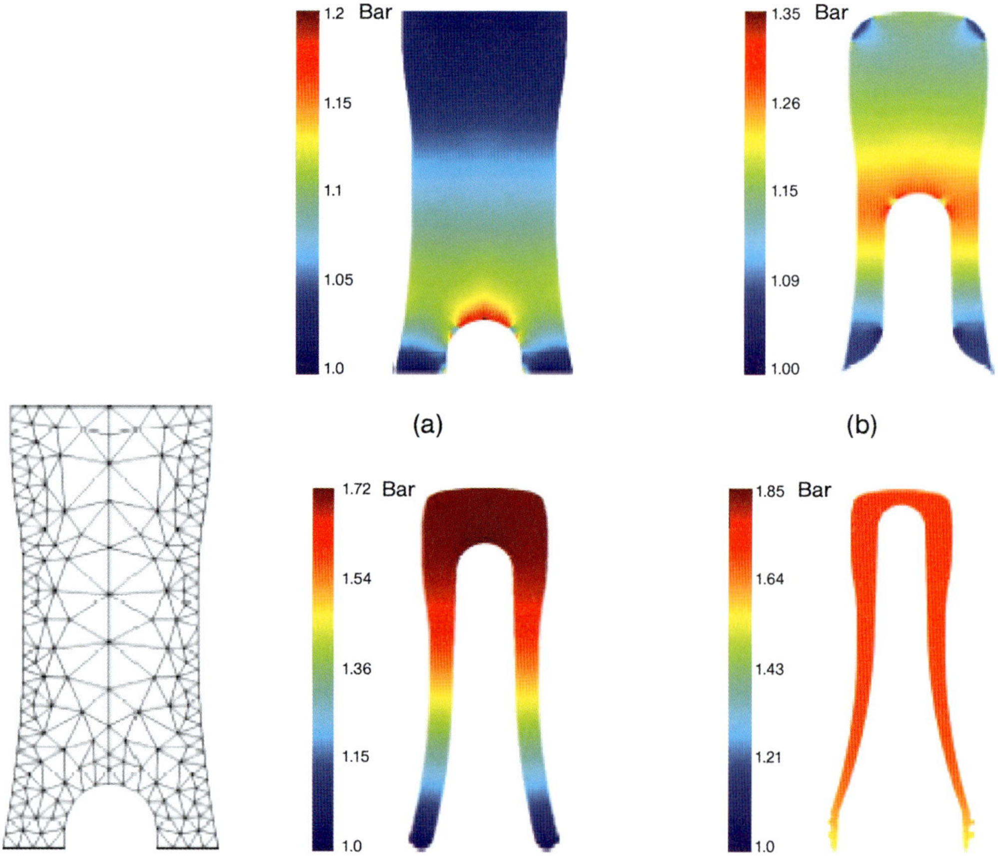

Fig. A11.1 FEM modelling of the pressing of a glass parison; mesh and pressure field (Laevsky, 2003).

moulds are curved to achieve the final lenses shape. The blank flows progressively while stress and stress gradients increase. Stress gradient is more intense at mould folding where strain gradient is expected as well. One should consider friction between the tools and the lens, which FEM manages. Also, FEM allows adjustment of mould geometric characteristics and tool loading carrying out numerical iterations. In order to improve the prediction precision, Jain et al. (2005) recommend determining the viscosity regime under similar compressive conditions and using these inputs so that the FEM will achieve even more precise predictions of the moulding operation.

A11.3 FEM of Fracture

Moving to the mechanical response at fracture, numerical simulations are of great interest since fracture mechanics requires statistics to be considered. Modelling

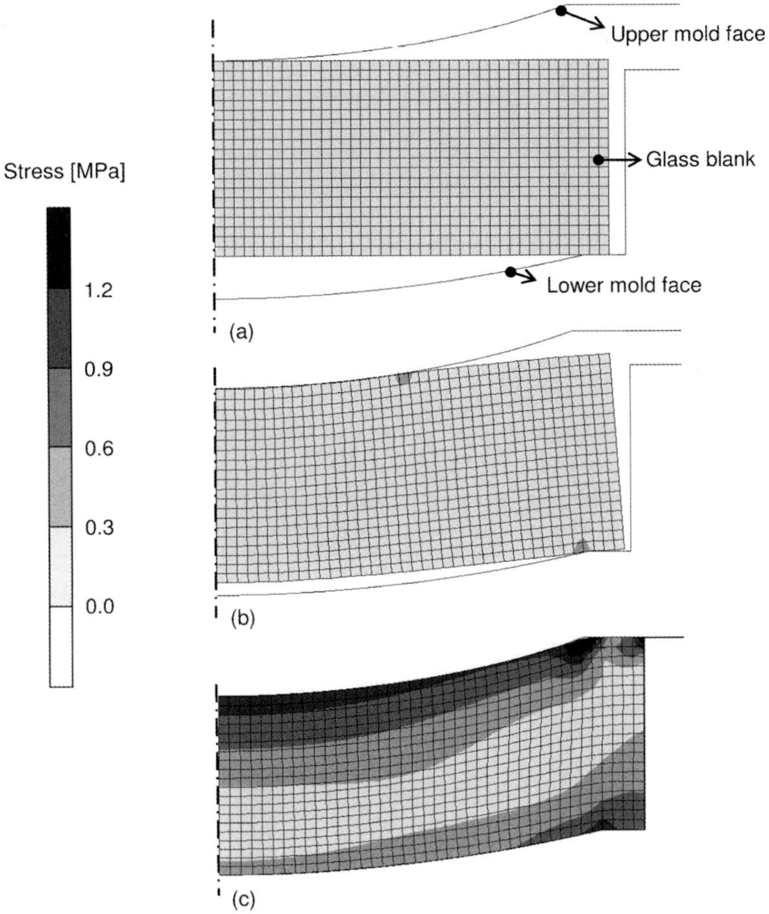

Fig. A11.2 FEM modelling of the compression moulding of a BK7 glass lens at a forming temperature of 671 °C and with a transition temperature of 557 °C. Predicted stresses are shown. (a) Initial glass blank and mould set-up; (b) compression of the glass blank; (c) final lens shape. (Reprinted from Jain et al., 2005, with permission from Blackwell Publishing.)

saves expensive and time-consuming experiments. Also, glazing may involve complex assemblies of several sheets of glass having different strengthening treatments and related interfaces. FEM yields helpful and valuable information in this domain. The study of the resistance of a laminated glass presented in Fig. A11.3 is difficult because of interfaces and sliding. Both glass sheets and polyvinylbutyral (PVB) bond have to be taken into account. Moreover, an impact is a dynamic loading and thanks to FEM great insight into glazing crack propagation is obtained (Du Bois et al., 2003). The dynamic impact of a ball (diameter 300 mm, mass 70 kg and initial velocity 10 m s^{-1}) on a laminated safety glass (length 1500 mm, thickness 5 mm) is shown in Fig. A11.3, the

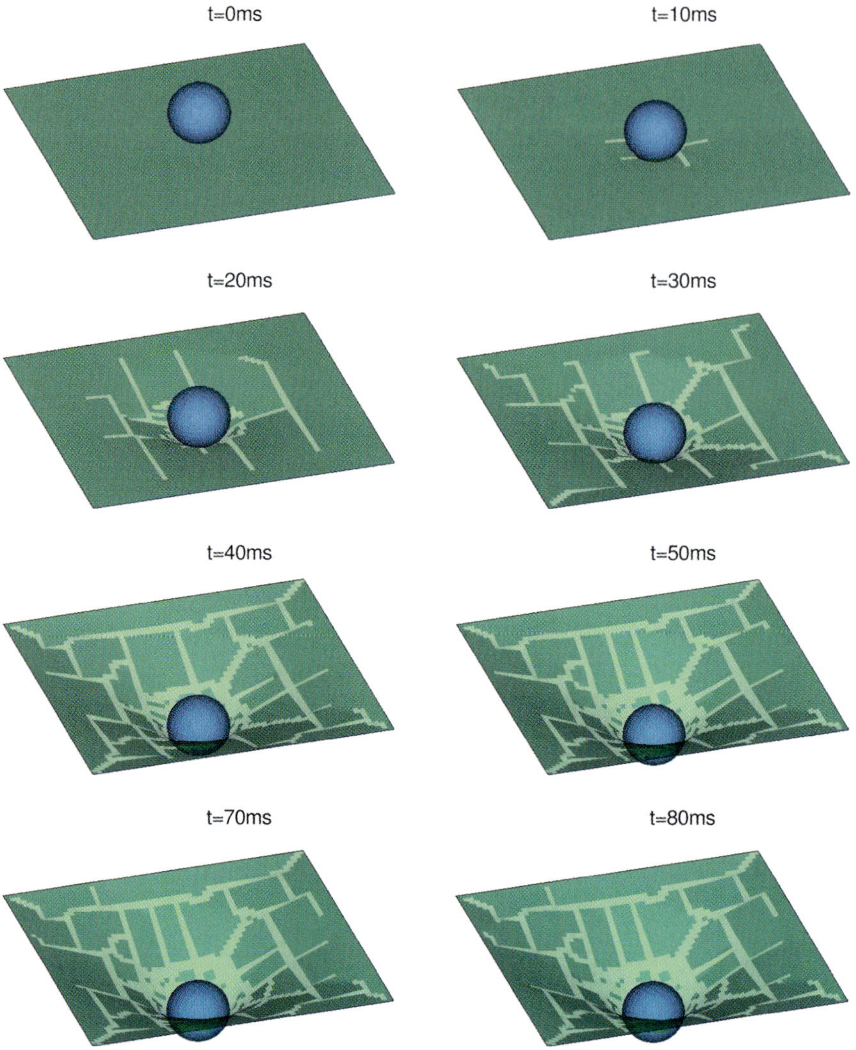

Fig. A11.3 Temporal evolution of an impact on a laminated safety glass. (Reprinted from Du Bois et al., 2003, with permission from Elsevier.)

resistance with time being observed. Du Bois et al. 2003 show very a realistic fracture pattern of the laminated glass, and indicate that the fracture pattern is sensitive to mesh refinement and regularity. The authors assume a failure strain at 0.1 % for the glass and no failure for the PVB interlayer. At 10 ms, the first elements fail and cracks start to propagate in different directions (bright areas are failed elements where only the PVB interlayer carries the load). Between 30 and 40 ms, some larger glass fragments are formed, which are still joined by the membrane. This can be observed until the end of the impact at 80 ms. These studies are also helpful in understanding how fracture propagates. Experimentally, high-speed photography is required (Chapters 7 and 8). Instead, FEM can be carried out at the appropriate time scale.

A11.4
FEM of Contact Loading

As detailed extensively in Chapter 8, glass response to contact loading is complex, in particular once plasticity and fracture are operating. Stress and strain fields are

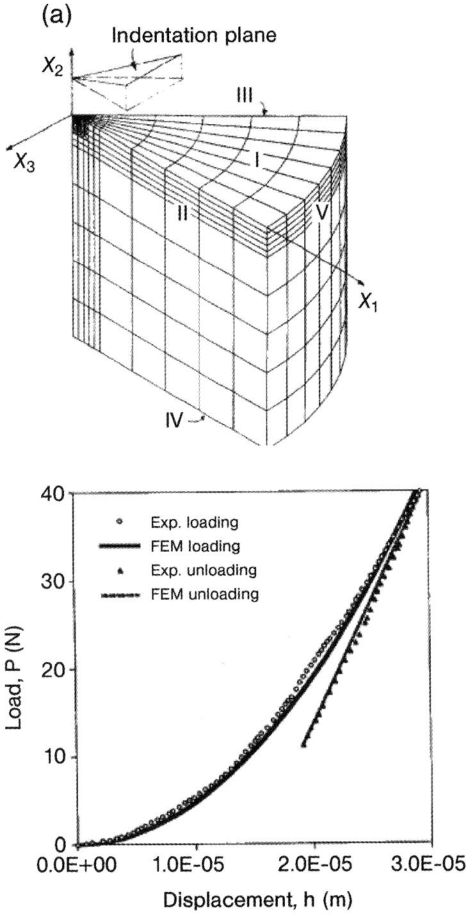

Fig. A11.4 FEM mesh used in the three-dimensional analysis of the Vickers indentation experiment by Zeng et al. (1995). (a) General view. (b) Detail of the mesh at the contact region (8524 eight-noded elements, 9914 nodes). (c) Loading–unloading curves of Vickers indentation on soda-lime glass obtained from the FEM as compared to experimental data reported by Cook and Pharr (1990). $H = 5.9$ GPa, $Y = 2.8$ GPa, $E = 74.3$ GPa. (Reprinted from Zeng et al., 1995, with permission from Elsevier.)

complex (see also Appendix 8). Again FEM is shown to be valuable and allows one to get great insight into glass response. As shown in Fig. A11.4, contact mechanics FEM is shown to account well for plasticity. Zeng et al. (1995) used a mesh with higher resolution at the indenter proximity, where plasticity is important and where stress and strain fields change rapidly. Using elastic and plastic behaviour inputs, they were able to predict quite well the experimental curves obtained for glass by Cook and Pharr (1990). The next step is to take into account fracture and have a predictable model for contact fracture.

Appendix 12
X-Ray Diffraction Analysis of Thin-Film Residual Stresses

A12.1
Thin-Film Stress and Strain

Let us consider a film under simple equibiaxial stresses. The stress tensor is (Appendix 9)

$$\begin{pmatrix} \sigma & 0 & 0 \\ 0 & \sigma & 0 \\ 0 & 0 & 0 \end{pmatrix}$$

and the strain tensor is

$$\begin{pmatrix} \varepsilon_{\|} & 0 & 0 \\ 0 & \varepsilon_{\|} & 0 \\ 0 & 0 & \varepsilon_{\perp} \end{pmatrix}$$

$\varepsilon_{\|}$ and $\varepsilon_{\perp}$ being the in-plane and out-of-plane strains respectively.

Most deposited FCC metallic thin films show two particular orientations or textures: either (0 0 1) or (1 1 1). Crystallites show a <0 0 1> or <1 1 1> preferential direction along the growth direction (surface normal). In such cases, stress–strain relationships are written

$$\sigma = M_{hkl}\varepsilon_{\|}$$

and

$$\eta_{hkl} = -\varepsilon_{\perp}/\varepsilon_{\|}$$

M_{hkl} is defined as the biaxial modulus and η_{hkl} the ratio of out-of-plane to in-plane strains depending on crystallographic direction $[hkl]$. M_{hkl} and η_{hkl} are listed for the two respective films texture in Table A12.1

We define now anisotropy ratio A of cubic crystals as

$$A = \frac{2C_{44}}{C_{11} - C_{12}}$$

Glass; Mechanics and Technology. Eric Le Bourhis

ISBN: 978-3-527-31549-9

Tab. A12.1 Biaxial modulus M_{hkl} and the ratio of out-of-plane to in-plane strains η_{hkl} for the two most common textures of thin FCC films

	(001)	(111)
M_{hkl}	$\dfrac{(MC_{11} - C_{12})(C_{11} + 2C_{12})}{C_{11}}$	$\dfrac{6C_{44}(C_{11} + 2C_{12})}{C_{11} + 2C_{12} + 4C_{44}}$
η_{hkl}	$2\dfrac{C_{12}}{C_{11}}$	$2\dfrac{C_{11} + 2C_{12} - 2C_{44}}{C_{11} + 2C_{12} + 4C_{44}}$

Most metals (Table A12.2) show elastic anisotropy. Note that a thin film is polycrystalline, that is, formed of monocrystalline grains, the size and orientation of which depend on deposition conditions. They will exhibit elastic macroscopic anisotropy only when texture is present, that is, when crystallites show a preferential orientation (001 or 111 for instance). Tungsten is an archetype since the crystallites are elastically isotropic. Hence tungsten films show microscopic and macroscopic elastic isotropy. For $A = 1$ the elastic properties are completely defined by two elastic constants E and ν and we have

$$M = E/(1 - \nu)$$

$$\eta = 2\nu/(1 - \nu)$$

Note that for $\nu = 0.33$, we have $\eta = 1$, that is, out-of-plane strain is exactly opposite of the in-plane strain.

Tab. A12.2 Biaxial modulus M_{hkl} and the ratio of out-of-plane to in-plane strains η_{hkl} for various elements

Element	C_{11} (GPa)	C_{12} (GPa)	C_{44} (GPa)	A	M_{001} (GPa)	η_{001}	M_{111} (GPa)	η_{111}
Al	108	61	28	1.2	100	1.1	114	1
V	228	119	42	0.8	222	1	187	1.2
Nb	246	134	28	0.5	234	1.1	140	1.5
Ta	267	161	82	1.6	233	1.2	317	0.9
Cr	350	57	101	0.7	388	0.3	324	0.6
Mo	460	176	110	0.8	501	0.8	428	0.9
W	501	198	151	1	542	0.8	542	0.8
Fe	238	133	110	2.1	222	1.1	352	0.6
Ni	246	147	124	2.5	217	1.2	389	0.6
Pd	227	176	71	2.8	130	1.5	287	1
Pt	346	250	76	1.6	234	1.4	337	1.2
Cu	168	121	75	3.2	114	1.4	261	0.7
Ag	124	93	46	3	76	1.5	173	0.9
Au	186	157	42	2.9	78	1.7	188	1.2

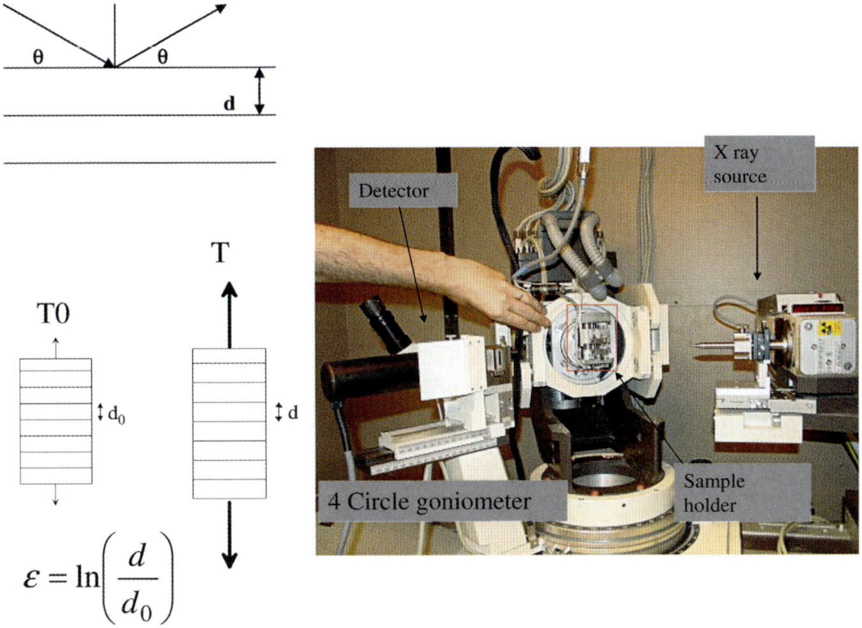

Fig. A12.1 X-ray diffraction principle and experimental set-up (four-circle goniometer) with, from right to left, the X-ray source, the sample holder and the detector.

A12.2 X-Ray Diffraction Method

X-ray diffraction (XRD) allows for the precise measurement of diffraction peak positions from which we obtain the inter-planar spacing d according to Bragg's law:

$$2d \sin\theta = k\lambda$$

where λ is the wavelength of the radiation and θ the incident angle. Calling d_o the strain-free spacing one can derive the corresponding strain using (Fig. A12.1)

$$\varepsilon = (d - d_o)/d_o$$

or in its rational form (this form is much helpful in the XRD frame; see Appendix 9)

$$\varepsilon = \ln(d/d_o)$$

Once in-plane $d_{||}$ and out-of-plane $d_{\perp}$ lattice parameters can be measured, one can determine the stress- and strain-free parameter using the biaxial modulus M, and the ratio of out-of-plane to in-plane strains η since we have

$$\sigma = M \frac{d_{\|} - d_o}{d_o}$$

and

$$d_o = \frac{d_{\perp} + \eta d_{\|}}{1 + \eta}$$

Stress-free parameter d_o informs on the structure of films, in particular the presence of impurities or vacancies that yield a different value of the stress-free parameter as compared to that of the bulk reference. Residual stress amplitude σ is of utmost importance for film mechanical stability as detailed in Chapter 12.

A12.3 The ε–$\sin^2\psi$ Method

The X-ray diffraction method can be generalized in the so-called ε–$\sin^2\psi$ method (Clemens and Bain, 1992; Hauk, 1997; Faurie et al., 2004; Thomas et al., 2004). It encompasses the strain ε measurement as a function of angle ψ defined as the angle between surface normal and diffracting plane normal. A series of lattice spacing measurements is carried out for different orientations with respect to the surface normal (Fig. A12.2).

For an equibiaxial stress state we have

$$\varepsilon_{\psi} = \varepsilon_{\|} \sin^2\psi + \varepsilon_{\perp} \cos^2\psi$$

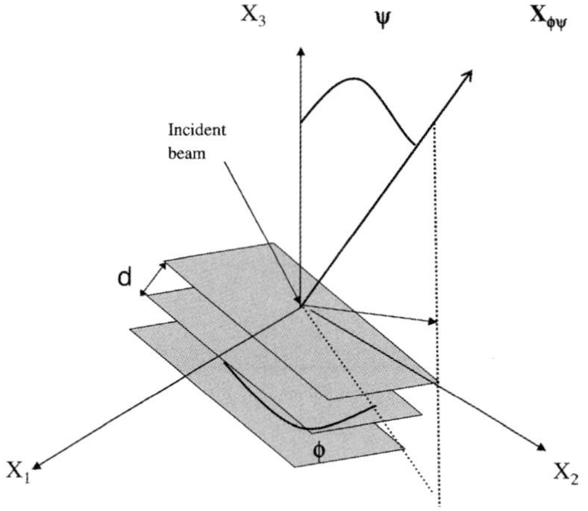

Fig. A12.2 X-ray diffraction method.

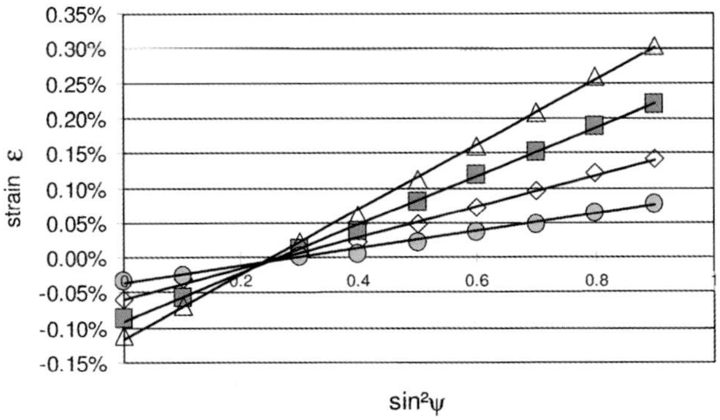

Fig. A12.3 Strain ε versus $\sin^2\psi$ for a polycrystalline tungsten film. Circles: as-deposited; diamonds, squares and triangles: increasing loading states (the specimens being tensile stressed). As stress state changes, the slope of the line, that is, stress varies. (Reprinted from Faurie et al., 2004, with permission from Elsevier.)

that shows only ψ angle dependence (shown in Fig. A12.2, independent of angle ϕ) and which can be rewritten

$$\varepsilon_\psi = \frac{1+\eta}{M}\sigma\sin^2\psi - \frac{\eta}{M}\sigma$$

This latter equation shows that plotting ε_ψ–$\sin^2\psi$ one should obtain a straight line the slope of which yields the stresses in the film (Faurie et al., 2004; Fig. A12.3).

For a particular direction ψ_o strain is zero as observed in Fig. A12.3. The zero strain relationship is written

$$\sin^2\psi_o = \eta/(1+\eta)$$

that for an isotropic material like tungsten is

$$\sin^2\psi_o = 2\nu/(1+\nu)$$

allowing for Poisson ratio to be determined.

Appendix 13
Diffusion

Diffusion plays an important role in glass science and technology since it governs electrical conductivity, corrosion resistance, container permeability, ion exchange used for inducing index gradient or residual stresses, and fining of glass. The diffusion of gas out of a bubble can control its rate of contraction and is then of utmost importance in bubble removal from glass melt. The rate of shrinkage can be determined from diffusion equations. Diffusion controls the atomic movement and hence the melt homogenization.

A13.1
Diffusion Laws

Diffusion is a time-dependent process. Without external force (electrical field, temperature gradient, stress) the diffusion flux J_x along the x direction is proportional to concentration gradient along this direction (we suppose concentration is only a function of x and t):

$$J_x = -D\frac{\partial c}{\partial x}$$

The constant of proportionality D is called the diffusion coefficient which is expressed in $m^2\,s^{-1}$, c is the concentration in m^{-3} and hence J_x is expressed in $m^{-2}\,s^{-1}$. The negative sign in this expression indicates that the direction of diffusion is opposite to the concentration gradient. This means that diffusion happens from a high- to a low-concentration region. This relationship between concentration gradient and flux is called Fick's first law and is formally identical to the Fourier law that relates thermal flux to temperature gradient.

Consider now matter conservation in an infinitesimal volume of size dx (Fig. A13.1). We derive

$$\frac{\partial c}{\partial t}\mathrm{d}x\mathrm{d}t = (J_x(x) - J_x(x + \mathrm{d}x))\mathrm{d}t = -\frac{\partial J_x}{\partial x}\mathrm{d}x\mathrm{d}t = -\frac{\partial}{\partial x}\left(-D\frac{\partial c}{\partial x}\right)\mathrm{d}x\mathrm{d}t$$

Glass; Mechanics and Technology. Eric Le Bourhis

ISBN: 978-3-527-31549-9

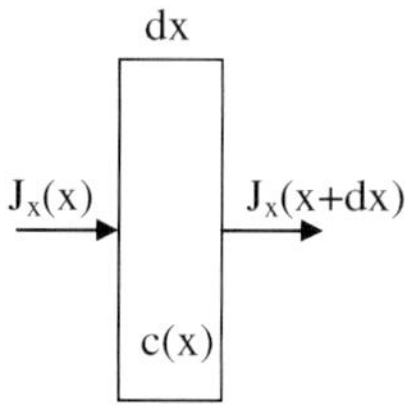

Fig. A13.1 Elemental volume subjected to diffusion along *x*-axis.

$$\frac{\partial c}{\partial t} = \frac{\partial}{\partial x}\left(D\frac{\partial c}{\partial x}\right)$$

When D is independent of concentration, Fick's second law is derived:

$$\frac{\partial c}{\partial t} = D\frac{\partial^2 c}{\partial^2 x}$$

This equation is to be solved to extract concentration as a function of time and position. The solution to extract $c(x, t)$ is complex and we shall review in Section A13.3 two important solutions, namely for constant surface concentration and for instantaneous plan source.

Diffusion coefficient D increases with temperature and follows generally an Arrhenius-type law:

$$D = D_o\exp\left(-\frac{\Delta H}{RT}\right)$$

where ΔH is the energy barrier and D_o the pre-exponential factor. Let us consider first the case of neutral molecule diffusion (rare gas) where the molecules do not interact with the glass network (Fig. A13.2). The molecules diffuse through the network excess volume. Values of D and ΔH are given in Table A13.1 and show that higher diffusion of atomic species is obtained for smaller species. The pre-exponential factor is often taken as a constant in a given range of temperatures. Studies over a wide range of temperatures in particular in silica showed that this factor changes with temperature almost linearly as expected from Fig. A13.2 that plots D/T as a function of reciprocal temperature (Doremus, 1994).

Diffusion of water, hydrogen and oxygen is complicated by reactions of these gases with the silica lattice. Indeed corrosion of glass is known to result from the reaction of water with silica (with formation of –OH groups; Chapters 5 and 9). Therefore, molecular diffusion of such species is to be measured and considered carefully. A term has to be added to the diffusion equation:

$$\frac{\partial c}{\partial t} = D\frac{\partial^2 c}{\partial^2 x} - \frac{\partial r}{\partial t}$$

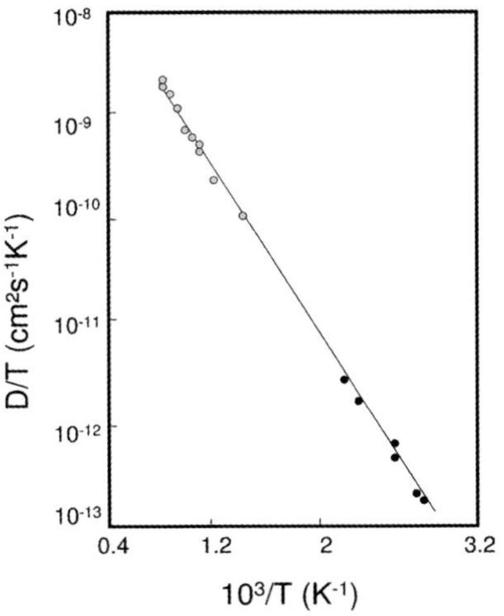

Fig. A13.2 Diffusion of neon in fused silica (open points from Frank et al. (1961); dark points from Perkins and Begeal (1971)).

where r is the concentration of reacted or immobilized species. As shown again in Table A13.1, molecular diffusion of a species is observed to depend on its size.

When the diffusing species appear also in the glass network, one expects network diffusion to happen in addition to molecular diffusion and viscosity should be closely linked to diffusion when network diffusion is predominant. For instance viscous activation energy for silica is about 580 kJ mol^{-1} and compares favourably with that measured for Si diffusion in silica (about 700 kJ mol^{-1}; Table A13.2). Moreover,

Tab. A13.1 Molecular diffusion coefficients in silica.

Species	Diameter (nm)	D (cm^2 s^{-1}) at 25 °C	D (cm^2 s^{-1}) at 1000 °C	ΔH (kJ mol^{-1})	Authors
He	0.2	2.4×10^{-8}	5.5×10^{-5}	1.1	Swets et al. (1961)
Ne	0.24	5×10^{-12}	2.5×10^{-6}	2.1	Frank et al. (1961); Perkins and Begeal (1971)
Ar	0.32		1.4×10^{-9}	0.6	Perkins and Begeal (1971); Nakayama and Shackelford (1990)
H_2	0.25	2.2×10^{-11}	7.3×10^{-6}	2.0	Lee (1963); Perkins and Begeal (1971)
O_2	0.32		6.6×10^{-9}	6.0	Norton (1961)
H_2O	0.33		2×10^{-7}	4.0	Shelby (1961); Doremus (1969)

Tab. A13.2 Diffusion coefficients.

	D_o ($m^2\,s^{-1}$)	ΔH (kJ mol^{-1})	Range of temperature (°C)	Authors
Na^+ in silica		90–125	200–1000	Frischat (1968)
O^{2-} in silica	2.6×10^{-4}	453	1200–1400	Mickelsen (1984)
Si^{4+} in silica	328×10^{-4}	579	1100–1400	Brebec et al. (1980)

applying Stokes' law ($k = 6\pi\eta r$) and calling r the radius of the ion, one gets the Stokes–Einstein law:

$$D = \frac{kT}{6\pi r\eta}$$

This allows determination of a $\ln(D)$ versus $1/T$ plot shown as a continuous line in Fig. A13.3. It applies well for O and Si that form the glass network. Below the transition temperature diffusion mechanisms are still debated.

The variation of D is shown for alkali metal ions (Na^+) in Fig. A13.4 and shows a good fit with a constant pre-exponential D_o. Interestingly when the concentration of ions increases the diffusion rate increases as well. This is attributed to more lattice network discontinuities that facilitate ion transport. Alkali diffusion is important to consider since electrical conduction of most commercial glasses results from sodium ions. While in many applications glass serves as an electrical insulator, ionic transport

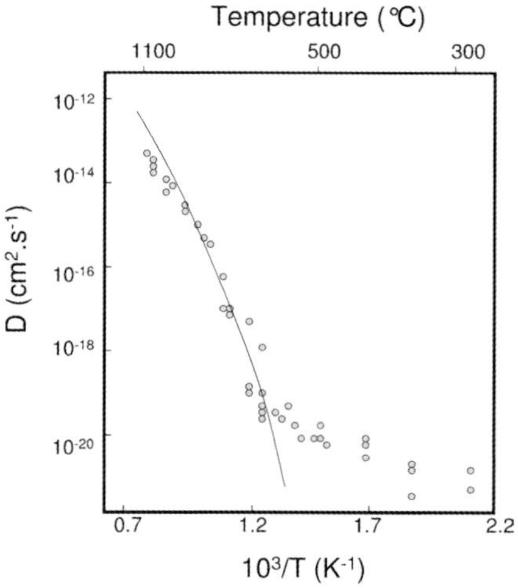

Fig. A13.3 Diffusion coefficient D for oxygen in SLS glass as a function of temperature (according to Yinnon and Cooper, (1980)). We observe two regimes with $Q = 57.8$ and 208 kJ mol^{-1} below and above T_g respectively. Curvature is observed above T_g as suggested by the Stokes–Einstein law. Note the change in behaviour at T_g.

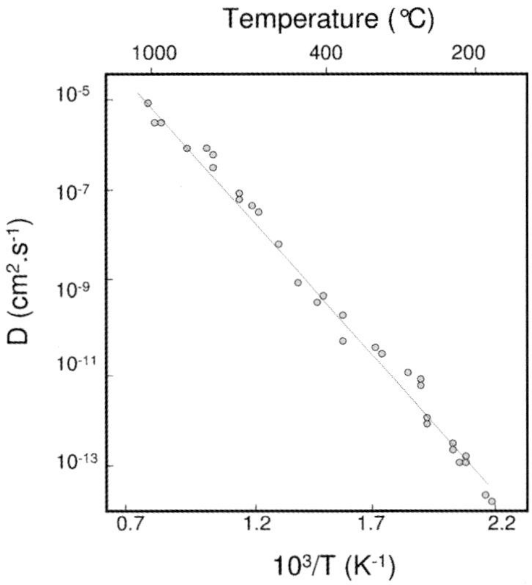

Fig. A13.4 Diffusion coefficient D for Na^+ in silica as a function of temperature. The line is a guide for the eye. (After Frischat, 1968).

and hence conduction plays an important role in the manufacture of glass. Ionic conduction can be approximately estimated from Einstein's law

$$\sigma = \frac{Z^2 F^2 Dc}{RT}$$

where Z is the ionic charge, F the Faraday $= Ne$, N: Avogadro number, D the diffusion coefficient, c the concentration, R the gas constant and T the temperature.

A still debated issue is the so-called mixed alkali effect that is one of the classical anomalies of glass science and has important technological consequences. Curiously when a glass contains two alkalis, for instance Na^+ and K^+, diffusion is faster for the one in larger concentration. When chemically tempering a glass, this effect can be used to increase diffusion.

A13.2 Steady-State Diffusion

If diffusion flux does not change with time, a steady-state condition exists. One common example is the diffusion of atoms of a gas through a vessel or tube for which the concentrations (or pressures) of diffusing species on both surfaces (external and internal) are held constant. This is represented in Fig. A13.5. Then

$$J = D\frac{c_i - c_e}{t_H}$$

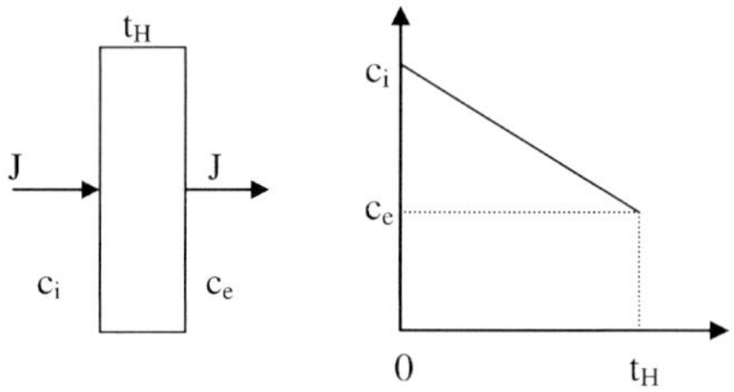

Fig. A13.5 Steady-state diffusion.

This simple relationship is applied to determine the diffusion coefficient (so-called membrane technique) or the lost volume of gas when a tube is used to transport gaseous reactive species. Consider a tube (Fig. A13.6) of length L, radius r and thickness t_H ($\ll r$). The lost volume of gas is $V = V_m J \times 2\pi r L = V_m D(c_i - c_e) \times 2\pi r L / t_H$, where V_m is the molar volume ($22.4\,\mathrm{l\,mol^{-1}}$ in standard conditions).

Also, air diffusion is to be prevented in electrical lamp envelopes and television tubes. In practice the concentration gradient is unknown in the glass so one has to use Henry's law that relates internal and external pressures and concentrations:

$$c = SP$$

where S is the solubility. Alternatively permeability $K = DS$ is employed. Combining the above equations yields

$$J = DS\frac{P_i - P_e}{t_H} = K\frac{P_i - P_e}{t_H}$$

When a vacuum is maintained inside the bulb or tube we have

$$-K = \frac{t_H J}{P_e}$$

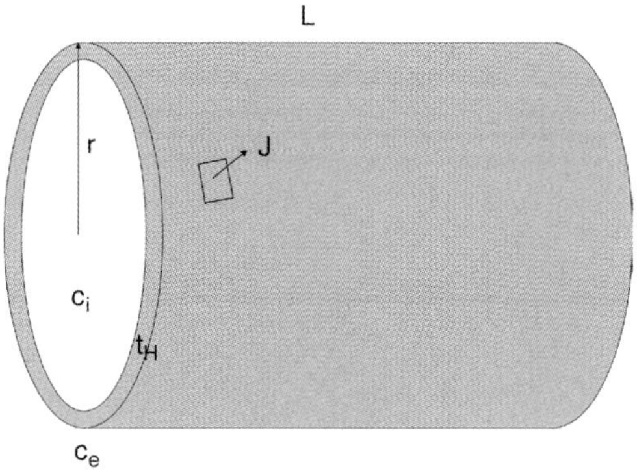

Fig. A13.6 Steady-state diffusion in liners.

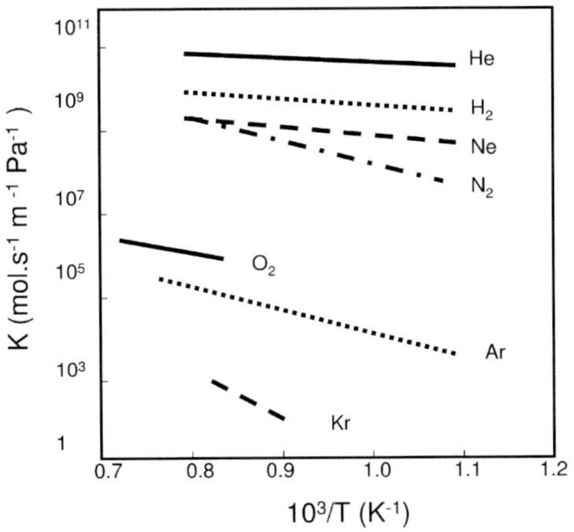

Fig. A13.7 Permeability K variation for different gases in silica (according to Shelby, 1996).

One observes that permeability increases when the atom or molecule size decreases (Fig. A13.7). Since permeability K is proportional to D one observe an Arrhenius evolution of K in the studied domains.

A13.3 Non-steady-State Diffusion

Most practical diffusion situations are non-steady-state ones. The diffusion flux and concentration gradient at some point in a solid vary with time, with a net accumulation or depletion of the diffusing species. This is illustrated in Fig. A13.8.

Solutions to Fick's second law can be extracted using boundary conditions. One practically important solution is for a semi-infinite solid in which surface concentration c_s is held constant. These conditions are met for chemical tempering when the

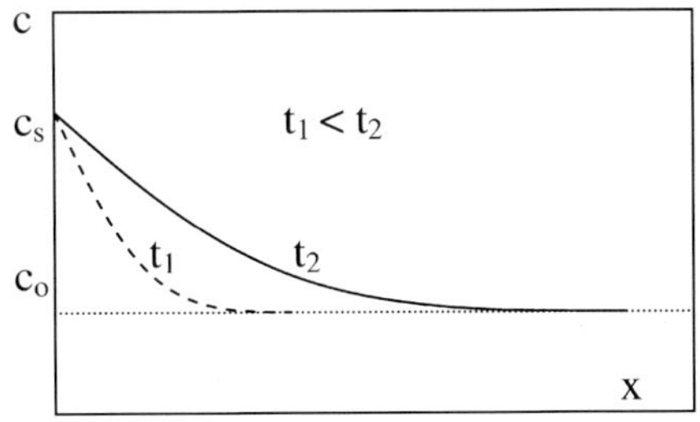

Fig. A13.8 Non-steady-state diffusion under constant surface concentration.

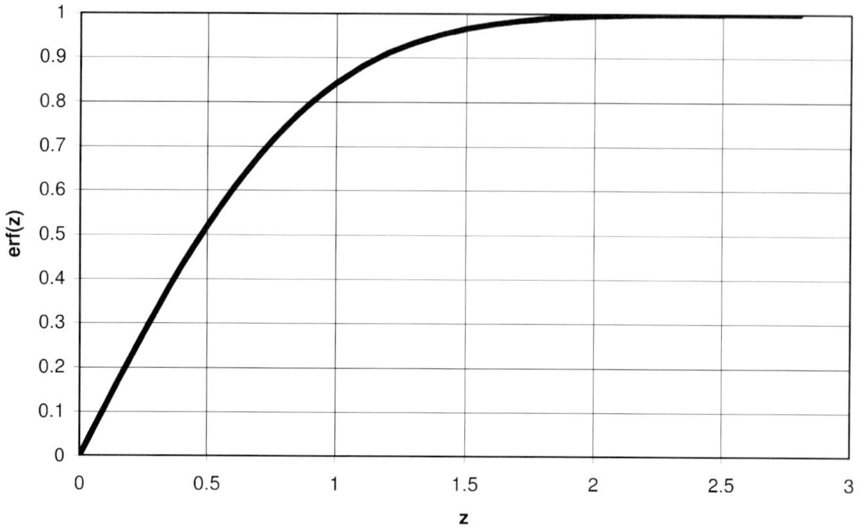

Fig. A13.9 Erf(z) function.

glass is immersed in a molten salt solution (Chapter 6) as well as for glass surface hydration (Chapter 5; diffusion depth being much less than the thickness, the two opposite surfaces can be considered separately). We shall assume that before diffusion, solute atoms are uniformly distributed with concentration c_o. The solution is written

$$\frac{c(x,t) - c_o}{c_s - c_o} = 1 - \operatorname{erf}\left(\frac{x}{2\sqrt{Dt}}\right)$$

The erf(z) function is plotted in Fig. A13.9 ($\operatorname{erf}(z) = (2/\sqrt{\pi}) \int_0^z \exp(-y^2)dy$). When initial concentration is zero or negligible ($c_o \ll c$ or c_s) the equation reduces to

$$c(x,t) = c_s\left[1 - \operatorname{erf}\left(\frac{x}{2\sqrt{Dt}}\right)\right]$$

Note that for the same Dt value the diffusion profile is the same. Hence, the same diffusion profile can be obtained more rapidly using a higher temperature and maintaining the same Dt product (decreasing time accordingly).

Another important solution is for an instantaneous plan source of diffusing species. This is used to determine diffusion coefficient, using radioactive tracers and plotting log c as a function of x^2. This also corresponds to the application of thin coatings with the diffusion element at a plane surface of the studied material (in practice it can be sandwiched, concentration being then taken as half the total value).

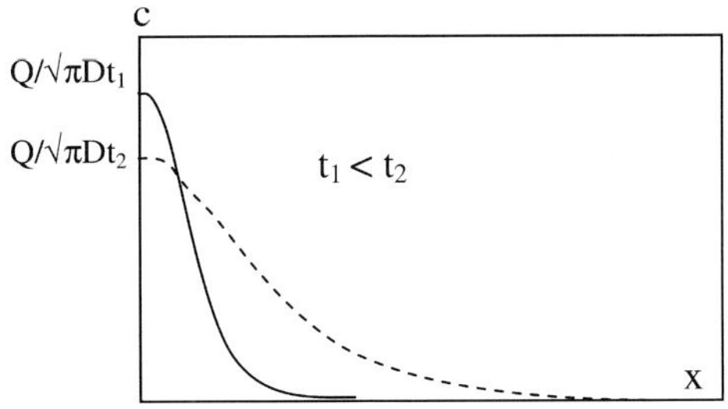

Fig. A13.10 Non-steady-state diffusion for an instantaneous plan source.

Diffusion solution is written

$$c(x,t) = \frac{Q}{\sqrt{\pi Dt}} \exp\left(-\frac{x^2}{4Dt}\right)$$

where Q is the amount of substance per unit area at time zero. It is half a Gaussian curve and is shown in Fig. A13.10 (at $x = 0$, $\partial c/\partial x|_{x=0} = 0$).

Glossary

Annealing: Thermal treatment to remove stress from glassware
Batch: Raw materials, properly proportioned and mixed for delivery to the furnace
Batch charger: Metallic device for mechanically conducting batch to furnace
Batch house: Place where batch materials are received, weighed and mixed before being delivered
Blister: Bubble or inclusion in glass
Bloom: Surface film resulting from attack by moisture or other vapour
Blow mould: Metal mould for blowing glass to its final shape
Blower: Person who forms the glass by blowing
Blowpipe: Pipe that the glassmaker uses for gathering glass and blowing
Borosilicate glass: Glass containing boron oxide
Chain marks: Marks left on the bottom of glass articles by the chain belt
Check: Surface crack at glass article surface
Chemical durability: Chemical resistance of glass to moisture
Chip: Abraded and removed glass piece
Cold end: Operations on cold glass (polymer coating, cutting, scoring)
Continuous tank: Glass furnace allowing continuous delivery
Cord: Glass inhomogeneity
Cullet: Waste and recycled broken glass used at furnace input
Devitrification: Crystallization in glass
Etch: Glass surface chemical attack for marking and decorating
Fining: Operation consisting in homogenizing and eliminating bubbles from glass melt
Float glass: Flat glass used in house, building and vehicle windows
Forming: Shaping hot glass
Gather: Hot glass collected at blowpipe extremity
Glass blowing: Shaping glass by air pressure
Gob: Hot glass delivered by the feeder
Hackle marks: Marks on fracture path
Hot end: Operations on hot glass (melting, forming, annealing)

Glass; Mechanics and Technology. Eric Le Bourhis

ISBN: 978-3-527-31549-9

Lehr: Long tunnel oven for annealing glass
Melting temperature: Temperature used by commercial furnace for melting raw materials. This is well above the melting temperature defining liquid-crystal phase change often called liquidus temperature
Mould: metallic mould used to shape glass
Mould mark: mark at mould joint
Oven glass: Borosilicate glass
Pressed glass: Glassware formed by pressure between a mould and plunger
Quartz: Most common form of silica
Residual stresses: Permanent stresses reinforcing glass
Recycling: Using glass (cullet) as raw material to produce new glass
Sagging: Glass deformation under gravity
Silica glass: Pure SiO_2 glass
Silico-soda-lime glass: Glass containing silica associated with soda and lime
Soda: Sodium oxide Na_2O
Stria: Cord
Tank: Melting unit
Temper: Strengthen glass (build in stress)
Tempered glass: Strengthened glass
Thermal shock resistance: Resistance to temperature variation
Weathering: Attack of glass surface by atmosphere

References

Aben, H. and Guillemet, C. (1993) *Photoelasticity of Glass*, Springer-Verlag, Berlin.

Abrams, M.B., Green, D.J., Glass, S.J. (2003) *J. Non-Cryst. Solids*, **321**, 10.

Acloque, P. (1951) *Verres Refract*, **5**, 247.

Adams, G. and Gibbs, J.H. (1965) *J. Chem. Phys*, **43**, 139.

Adams, L.H. and Williamson, E.D. (1920) *J. Franklin Inst*, **190**, 597.

Akeyoshi, K., Kanai, E., Yamamoto, K., Shima, S. (1967) *Rep. Res. Lab. Asahi Glass.*, **17**, 23.

Alliche, A. and Le Bourhis, E. (2000) *Mater. Sci. Eng. A*, **278**, 255.

Angell, C.A. (1995) *Science*, **267**, 1924.

Anstis, G.R., Chantikul, P., Lawn, B.R., Marshall, D.B. (1981) *J. Am. Ceram. Soc.*, **64**, 533.

Arbad, M., Shelestak, L.J., Harris, C.S. (2005) *Am. Ceram. Soc. Bull*, **84**, 34.

Argon, A.S. (1979) *Acta Metall.*, **27**, 47.

Arora, A., Marshall, D.B., Lawn, B.R., Swain, M.V. (1979) *J. Non-Cryst. Solids*, **31**, 415.

Ashby, M.F. (2001) *Materials Selection in Mechanical Design*, Butterworth-Heinemann, Oxford, UK.

Ashby, M. and Johnson, K. (2003) *Mater. Today.*, **12**, 24.

Ashby, M.F. and Jones, D.R.H. (1991) *Matériaux 1* & 2, Dunod, France.

Askeland, D.R. (1989) *The Science and Engineering of Materials*, PWS-Kent Publishing, Boston, MA.

Atanacio, A.J., Latella, B.A., Barbé, C.J., Swain, M.V. (2005) *Surf. Coat. Technol.*, **192**, 254.

Auerbach, F. (1891) *Ann. Phys. Chem.*, **43**, 61.

Avramov, I. (2005) *J. Non-Cryst. Solids*, **351**, 3165.

Barry, F., Chaunac, M., Chiale, D., Cordier, A., Criaud, A., De La Rochefoucauld, C., Deschanet, P., Désert, C., De Zorzi, M., Faivre Delord, C., Grell, P., Holle, R., Lafarge, M., Levet, J., Macocco, B., Marozeau, D., Thibault, P. (1998) *Verre.*, **12**, 12.

Bartoe, R.D. and McDanel, V. (2003) *Industrie Céram.* & Verrière, **986**, pp. 58.

Barton, J. and Guillemet, C. (2005) *Le verre, science et technologie*, EDP Sciences, Les Ulis, France.

Bather, J.M. and Gray, R.A.A.C. (1976) *J. Chromatog*, **122**, 159.

Beegan, D. and Laugier, M.T. (2005) *Surf. Coat. Technol.*, **199**, 32.

Bergman, B. (1984) *J. Mater. Sci. Lett.*, **3**, 689.

Bergonnier, S., Hild, F., Rieunier, J.B., Roux, S. (2005) *J. Mater. Sci.*, **40**, 5949.

Berkovich, E.S. (1951) *Ind. Diamond Rev.*, **11**, 129.

Bertin, M., Faroux, J.P., Renault, J. (1984) *Electromagnétisme 4*, Dunod, Paris.

Bertoldi, M. and Sglavo, V.M. (2004) *J. Non-Cryst. Solids*, **344**, 51.

Bhattacharya, A.K. and Nix, W.D. (1988) *Int. J. Solids Struct.*, **24**, 1287.

Boccaccini, A.R. (1997) *J. Mater. Sci. Proc. Tech.*, **65**, 302.

Bouyne, E. and Gaume, O. (2001) *Proc. Int. Cong. Glass*, **2**, 106.

Brebec, G., Seguin, R., Sella, C., Bevenot, J., Martin, J.C. (1980) *Acta Metall*, **28**, 327.

Briard, R. (2004) Renforcement du verre par un revetement hybride en formulation aqueuse, Thesis, Rennes.

Glass; Mechanics and Technology. Eric Le Bourhis

ISBN: 978-3-527-31549-9

Brinker, C.J. and Scherrer, G.W. (1990) *Sol-Gel Science: The Physics and Chemistry of Sol-Gel Processing*, Academic Press, San Diego, CA.

Brown, J.T., Kobayashi, H., (1998) *Am. Ceram. Soc. Bull.*, **77**, 81.

Bruck, H.A., Christman, T., Rosakis, A.J., Johnson, W.L. (1994) *Scripta Mater*, **30**, 429.

Buehl, W.M. and Ryszyticoskyj, W.P. (1991) *SID91 Digest.*, 667.

Buijs, M. (1994) *J. Am. Ceram. Soc*, **77**, 1676.

Cable, M. and Smedley, J.W. (1989) *Glass Sci. Technol*, **30**, 39.

Caillot, E., Barlier, P., Moffatt-Fairbanks, D.M., Tennent, D.L. (1997) *Verre.*, **3**, 3.

Carré, H. (1996) Etude du comportement à la rupture d'un matériau fragile précontraint: le verre trempé, Thesis, Paris.

Célarié, F., Prades, S., Bonamy, D., Ferrero, L., Bouchaud, E., Guillot, C., Marlière, C. (2003) *Phys. Rev. Lett.*, **90**, 0755504.

Chandrasekar, S. and Chaudhri, M.M. (1993) *Phil. Mag. A*, **67**, 1187.

Chang, H.J., Kim, D.H., Kim, T.M., Kim, T.J., Chattopadhyay, K. (2006) *Scripta Mater*, **55**, 509.

Charles, R.J., Hillig, W.B. (1962) *The Kinetics of Glass Failure by Stress Corrosion*, Union Scientifique Continentale du Verre, Belgium.

Chartier, P. (1997) *Verre*, **3**, 5.

Chaudhri, M.M. (1998) *Acta Mater*, **46**, 3047.

Chaudhri, M.M. (2004) *in FRN Nabarro Collection*, chapter 70.

Chaudhri, M.M. and Kurkjian, C.R. (1986) *J. Am. Ceram. Soc.*, **69**, 404.

Chaudhri, M.M. and Liangyi, C. (1986) *Nature*, **320**, 48.

Chaudhri, M.M. and Phillips, M.A. (1990) *Phil. Mag. A*, **62**, 1.

Chemin, N. (2007) Relation entre la structure et les propriétés mécaniques de films minces hybrides sol-gel, Thesis, Paris VI.

Chen, J. and Bull, S.J. (2006) *J. Mater. Res.*, **21**, 2617.

Cheng, Y.T. and Cheng, C.M. (2005) *J. Mater. Res.*, **20**, 1046.

Clemens, B.M. and Bain, J.A. (1992) *Mater. Res. Soc. Bull*, **17**, 46.

Coes, L. (1971) *Abrasives*, Springer Verlag, New York.

Cohen, M.H. and Turnbull, D. (1959) *J. Chem. Phys*, **31**, 1164.

Cook, R.F. (2006) *J. Mater. Sci.*, **41**, 841.

Cook, R.F. and Pharr, G.M. (1990) *J. Am. Ceram. Soc.*, **73**, 787.

Creuzet, F., Guilloteau, E., Arribart, H. (1993) Fundamentals of Glass Science and Technology, *Proc. Second Conf. ESG*, 163.

Dally, J.W. and Riley, W.F. (1965) *Experimental Stress Analysis*, McGraw-Hill, New York.

DeBast, J. and Gilard, P. (1963) *Phys. Chem. Glasses*, **4**, 117.

Dejean-Arrecgros, J. (1978) *Petit guide panoramique des roches*, Delachaux & Niestlé, Paris.

Dériano, S. (2002) Conception chimique de verres silicatés à hautes performances mécaniques, Thesis, Rennes.

Dériano, S., Jarry, A., Rouxel, T., Sangleboeuf, J.-C., Hampshire, S. (2004) *J. Non-Cryst. Solids*, **344**, 44.

Dietzel, A. (1942) *Z. Electrochem*, **48**, 3.

Doremus, R.H. (1969) *Reactivity of Solids* eds J.W.Mitchell, R.C. Devries, R.W. Roberts, P. Cannon, Wiley, New York. pp. 667.

Doremus, R.H. (1994) *Glass Science*, 2nd edn, John Wiley, New York.

Doremus, R.H. (2003) *Am. Ceram. Soc. Bull*, **82**, 1.

Douglas, R.W. (1958) *J. Soc. Glass Technol*, **42**, 145T.

Du Bois, P.A., Kolling, S., Fassnacht, W. (2003) *Comput. Mater. Sci*, **28**, 675.

Dubois Petroff, M.P. (2001) *Recettes d'Architecture-la cuisine*, Massin, Paris.

Dugdale, D.S. (1960) *J. Mech. Phys. Solids*, **8**, 100.

Durville, D. (2005) *J. Mater. Sci.*, **40**, 5941.

Elmustafa, A.A. and Stone, D.S. (2003) *J. Mech. Phys. Solids.*, **51**, 357.

Entwistle, K.M., (1993) *J. Mater. Sci.*, **28**, 2007.

Ernout, A. *et al.* (eds) (1947) *Pline l'Ancien, Histoire naturelle.*

Faraday, M. (1830) *Phil. Trans. R. Soc*, 49.

Faraday, M. (1857) *Phil. Mag*, **14**, 401.

Faurie, D., Renault, P.O., Le Bourhis, E., Villain, P., Goudeau, P., Badawi, F. (2004) *Thin Solid Films*, **469/470**, 201.

Ferry, J.D. (1970) *Viscoelastic Properties of Polymers*, John Wiley, New York.

Feynmann, R.P., Leighton, R.B., Sands, M. (1979) *Physique*, **Volume 1**, Interedition.

Field, J.S. and Swain, M.V. (1993) *J. Mater. Res.*, **8**, 297.

Fischer-Cripps, A.C. (2004a) *Nanoindentation*, Springer.

Fischer-Cripps, A.C. (2004b) *Mater. Sci. Eng. A*, **385**, 74.

Flessel, J.M. (2003) *Bull. Soc. Fran. Phys.*, **140**, 28.

Fokin, V.M., Zanotto, E.D., Yuritzyn, N.S., Schmelzer, J.W.P. (2006) *J. Non-Cryst. Solids*, **352**, 2681.

Fouquet, V., Le Bourhis, E., Pichon, L., Drouet, M., Straboni, A. (2004) *Scripta Mater*, **51**, 899.

François-Baillet and Marchi, 2006 François-Baillet, B. and Marchi, A. (2006) *Est Magazine*, **3**, 6.

Frank, F.C., Lawn, B.R. (1967) *Proc. R. Soc. A*, **229**, 291.

Frank, R.C., Swets, D.E., Lee, R.W. (1961) *J. Chem. Phys*, **35**, 1451.

Freiman, S.W. (1985) *Strength of Inorganic Glass* (ed C.R.Kurkjian), Plenum, New York, 197.

Friedman, G.S., Cushman, K., Doremus, R.H. (1982) *J. Mater. Sci.*, **17**, 994.

Frischat, G.H. (1968) *J. Am. Ceram. Soc*, **51**, 528.

Frischat, G.H. (1987) in *Glas-Struktur und Eigenschaften*, (ed. Lohmeyer et al.), Werkstoff Glas I. Ehningen: ex-pert, p. 47.

Frischat, G.H. (2002) *C.R. Chimie*, **5**, 759.

Frischat, G.H., Heide, G., Muller, B., Weeks, R.A. (2001) *Phys. Chem. Glasses*, **42**, 179.

Frost, H.J. and Ashby, M.F. (1982) *Deformation-Mechanism Maps*, Pergamon, New York, (electronic version available at http://thayer.dartmouth.edu/~defmech/).

Gadaud, P. and Pautrot, S. (2003) *J. Non-Cryst. Solids*, **316**, 146.

Gao, H., Huang, Y., Nix, W.D., Hutchinson, J.W. (1999) *J. Mech. Phys. Solids.*, **47**, 1239.

Garcia-Vallès, M., Gimeno-Torrente, D., Martinez-Manent, S., Fernandez-Turiel, J.L. (2003) *Am. Miner*, **88**, 1996.

Gardon, R. (1978) *J. Am. Ceram. Soc*, **61**, 143.

Gay, D. (1997) *Matériaux composites*, Hermes, France.

Geandier, G., Denis, S., Mocellin, A. (2003) *J. Non-Cryst. Solids*, **318**, 284.

Gilman, J.J. (1949) *Rep. Prog. Phys*, **12**, 48.

Gilman, J.J., (1973) *The Science of Hardness Testing and Its Research Applications*, in ed. J.H.Westbrook, H.Conrad, American Society for Metals, Metal Park, OH, chapter 4.

Goodier, J.N. (1936) *Phil. Mag*, **22**, 678.

Grau, P., Berg, G., Meinhard, H., Mosch, S. (1998) *J. Am. Ceram. Soc*, **81**, 1557.

Greaves, G.N. (1985) *J. Non-Cryst. Solids*, **71**, 203.

Green, D.J., Tandon, R., Sglavo, V.M. (1999) *Science*, **283**, 1295.

Griffith, A. (1920) *Phil. Trans. R. Soc.*, **A221**, 163.

Griffith, A. (1924) *Int. Congr. Appl. Mech.*, **61**, 55.

G'Sell, C. (2001) *Mater. Sci. Eng. A*, **309/310**, 539.

Guillemet, C. (1990) *J. Non-Cryst. Solids*, **123**, 415.

Guillemet, C. and Gy, R. (1996) *Rev. Metall*, 701.

Guin, J.P. and Wiederhorn, S.M. (2004) *Phys. Rev. Lett.*, **92**, 215502.

Guin, J.P., Rouxel, T., Sangleboeuf, J.C., Melscoëy, I., Lucas, J. (2002) *J. Am. Ceram. Soc*, **85**, 1545.

Gy, R. (1997) *Verre*, **3**, 21.

Gy, R. (1999) *Proceedings of Glass Processing Days, Tempere*, Finland.

Gy, R. (2003) *J. Non-Cryst. Solids.*, **316**, 1.

Gy, R., Duffrène, L., Labrot, M. (1994) *J. Non-Cryst. Solids*, **175**, 103.

Hagan, J.T. (1979) *J. Mater. Sci.*, **14**, 462.

Hainsworth, S.V., Chandler, H.W., Page, T.F. (1987) *J. Mater. Res.*, **11**, 2283.

Han, W.T. and Tomozawa, M. (1990) *J. Am. Ceram. Soc.*, **73**, 3626.

Hand, R., Ellis, B., Whittle, B., Wang, F. (2003) *J. Non-Cryst. Solids.*, **315**, 276.

Harvey, S., Huang, H., Venkataraman, S., Gerberich, W.W. (1993) *J. Mater. Res.*, **8**, 1291.

Hauk, V. (1997) *Structural and Residual Stress Analysis by Non Destructive Methods: Evaluation, Application, Assessment*, Elsevier, Amsterdam.

Heide, K., Kletti, H., Völksch, G. (2003) *Glass Sci. Technol*, **76**, 118.

Heilmaier, M. and Eckert, J. (2005) *Adv. Eng. Mater*, **7**, 833.

Heller, A. (1995) *Acc. Chem. Res.*, **28**, 503.
Hertz, H.H. (1881) *J Reine Angew. Math*, **92**, 156. Translated and reprinted in English in Hertz's Miscellaneous Papers, Macmillan, London, 1896.
Hillig, W.B. (1962) *Modern Aspects of the Vitreous State, vol. II* (ed J.D.Mackenzie), Butterworths, London, pp. 152.
Höhne, G., Hemminger, W., Flammershein, H.J. (1996) *DSC: An Introduction to Practitioners*, Springer-Verlag, Berlin.
Hollomon, J.H. (1945) *AIME Trans.*, **162**, 268.
Hufnagel, T.C. (ed) (2006) *Scripta Mater*, **54**, pp. 317.
Inglis, C. (1913) *Trans. Inst. Naval. Archit.*, **55**.
Inoue, A. (2000) *Acta Mater*, **48**, 279.
Inoue, A. and Takeuchi, A. (2002) *Mater. Trans. JIM*, **43**, 1892.
Inoue, A., Kawamura, Y., Shibata, T., Sasamori, K. (1996) *Mater. Trans*, **37**, 1337.
Irwin, G.R. (1958) Fracture, *Handbuch der Physik*, Springer-Verlag, Berlin.
Ishikawa, H. and Shinkai, N. (1982) *J. Am. Ceram. Soc.*, **65**, C124.
Ito, S. and Taniguchi, T. (2004) *J. Non-Cryst. Solids*, **349**, 173.
Jain, A., Firestone, G.C., Yi, A.Y. (2005) *J. Am. Ceram. Soc*, **88**, 2409.
James, F.A.J.L. (1991) *Phys. Educ*, **26**, 296.
Jiang, W.H. and Atzmon, M. (2003) *J. Mater. Res*, **18**, 755.
Johnson, K.L. (1985) *Contact Mechanics*, University Press, Cambridge.
Johnson, W.L. (1996) *Mater. Sci. Forum*, **225/227**, 35.
Jones, G.O. (1956) *Glass, Methuen's Monographs on Physical Subjects*, Methuen, London.
Joslin, D.L. and Oliver, W.C. (1990) *J. Mater. Res.*, **5**, 123.
Kawamura, Y., Nakamura, T., Inoue, A. (1998) *Scripta Mater*, **39**, 301.
Kennedy, C.R., Bradt, R.C., Rindone, G.E. (1980) *Phys. Chem. Glasses*, **21**, 99.
Keulen, N.M. (1993) *J. Am. Ceram. Soc*, **76**, 904.
Kieffer, J. (2002) *Am. Ceram. Soc. Bull*, **81**, 1.
Kim, J.J., Choi, Y., Suresh, S., Argon, A.S. (2002) *Science*, **25**, 654.
Kittel, C. (1983) *Physique de l'état solide*, Dunod, France.
Kittl, P. and Diaz, G. (1988) *Int. J. Struc. Mech. Mater. Sci.*, **24**, 99.
Klement, W., Willens, R.H., Duwez, P. (1960) *Nature*, **187**, 869.
Koike, A. and Tomozawa, M. (2006) *J. Non-Cryst. Solids*, **352**, 3787.
Kramer, D., Huang, H., Kriese, M., Robach, J., Nelson, J., Wright, A., Bahr, D., Gerberich, W.W. (1999) *Acta Mater*, **47**, 333.
Krause, J.T., Testardi, L.R., Thurston, R.N. (1979) *Phys. Chem. Glasses.*, **20**, 135.
Kurkjian, C.R. (1963) *Phys. Chem. Glasses*, **4**, 128.
Kurkjian, C.R. and Prindle, W.R. (1998) *J. Am. Ceram. Soc*, **81**, 795.
Kurkjian, C.R., Kammlot, G.W., Chaudhri, M.M. (1995) *J. Am. Ceram. Soc.*, **78**, 737.
Laevsky, K.Y. (2003) Pressing of glass in bottle and jar manufacturing: numerical analysis and computation, PhD thesis, Eindhoven University of Technology.
Landau, L. and Levich, B. (1942) *Acta Physiochim.*, **17**, 42.
Langmuir, I. (1933) *J. Chem. Phys.*, **1**, 756.
Largeau, L., Patriarche, G., Le Bourhis, E. (2002) *J. Mater. Sci. Lett*, **21**, 401.
Lawn, B. (1993) *Fracture of Brittle Solids*, Cambridge Solid State Science Series, Cambridge.
Lawn, B. (1998) *J. Am. Ceram. Soc.*, **81**, 1977.
Lawn, B.R. and Evans, A.G. (1977) *J. Mater. Sci.*, **12**, 2195.
Lawn, B.R. and Marschall, D.B. (1979) *J. Am. Ceram. Soc.*, **62**, 347.
Le Bourhis, E. (1996) Vitrified bonded abrasives: post-annealing influence on residual stresses and wheel strength, Saint-Gobain internal report.
Le Bourhis, E. (1998a) Endommagement du verre par contact, Saint-Gobain internal report.
Le Bourhis, E. (1998b) *Verre*, **4**, 5.
Le Bourhis, E. and Le Guiner, F. (1986) *Report*, Laboratoire de recherche des Musées de France, Paris.
Le Bourhis, E. and Metayer, D. (2000) *J. Non-Cryst. Solids*, **272**, 34.
Le Bourhis, E. and Patriarche, G. (2003) *Prog. Cryst. Growth Charac. Mater*, **47**, 1.
Le Bourhis, E. and Rouxel, T. (2003) *J. Non-Cryst. Solids*, **316**, 153.

Le Bourhis, E., Gadaud, P., Guin, J.-P., Tournerie, N., Zhang, X.H., Lucas, J., Rouxel, T. (2001a) *Scripta Mater,* **45**, 317.

Le Bourhis, E., Largeau, L., Patriarche, G., Rivière, J.P. (2001b) *J. Mater. Sci. Lett,* **20**, 1361.

Le Bourhis, E., Patriarche, G., Largeau, L., Rivière, J.P. (2004) *J. Mater. Res,* **9**, 131.

Le Houérou, V., Sangleboeuf, J.C., Dériano, S., Rouxel, T., Duisit, G. (2003) *J. Non-Cryst. Solids,* **316**, 54.

Lee, E.H. and Radok, J.R.M. (1960) *Trans. ASME Ser. E, J. Appl. Mech.,* **27**, 438.

Lee, E.H., Rogers, T.G., Woo, T.C. (1965) *J. Am. Ceram. Soc,* **48**, 480.

Lee, R.W. (1963) *J. Chem. Phys,* **38**, 448.

Lehmann, J.C. (2005) *Bull. Soc. Fran. Phys.,* **150**, 4. English version: Europhysicsnews 2006, 37, 23.

Levelut, C., Faivre, A., Le Parc, R., Champagnon, B., Hazemann, J.-L., David, L., Rochas, C., Simon, J.P. (2002) *J. Non-Cryst. Solids,* **307/310**, 426.

Lilleodden, E.T., Bonin W, W., Nelson, J., Wyrobek, J.T., Gerberich, W.W.(1995) *J. Mater. Res.,***10**,2162.

Lim, Y.Y. and Chaudhri, M.M. (2004) *Phil. Mag.,* **84**, 2877.

Lim, Y.Y. and Chaudhri, M.M. (2006) *Mech. Mater.,* **38**, 1213.

Livshits, V.Y., Tennison, D.G., Gukasyan, S.B. (1982) *Sov. J. Glass Phys. Chem.,* **8**, 463.

Lornage, T. (2004) Optimisation des bouteilles par modélisation, Lecture, Réunion annuelle GDR Verre.

Loubet, J.L., Georges, J.M., Meille, J. (1986) *Microindentation Techniques in Materials Science and Engineering* (eds P.J.Blau and B.R.Lawn), ASTM International, Philadelphia, PA, pp. 72.

Lu, J., Ravichandran, G., Johnson, W.L. (2003) *Acta Mater,* **51**, 3429.

Lucas, J. (1999) *Curr. Opin. Solid State Mater Sci.* **4**, 181.

McCurty, C. (2000) *Proceedings of Coating on Glass Workshop, Livermore, CA,* p. 5.

Macedo, P.B. and Litovitz, T.A. (1965) *J. Chem. Phys,* **42**, 245.

Mackenzie, J.D. (1963) *J. Am. Ceram. Soc.,* **46**, 461.

McKown, C. (2000) *Proceedings of Coating on Glass Workshop, Livermore, CA,* p. 8.

McMillan, P.W. (1979) *Phys. Educ,* **14**, 441.

Macosco, C.W. (1994) *Rheology: Principles, Measurements and Applications,* VCH, New York.

Malkin, S. (1989) *Grinding Technology, Theory and Applications of Machining with Abrasive,* Ellis Horwood, Chichester, UK.

Malzbender, J., de With, G., den Toonder, J. (2000) *J. Mater. Res.,* **15**, 1209.

Mammeri, F. (2003) Relation entre la structure et les propriétés mécaniques de films minces hybrides organiques-inorganiques préparés par voie sol-gel, Thesis, Paris VI.

Mammeri, F., Le Bourhis, E., Rozes, L., Sanchez, C., Huignard, A., Lefevre, D. (2004) *J. Non-Cryst. Solids,* **345/346**, 610.

Mammeri, F., Le Bourhis, E., Rozes, L., Sanchez, C. (2005) *J. Mater. Chem.,* **15**, 3787.

Mamor, M., Dufour-Gergam, E., Finkman, L., Tremblay, G., Meyer, F., Bouziane, K. (1995) *Appl. Surf. Sci.,* **91**, 342.

Markovsky, A. and Soules, T.F. (1984) *J. Am. Ceram. Soc,* **67**, C56.

Marot, L., Le Bourhis, E., Straboni, A. (2002) *Mater. Lett.,* **56**, 76.

Marschall, D.B., Lawn, B.R., Evans, A.G. (1982) *J. Am. Ceram. Soc.,* **65**, 561.

Marsh, D.M. (1964) *Proc. R. Soc,* **A279**, 420.

Mattoni, A., Colombo, L., Cleri, F. (2005) *Phys. Rev. Lett.,* **95**, 115501.

Mencik, J., Munz, D., Quandt, E., Weppelmann, E.R., Swain, M.V. (1997) *J. Mater. Res.,* **12**, 2475.

Meyerhofer, D. (1978) *J. Appl. Phys.,* **49**, 3993.

Mickelsen, J.C. (1984) *Appl. Phys. Lett,* **45**, 1187.

Mills, J.J. (1974) *J. Non-Cryst. Solids,* **14**, 255.

Morris, D.J., Myers, S.B., Cook, R.F. (2004) *J. Mater. Sci.,* **39**, 2399.

Moseler, D., Heide, G., Frischat, G.H. (2002) *Glass Sci. Technol.,* **75**, 174.

Moynihan, C.T. (1996) Structural relaxation and the glass transition, *Structure, Dynamics and Properties of Silicate Melts* (eds J.F. Stebbins, P.F. McMillan, D.B. Dingwell), Mineral Society of America.

Mozdierz, N., Le Bourhis, E., Peyrot, A., Rivière, J.P. (1993) *Céramiques, polymères et matériaux composites, Encyclopédie CLARTE,* **1**, 3390.

Murphy, D.L. (2005) *Am. Ceram. Soc. Bull*, **84**, 41.
Musil, J. (2000) *Surf. Coat. Technol.*, **125**, 515.
Nakayama, G.S. and Schackelford, J.F. (1990) *J. Non-Cryst. Solids*, **126**, 249.
Narayanaswamy, O.S. (1978) *J. Am. Ceram. Soc*, **61**, 146.
Neely, J.E. and Mackenzy, J.D. (1968) *J. Mater. Sci.*, **3**, 603.
Nghiêm, B. (1998) Thesis, Université Paris-VI.
Ngô, H. and Ngô, C. (1988) *Physique statistique*, Masson, Paris.
Nieh, T.G., Schuh, C., Wadsworth, J., Li, Y. (2002) *Intermetallics*, **10**, 1177.
Norton, F.J. (1961) *Nature*, **191**, 701.
Novikov, V.N. and Sokolov, A.P. (2004) *Nature*, **431**, 961.
Oliver, W.C. and Pharr, G.M. (1992) *J. Mater. Res.*, **7**, 1564.
Orowan, E. (1949) *Rep. Prog. Phys.*, **12**, 185.
Oyen, M.L. (2005) *J. Mater. Res.*, **20**, 2094.
Oyen, M.L. and Cook, R.F. (2003) *J. Mater. Res.*, **18**, 139.
Palucka, T. (2006) *Mater. Res. Soc. Bull*, **31**, 697.
Pampillo, C.A. and Chen, H.S. (1974) *Mater. Sci. Eng.*, **13**, 181.
Pantano, C.G., Bojan, V., Verita, M., Geotti-Bianchini, F., Hreglich, S. (1993) *2nd Conf. European Soc. Glass Sci. Techn., Venice.*
Paris, P.C. and Sih, G.C. (1965) *ASTM Spec. Tech. Publ.*, **381**, 30.
Patriarche, G., Le Bourhis, E., Khayyat, M.M., Chaudhri, M.M. (2004) *J. Appl. Phys.*, **96**, 1464.
Pavelchek, E.K. and Doremus, D.H. (1974) *J. Mater. Sci.*, **9**, 1803.
Pérez, J.P., Carles, R., Fleckinger, R. (1997) *Electromagnétisme*, Masson, Paris.
Perkins, W.G. and Begeal, D.R. (1971) *J. Chem. Phys*, **54**, 1683.
Perriot, A. (2005) Nanoindentation de couches minces deposes sur substrat de verre de silice, Thesis, Paris VI.
Peter, K.W. (1964) *Glasstech. Ber.*, **37**, 333.
Piellard, M., 2005 Etude de la liaison Platine-Alumine, Thesis, Poitiers.
Pilkington, A. (1969) *Proc. R. Soc. Lond. A*, **314**, 1.
Priven, A.I. (2004) *Glass Technol*, **45**, 244.
Puchi-Cabrera, E.S. (2002) *Surf. Coat. Technol*, **160**, 177.
Pugno, N.M. and Ruoff, R.S. (2006) *J. Appl. Phys.*, **99**, 024301.
Radok, J.R.M. (1957) *Q. Appl. Math*, **15**, 198.
Ramamurty, U., Jana, S., Kawamura, Y., Chattopadhyay, K. (2005) *Acta Mater*, **53**, 705.
Rawlings, R.D., Wu, J.P., Boccaccini, A.R. (2006) *J. Mater. Sci*, **41**, 733.
Redner, A.S. (1998) Measurement of Residual Strain in Glass Products, Short course lecture notes, Strainoptics Technologies.
Rehren, T. (2006) *GEO*, **334**, 118.
Rekhson, S.M. (1986) (eds D.R.Uhlmann and N.J.Kreidl), *Glass Science and Technology*, Academic, Orlando, FL, vol. **3**, chap. 1.
Reuss, R.H., Hopper, D.G., Park, J.G. (2006) *Mater. Res. Soc. Bull.*, **31**, 447.
Ridealgh, J.A. (2006) *Mater. Res. Soc. Symp. Proc.*, **890**, Y01-10.1.
Rizkalla, A.S., Jones, D.W., Miller, R.P. (1996) *Brit. Ceram. Trans.*, **95**, 151.
Rouxel, T. (2006) *C. R. Mecanique*, in press.
Rouxel, T. and Sangleboeuf, J.C. (2000) *J. Non-Cryst. Solids*, **271**, 224.
Rouxel, T., Yoshida, S., Shang, H., Sangleboeuf, J.C. (2006) *Mater. Res. Soc. Proc.*, **904**, BB02-03.1.
Ryu, S.R. and Tomozawa, M. (2006) *J. Non-Cryst. Solids*, **352**, 3929.
Sado, G. and Sado, M.C. (1991) *Les plans d'expériences*, AFNOR Technique, France.
Sakaï, M. (1999) *J. Mater. Res*, **14**, 3630.
Sakaï, M. (2002) *Phil. Mag. A*, **82**, 1841.
Sakaï, M., Sasaki, M., Matsuda, A. (2005) *Acta Mater*, **53**, 4455.
Sangwal, K.J. (2000) *Opto Adv. Mater.*, **2**, 105.
Scherer, G.W. (1982) *J. Am. Ceram. Soc*, **65**, 352.
Scherer, G.W. (1996) *Relaxation in Glass and Composites*, Wiley Interscience, New York.
Scholze, H. (1991) *Glass: Nature, Structure, and Properties*, Springer-Verlag, New York.
Schottner, G. (2001) *Chem. Mater.*, **13**, 3422.
Schuh, C.A., Lund, A.C., Nieh, T.G. (2004) *Acta Mater*, **52**, 5879.
Schzalke, K. (1981) *Test Technique in Fracture*, Pergamon Press.
Sehgal, J., Nakao, Y., Takahashi, H., Ito, S. (1995) *J. Mater. Sci. Lett.*, **14**, 167.

Sehgal, J. Nakao, Y. Takahashi, H. Ito, S. Patent EP 07570020A1.

Sglavo, V.M. and Green, D.J. (2001) *J. Eur. Ceram. Soc.*, **21**, 561.

Shang, H. and Rouxel, T. (2005) *J. Am. Ceram. Soc.*, **88**, 2625.

Shelby, J.E. (1961) *J. Am. Ceram. Soc*, **54**, 125.

Shelby, J.E. (1996) *Handbook of Gas Diffusion in Solids and Melts*, ASM International, Materials Park, OH.

Shen, J. and Green, D.J. (2004) *J. Non-Cryst. Solids*, **344**, 79.

Sheng, H.W., Luo, W.K., Alamgir, F.M., Bai, J.M., Ma, E. (2006) *Nature*, **439**, 419.

Shimizu, S., Yanagimoto, T., Sakai, M. (1999) *J. Mater. Res.*, **14**, 4075.

Shiue, Y.S., Matthewson, M.J., Kurkjian, C.R., Biswas, D.R. (1996) *SPIE*, **2611**, 117.

Sih, G. (1972) *Mechanics of Fracture*, Wolters Noordhoff.

Simmons, J.H., Ochoa, R., Simmons, K.D., Mills, J.J. (1988) *J. Non-Cryst. Solids*, **105**, 313.

Smith, J.F. and Zheng, S. (2000) *Surf. Eng.*, **16**, 143.

Sneddon, I.N. (1965) *Int. J. Eng. Sci.*, **3**, 47.

Spaepen, F. (1977) *Acta Metall.*, **25**, 407.

Steif, P.S., Spaepen, F., Hutchinson, J.W. (1982) *Acta Metall*, **30**, 447.

Stella, A. and Verita, M. (1993) *2nd Conf. European Soc. Glass Sci. Techn., Venice.*

Stoney, G.G. (1909) *Proc. R. Soc. Lond.*, **A82**, 172.

Sundar, V.C., Yablon, A.D., Grazul, J.L., Ilan, M., Aizenberg, J. (2003) *Nature*, **424**, 899.

Suscavage, M.J. and Pantano, C.G. (1983) *Glastech. Ber.*, **56K**, 498.

Swain, M.V. (1981) *J. Mater. Sci.*, **16**, 151.

Swets, D.E., Lee, R.W., Frank, R.C. (1961) *J. Chem. Phys*, **34**, 17.

Tabor, D. (1951) *The Hardness of Metals*, Oxford at Clarendon Press, London.

Tabor, D. (1996) *Phil. Mag A.*, **4**, 1207.

Tammann, G. (1925) *The States of Aggregation*, Van Nostrand, New York.

Tanaka, H. (2005) *J. Non-Cryst. Solids*, **351**, 3371, 3385, 3396.

Tandon, R. and Cook, R.F. (1992) *J. Am. Ceram. Soc.*, **75**, 2877.

Tandon, R. and Cook, R.F. (1993) *J. Am. Ceram. Soc.*, **76**, 885.

Tavares, C.J., Vidrago, C., Rebouta, L., Rivière, J.P., Le Bourhis, E., Denanot, M.F. (2005) *Surf. Coat. Technol*, **200**, 288.

Tawakoli, T. (1993) *High Efficiency Deep Grinding*, VDI-Verlag, Germany.

Thomas, O., Labat, S., Bigault, T., Gergaud, P., Bocquet, F. (2004) *J. Meta. Nanocryst. Mater*, **19**, 129.

Thonggoom, R. and Funkenbusch, P.D. (2005) *J. Mater. Sci.*, **40**, 4279.

Thornston, J.A. and Hoffman, D.W. (1989) *Thin Solid Films.*, **171**, 5.

Tillett, J.P.A. (1956) *Proc. Phys. Soc. B*, **69**, 47.

Timoshenko, M. (1951) *Théorie des plaques et des coques*, Librairie Polytechnique, France.

Ting, T.C.T. (1966) *J. Appl. Mech*, **33**, 845.

Tomozawa, M. (1998) *Phys. Chem. Glasses.*, **39**, 65.

Tool, A.Q. (1946) *J. Am. Ceram. Soc*, **29**, 240.

Turnbull, D. and Cohen, M.H. (1961) *J. Chem. Phys*, **34**, 120.

Tyagi, V. and Varshneya, A.K. (1998) *J. Non-Cryst. Solids*, **238**, 186.

Van Mol, A.M.B., Alcott, G.R., Allendorf, M.D. (2005) *Am. Ceram. Soc. Bull.*, **84**, 37.

Veprek, S., Veprek-Heijman, M.G.J., Karvankova, P., Prochazka, J. (2005) *Thin Solid Films.*, **476**, 1.

Wang, T.J. (1997) *Glass Technol*, **38**, 104.

Wang, W.H., Dong, C., Shek, C.H. (2004) *Mater. Sci. Eng. R*, **44**, 45.

Warren, P.D. (1995) *J. Eur. Ceram. Soc.*, **15**, 201.

Watanabe, T., Puratsubaki, K., Benino, Y., Saitoh, H., Komatsu, T. (2001) *J. Mater. Sci.*, **36**, 2427.

Weibull, W.A. (1951) *J. Appl. Phys.*, **18**, 293.

Weibull, W.A. (1952) *Appl. Mech. Rev.*, **5**, 11.

Westbrook, J.H. (1960) *Phys. Chem. Glasses*, **1**, 32.

Wiederhorn, S.M. (1967) *J. Am. Ceram. Soc.*, **50**, 407.

Wiederhorn, S.M. and Bolz, L.H. (1970) *J. Am. Ceram. Soc.*, **53**, 543.

Wiederhorn, S.M. and Hockey, B.J. (1980) *J. Non-Cryst. Solids*, **38/39**, 433.

Williams, K.F.E., Johnson, C.E., Greegrass, J., Tilley, B.P., Gelder, D., Johnson, J.A. (1997) *J. Non-Cryst. Solids*, **211**, 164.

Williams, M.L., Landel, R.F., Ferry, J.D. (1955) *J. Am. Chem. Soc*, **77**, 3701.

Woirgard, J. and Dargenton, J.C. (1997) *J. Mater. Res.*, **12**, 2455.

Wu, D., Lu, G., Jiang, H., Li, Y. (2004) *J. Am. Ceram. Soc.*, **87**, 1799.

Wu, D., Zhou, J., Li, Y. (2006) *J. Mater. Sci*, **41**, 5630.

Xing, L.Q., Bertrand, C., Dallas, J.P., Cornet, M. (1998) *Mater. Sci. Eng. A*, **241**, 216.

Yinnon, H. and Cooper, A.R. (1980) *Phys. Chem. Glasses*, **21**, 204.

Yoshida, S., Isono, S., Matsuoka, J., Soga, N. (2001) *J. Am. Ceram. Soc.*, **84**, 2141.

Yoshida, S., Sangleboeuf, J.C., Rouxel, T. (2005) *J. Mater. Res.*, **20**, 3404.

Zachariasen, W.H. (1932) *J. Am. Chem. Soc*, **54**, 3841.

Zandonella, C. (2005) *Courrier Int*, **755**, 60.

Zarzycki, J. (1982) *Les verres et l'état vitreux*, Masson, Paris. [*Glasses and the Vitreous State*, Cambridge University Press, Cambridge, 1991.]

Zeng, K., Giannakopoulos, A.E., Rowcliffe, D.J. (1995) *Acta Metall. Mater*, **43**, 1945.

Zhou, J., Wang, Y., Xia, Y. (2006) *J. Mater. Sci.*, **41**, 5778.

Zinck, P., Pays, M.F., Rezakhnlou, R., Gerard, J.F. (1999) *Phil. Mag. A.*, **79**, 2103.

Zou, X. and Toratani, H. (2001) *J. Non-Cryst. Solids*, **290**, 180.

Web Sites

http://thayer.dartmouth.edu/~defmech/

http://www.musee.ensmp.fr/gm//lettres.html

http://www.saint-gobain.com/fr/html/groupe/verre_plat.asp

http://www.saint-gobain-glass.com/fr/b1107.htm

http://www.saint-gobain-glass.com/saint-just/

http://www.saint-gobain.com/fr/html/investisseurs/pdf/LeGuide2i.pdf

http://www.sharplesstress.com/

http://www.strainoptic.com/

http://www.tamglass.com/

http://www.webelements.com/webelements/

Video

'C'est pas sorcier se met au verre', France Télévision (FR3).

Index

Glass; Mechanics and Technology. Eric Le Bourhis

ISBN: 978-3-527-31549-9